A Review of Science and Technology During the 1977 School Year

Science Year

The World Book Science Annual

1978

Field Enterprises Educational Corporation

Chicago   Frankfurt   London   Paris   Rome   Sydney   Tokyo   Toronto

A subsidiary of
Field Enterprises, Inc.   fe

The publishers of *Science Year* gratefully
acknowledge the following for permission to use
copyrighted illustrations. A full listing of illustration
acknowledgments appears on pages 430 and 431.

140-141   Background photo: © California Institute of Technology and
Carnegie Institution of Washington, from Hale Observatories

151   © California Institute of Technology and Carnegie Institution
of Washington, from Hale Observatories

161   CLIMAP, International Decade of Ocean Exploration, National
Science Foundation, from *Science*. Copyright © 1976 by the
American Association for the Advancement of Science

228   E. A. Pillemer and W. M. Tingey, New York State College of
Agriculture and Life Sciences, from *Science*. Copyright ©
1976 by the American Association for the Advancement of Science

248   Drawing by H. Martin; © 1977 The New Yorker Magazine, Inc.

284   Drawing by Lorenz; © 1976 The New Yorker Magazine, Inc.

300   *Temporomandibular Joint Dysfunction and Occlusal Equilibration* by Nathan A. Shore, D.D.S. © 1976 J. B. Lippincott
Company

340   Reprinted with permission from *Science News,* the weekly news
magazine of science. Copyright 1977 by Science Service, Inc.

The Cover: The awesome fire of an erupting volcano on the island
of Heimaey silhouettes residents and their homes.

Preface

This edition of *Science Year* devotes an unusual amount of space to one subject–the continuing debate over whether recombinant DNA research should be encouraged, discouraged, or even permitted to proceed. The Special Report GENES: HANDLE WITH CARE describes the technology and the guidelines established by the National Institutes of Health (NIH) under which the research is currently being conducted. Following this Special Report, six scientists who have participated in the debate recall the highlights of its history–from the advent of the technique in the early 1970s to legislation pending in the summer of 1977. In the *Science Year* essay SCIENCE AND THE DECISION MAKERS, Daniel M. Singer, a lawyer and science layman, puts in historical perspective the risks that some segments of society have commonly imposed on other segments. He goes on to argue for the need for all nonscientists to learn what is going on in science and to participate in decisions about the speed and direction of its progress.

There are several reasons for the extended coverage of this subject, all involving scientific precedents to some degree. First, its potential for good or evil extends far beyond that of other new advances. In fact, some scientists believe that this powerful biological technique will have a greater impact on the human race than the unleashing of the atom. Second, when the molecular biologists realized that they had discovered how to splice together genes from different species, they sensed the awesome possibilities of this work and voluntarily imposed a number of restrictions on it. Finally, the public became deeply involved in the discussions and in one notable case–that of the Cambridge Experimental Review Board–undertook to learn enough about the subject to make reasonable decisions on how recombinant DNA research should proceed at the two research institutions in their Cambridge, Mass., community.

Many scientists and engineers have come to believe that science and technology need to be monitored more closely than in the past. More and more they realize that the public especially should be aware of what is coming out of the research laboratories and how it may affect them. In order to do this, people are going to have to educate themselves in a subject that many have considered to be too difficult to comprehend. The members of the Cambridge Experimental Review Board, however, proved that it is not necessary to be a scientist to make competent judgments about how certain research should proceed. All that an intelligent group of lay people needs is a reason to learn and the willingness to work.

Most nonscientists, of course, would not want or need to become as immersed in a scientific subject as did the Cambridge Board members. But if people are to play a meaningful role in the scientific decision-making process, they will need at least to keep abreast of the major achievements in science. This has been the purpose of *Science Year*–to provide the science-interested public with a current view of the progress of science and technology.   [Arthur G. Tressler]

Contents

Staff

Editorial Director
William H. Nault

Editorial
Executive Editor
Arthur G. Tressler

Managing Editor
Michael Reed

Chief Copy Editor
Joseph P. Spohn

Senior Editors
Robert K. Johnson
Edward G. Nash
Kathryn Sederberg
Darlene Stille
Foster P. Stockwell

Assistant Copy Editor
Irene B. Keller

Editorial Assistant
Madelyn Krzak

Art
Executive Art Director
William Dobias

Art Director
Alfred de Simone

Senior Artists
Roberta Dimmer
Mary-Ann Lupa

Artist
Wilma Stevens

Photography Director
Fred C. Eckhardt, Jr.

Photo Editing Director
Ann Eriksen

Senior Photographs Editors
Blanche Cohen
John S. Marshall
Carol A. Parden
Paul Quirico
Jo Anne M. Ticzkus

Research and Services
Head, Editorial Research
Jo Ann McDonald

Head, Research Library
Vera Busselle

Head, Cartographic Services
H. George Stoll

Senior Index Editor
Marilyn Boerding

Pre-Press Services
Director
Richard A. Atwood

Manager, Manufacturing Liaison
John Babrick

Supervisor, Keyboarding Section
Lynn Iverson

Manager, Film Separations
Alfred J. Mozdzen

Assistant Manager, Film Separations
Barbara J. McDonald

Manager, Production Control
J. J. Stack

Supervisor, Art Traffic
Joe Gound

Supervisor, Scheduling
Barbara Podczerwinski

Manufacturing
Executive Director
Philip B. Hall

Production Manager
Jerry R. Higdon

Manager, Research and Development
Henry Koval

Contributors

Adelman, George, M.S.
Editor & Librarian
Neurosciences Research Program
Massachusetts Institute of Technology
Neurology

Alderman, Michael H., M.D.
Assistant Professor of Medicine and
Public Health
Cornell University Medical Center
Medicine, Internal
Public Health

Anthes, Richard A., Ph.D.
Associate Professor of Meteorology
Pennsylvania State University
Meteorology

Araujo, Paul E., Ph.D.
Assistant Professor
Department of Food Science
University of Florida
Nutrition

Arehart-Treichel, Joan, B.A.
Medical Editor
Science News
Rita Levi-Montalcini

Auerbach, Stanley I., Ph.D.
Director, Environmental Sciences Division
Oak Ridge National Laboratory
Ecology

Bell, William J., Ph.D.
Professor of Biology
University of Kansas
Zoology

Belton, Michael J. S., Ph.D.
Astronomer
Kitt Peak National Observatory
Astronomy, Planetary

Boffey, Philip M., B.A.
Staff Writer
Science
Close-Up, Public Health

Bromley, D. Allan, Ph.D.
Henry Ford II Professor and Chairman
Department of Physics
Yale University
Physics, Nuclear

Brown, Barbara B., Ph.D.
Biofeedback Researcher
Veterans Administration Hospital
Signals from Inner Space

Cermak, Jack E., Ph.D.
Director, Fluid Dynamics
& Diffusion Laboratory
College of Engineering
Colorado State University
Taming the Winds

Chiller, Jacques M., Ph.D.
Associate Professor
National Jewish Hospital
& Research Center
Immunology

Cromie, William J., B.S.
Executive Director
Council for the Advancement of
Science Writing
*The International Science and
Engineering Fair*

Crumley, Carole L., Ph.D.
Assistant Professor
Department of Anthropology
University of Missouri
Archaeology, Old World

Davies, Julian, Ph.D.
Professor of Biochemistry
University of Wisconsin
Biochemistry

Dix, Donald M., M.S.
Research Staff Member
Institute for Defense Analyses
Engines for the Eighties

Doniach, Sebastian, Ph.D.
Professor of Applied Physics
Stanford University
Physics, Solid State

Eberhart, Jonathan
Space Sciences Editor
Science News
Viking's View of Mars
Space Exploration

Ensign, Jerald C., Ph.D.
Professor of Bacteriology
University of Wisconsin
Microbiology

Freedman, David Noel, Litt.D., Sc.D.
Professor of Biblical Studies
University of Michigan and
Director
Albright Institute of Archaeological
Research, Jerusalem
A City Beneath the Sands

Gates, W. Lawrence, Sc.D.
Professor and Chairman
Department of Atmospheric Sciences
Oregon State University
New Clues to Changing Climate

Giacconi, Riccardo, Ph.D.
Professor of Astronomy
Harvard University and
Associate Director
Center for Astrophysics
Astronomy, High Energy

Goldhaber, Paul, D.D.S.
Dean
Harvard School of Dental Medicine
Medicine, Dentistry

Goldstein, Allan L., Ph.D.
Professor and Director
Division of Biochemistry
University of Texas Medical Branch
Close-Up, Immunology

Griffin, James B., Ph.D.
Senior Research Scientist
Museum of Anthropology
University of Michigan
Archaeology, New World

Grobstein, Clifford, Ph.D.
Vice-Chancellor, University Relations
University of California, San Diego
The Scientists Respond

Gump, Frank E., M.D.
Professor of Surgery
College of Physicians and Surgeons
Columbia University
Medicine, Surgery

Hamilton, Warren, Ph.D.
Research Geologist
U.S. Geological Survey
Geoscience, Geology

Hammond, Allen L., Ph.D.
Staff Writer
Science
Energy

Hartl, Daniel L., Ph.D.
Professor of Biology
Purdue University
Genes: Handle With Care
Close-Up, Biochemistry
Genetics

Hayes, Arthur H., Jr., M.D.
Associate Professor of Medicine
and Pharmacology
Milton S. Hershey Medical Center
Pennsylvania State University
Drugs

Hillman, William S., Ph.D.
Senior Plant Physiologist
Biology Department
Brookhaven National Laboratory
How Plants Tune In the Sun

Holden, Constance, B.A.
Staff Writer
Science
Close-Up, Zoology

Jennings, Feenan D., Ch.E.
Head,
International Decade of Ocean Exploration
National Science Foundation
Oceanography

Kellermann, Kenneth I., Ph.D.
Scientist
National Radio Astronomy Observatory
A Sharper Focus on the Universe

Kessler, Karl G., Ph.D.
Chief, Optical Physics Division
National Bureau of Standards
Physics, Atomic and Molecular

Koshland, Daniel E., Jr., Ph.D.
Chairman
Department of Biochemistry
University of California, Berkeley
The Scientists Respond

Maran, Stephen P., Ph.D.
Astrophysicist
NASA-Goddard Space Flight Center
Astronomy, Stellar

March, Robert H., Ph.D.
Professor of Physics
University of Wisconsin
Physics, Elementary Particles

Maugh, Thomas H. II, Ph.D.
Staff Writer
Science
The Two Faces of Diabetes
Close-Up, Environment

May, Robert M., Ph.D.
Class of 1877 Professor of Zoology
Princeton University
The Scientists Respond

McGetchin, Thomas R., Ph.D.
Geologist
Invar Science Institute
Harnessing Earth's Fountains of Fire

Meade, Dale M., Ph.D.
Research Physicist
Princeton University
Physics, Plasma

Meier, Albin R.
Technical Editorial Director
Telephony
Communications

Merbs, Charles F., Ph.D.
Professor and Chairman
Department of Anthropology
Arizona State University
Anthropology

Moran, Edward, B.A.
Editor and Writer
Popular Science
Technology

Moseley, Michael Edward, Ph.D.
Associate Curator
South and Middle American Archaeology
and Ethnology
Field Museum of Natural History
Close-Up, Archaeology

Nix, J. Rayford, Ph.D.
Staff Member
Theoretical Division
Los Alamos Scientific Laboratory
Cores in Collision

Norman, Colin, B.Sc.
Washington Correspondent
Nature
Science Policy

Perlberg, Mark, B.A.
Free-Lance Writer
The Wolf's Last Stand

Price, Frederick C., Ch.E.
Correspondent
McGraw-Hill, Incorporated
Chemical Technology

Riddell, Frederick R., Ph.D.
Research Staff Member
Institute for Defense Analyses
Engines for the Eighties

Roblin, Richard O. III, Ph.D.
Head, Molecular Biology of Tumor
Cells Group
Basic Research Program
Frederick Cancer Research Center
The Scientists Respond

Salisbury, Frank B., Ph.D.
Professor of Plant Physiology
Plant Science Department
Utah State University
Botany

Shank, Russell, D.L.S.
Director of Libraries
Smithsonian Institution
Books of Science

Silk, Joseph, Ph.D.
Professor of Astronomy
University of California, Berkeley
Astronomy, Cosmology

Singer, Daniel M., LL.B.
Law Partner
Fried, Frank, Harris, Shriver & Kampelman
Science and the Decision Makers

Sinsheimer, Robert L., Ph.D.
Chancellor
University of California, Santa Cruz
The Scientists Respond

Snyder, Solomon H., M.D.
Professor of Psychiatry and Pharmacology
Department of Pharmacology
Johns Hopkins Medical School
Our Body's Own Narcotics

Sperling, Sally E., Ph.D.
Professor of Psychology
University of California, Riverside
Psychology

Stetten, DeWitt, Jr., M.D., Ph.D.
Deputy Director for Science
National Institutes of Health
The Scientists Respond

Thompson, Ida, Ph.D.
Assistant Professor of Geological and
Geophysical Sciences
Princeton University
Geoscience, Paleontology

Todaro, George J., M.D.
Chief, Laboratory of Viral Carcinogenesis
National Institutes of Health
Close-Up, Genetics

Verbit, Lawrence, Ph.D.
Professor of Chemistry
State University of New York at Binghamton
Chemistry

Wade, Nicholas
Staff Writer
Science
Close-Up, Psychology

Ward, Harold R., Ph.D.
Professor of Chemistry and
Associate Dean
Brown University
Environment

Wargo, J. R., B.A.
Washington Correspondent
Freed-Crown Publishing Company
Transportation

Weber, Samuel, B.S.E.E.
Executive Editor
Electronics Magazine
Electronics

Wetherill, George W., Ph.D.
Director
Department of Terrestrial Magnetism
Carnegie Institution of Washington
Geoscience, Geochemistry

Wittwer, Sylvan H., Ph.D.
Director
Agricultural Experimental Station
Michigan State University
Agriculture

Contributors not listed on
these pages are members of the
Science Year editorial staff.

Special Reports

The Special Reports give in-depth treatment to the major advances in science and technology. The subjects were chosen for their current importance and lasting interest.

Viking's View
Of Mars

By Jonathan Eberhart

**The most extensive reconnoitering mission
ever sent to another world is drastically
altering our picture of the red planet**

Mars will never seem the same. Gone are the fabled canals and
vegetation. Gone are the four-armed, copper-skinned natives of science fiction with their exotic civilizations of sword and sorcery. Their
fictional world has been replaced by a fantastic real one of trackless
dunes and icy crags. There is a volcano almost as big as New Mexico;
a canyon that would span the North American continent. Skies are the
color of terrestrial sunsets and sunsets recall the hue of an unpolluted
daytime sky on Earth.

The newest additions to the spectacular picture of Mars were
painted by the United States Viking mission, which sent two spacecraft to orbit Mars and put down landers in the summer of 1976 after a
journey of nearly a year. Each lander is only about 1.5 meters (5 feet)
wide, yet each is crammed with a battery of sophisticated instruments
to study surface characteristics, analyze the soil, measure the winds,
determine atmospheric composition, detect quakes, and search for

evidence of life. All these data, plus pictures of the surface, are radioed back to Earth via the Viking orbiters. These sister ships to the landers circle Mars, producing their own stunning pictures of the planet from above. In addition, the orbiters measure the moisture content of the atmosphere and take infrared temperature measurements of the Martian atmosphere and surface.

For years, a special team of researchers worked with the photos of Mars produced from earlier spacecraft missions to select the Viking landing sites, only to see much of their work go for nothing, however, when photos from orbit suggested the proposed terrain was too irregular for safe landing. Lander 1 finally touched down safely on July 20, 1976, more than two weeks behind schedule, on the carefully chosen upper slopes of a dry basin in Chryse Planitia (the Plains of Gold) about 1,400 kilometers (850 miles) north of the Martian equator. About seven weeks later, Lander 2 miraculously found a safe spot on a flat, but rock-strewn, region of Utopia Planitia, which is roughly halfway around the planet from Lander 1, and some 1,400 kilometers closer to the north pole.

Using data sent back to the Jet Propulsion Laboratory in Pasadena, Calif., during Lander 1's first day on the surface, Seymour L. Hess of Florida State University and the rest of Viking's meteorology team produced the first weather report from another world: "Light winds from the east in the late afternoon, changing to light winds from the southwest after midnight. Maximum wind was 25 kilometers [15 miles] per hour. Temperature range from −122°F. just after dawn to −22°F. . . . Pressure steady at 7.70 millibars."

Apart from the specific details, the ability to study the relatively simple weather on Mars can be valuable to researchers still struggling to decode the complex relationships that control weather on Earth. "Trying to understand the Earth's atmosphere," Hess said, "can be compared to trying to understand a manic-depressive paranoid who has schizophrenic tendencies. But when numerical studies are developed with a good observed data base on a simpler atmosphere such as [that on] Mars, one would expect that we would get a better feeling for what is important on Earth."

The Viking spacecraft confirmed that Mars does indeed behave in a more orderly meteorological fashion. That first-day weather report changed little over the following weeks, varying only a few degrees in temperature and just slightly in wind patterns. The landers touched down in the northern hemisphere during the Martian summer, but even the somewhat more tumultuous winter seems to have a regular behavior cycle. Lander 1 revealed a tiny, but extremely regular, daily drop in atmospheric pressure, which continued for about 100 days before it began to rise again on a similarly smooth curve. As soon as Lander 2 reached the surface, it began recording the same kind of pattern. The Viking meteorologists concluded that the pressure drop signaled the onset of winter in the *southern* hemisphere, theorizing that

The author:
Jonathan Eberhart is space sciences editor for *Science News.* He writes the Space Exploration article for *Science Year.*

Viking Reaches Mars

1. Orbiter releases lander

2. Lander turns over, leaves orbit, and enters Martian atmosphere

3. Lander opens parachute, releases aeroshell shield

4. Lander fires engines to slow final descent

Viking 1 Landing Site
Chryse Planitia

North Pole

60° W

Viking 2 Landing Site
Utopia Planitia

Valles Marineris

Olympus Mons

Approaching Mars after nearly a year en route, each Viking spacecraft released its lander. Retrorockets and a parachute slowed the lander's descent to its preselected site.

the carbon dioxide that forms about 95 per cent of the Martian atmosphere was freezing out onto the growing south polar cap.

Many researchers suspect that the Mars of a billion or more years ago may have been even more interesting than the planet today. One reason for this has to do with water, as do so many questions about the red planet. Mariner 9, visiting Mars in 1971-1972, sent back photos showing surface features resembling patterns that might have been cut by massive quantities of flowing water–rivers, floods, and meandering streams. But in order for liquid water to have existed on the planet in ancient times, the atmosphere must have been much thicker than it is today. The present surface pressure, which averages roughly 6 millibars compared with more than 1,000 millibars on Earth, is not enough to hold water in liquid form. And a thicker atmosphere would have

A wild and lonely dune field, littered with rocks, greeted the Viking 1 Lander, *above.* Halfway across Mars, under a salmon-colored sky, Lander 2 found a similar scene, *left.*

meant higher temperatures, altogether a more appealing environment – at least by Earth standards.

The Viking orbiters discovered other seemingly water-caused patterns. On Mars, these patterns are usually referred to as fluvial, or liquid-caused, features because it is not 100 per cent certain that water was the liquid responsible for carving them. Months after the Vikings arrived, a few scientists still thought that some other liquid, such as extremely runny lava, could have been at work instead of water.

But the two Viking orbiters revealed additional water-related features that extend far beyond dry riverbeds and flood plains. For example, an area crisscrossed by wide, flat-bottomed grooves similar to the tracks left by large-scale glacial flows on Earth lies in a region known as Nilosyrtis, roughly 2,000 kilometers (1,250 miles) north of the Martian equator. To the northwest, near the Cydonia region, is the crater Arandas, whose smooth and contoured flanks remind some researchers of test craters made by impacts in water-logged soil in laboratories on Earth. In another major revelation from Viking, summertime temperatures at the north pole measured from orbit proved much too high for frozen carbon dioxide to exist in solid form. But the temperatures were consistent with frozen water, strongly implying that the permanent polar caps, which remain through the summer, consist entirely of ordinary ice.

Some of the gases in the Martian atmosphere are also providing clues to the possible earlier existence of liquid water. Various Viking instruments, operating as the landers descended through the atmosphere and after they landed, detected not only carbon dioxide, but also small quantities of carbon monoxide, oxygen, nitrogen, nitric oxide, and several of the so-called "noble" or "inert" gases – argon, krypton, xenon, and neon.

Some of these gases tell their stories by the rates at which they are formed from other elements; argon 40, for example, results from the decay, at a precisely known rate, of potassium 40. Measuring the present ratio of argon 40 to potassium 40 helps researchers estimate how much potassium 40 may have been around initially. Other gases yield clues from the ratios between lighter and heavier isotopes – forms of the same element. Comparing present-day isotopic ratios on Mars with previously esti-

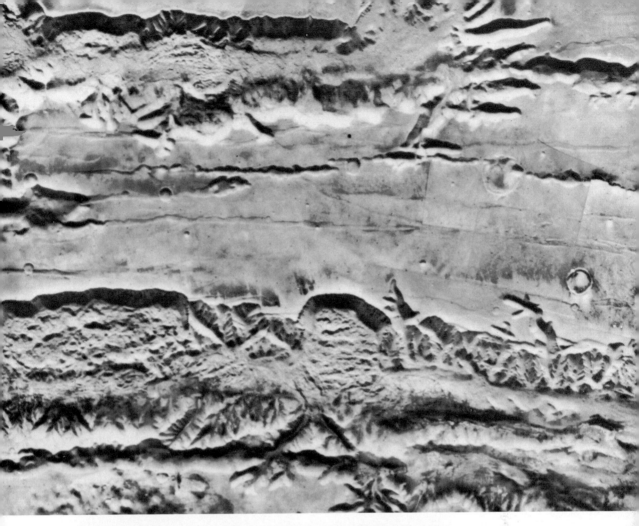

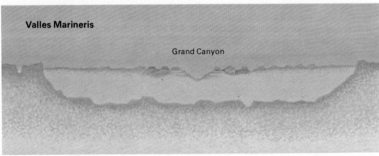

Valles Marineris

Grand Canyon

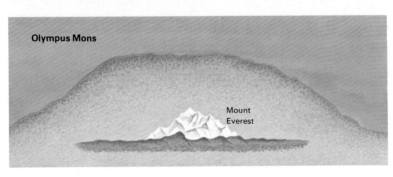

Olympus Mons

Mount
Everest

Geographic features on Mars dwarf their earthly counterparts. Valles Marineris, *above,* a huge canyon near the Martian equator, makes the Grand Canyon look like a ditch. Volcano Olympus Mons, *opposite page,* Mars's highest peak, is nearly three times as tall as Mount Everest.

mated ratios based on other solar system studies can help scientists deduce how much total gas may have been present in the early Martian atmosphere.

For example, the ratio of argon 36 to argon 40 on Mars is only about one-ninth the ratio found on Earth. From this, astrophysicist Tobias Owen of the State University of New York at Stony Brook and some of his colleagues concluded that the early atmosphere of Mars was never more than 10 times thicker than it is today. This would allow liquid water to exist, though the atmosphere might not be sufficiently thick to trap the heat needed to keep the water unfrozen long enough to cut all of those "riverbeds." On the other hand, physicist Michael B. McElroy of Harvard University in Cambridge, Mass., cited the amount of nitrogen–something close to 3 per cent of the atmosphere–as evidence of an early atmosphere that may have been some 70 times thicker than the present one.

Where did that early atmosphere go? Some has certainly escaped into space because Mars's relatively weak gravitational field could not hold it. More is chemically bound, or physically trapped, in the soil, and still more may be frozen beneath the surface as permafrost. One possibility is that the solar wind, the high-speed charged particles that stream out from the Sun, could have swept away a substantial fraction of the atmosphere while the planet was forming. However, Owen and his colleagues said that this is unlikely because it would have left a present ratio of argon 36 to total krypton much lower than that of the Earth. Viking data show that this is not the case.

A more likely possibility, Owen's team reported, is that much less gas escaped from the hot Martian interior to form the atmosphere than was the case on Earth. But again there are uncertainties. McElroy reported that the ratio of nitrogen 15 to nitrogen 14 is about 75 per cent greater on Mars than it is on Earth. This seems to suggest that a large amount of gas originally escaped from the interior, with the lighter nitrogen 14 escaping into space or combining into heavier molecules that fell back onto the Martian surface while the heavier nitrogen 15 remained in the atmosphere.

Most atmospheric-history calculations depend upon being able to determine the original mixture of materials from which the planet formed. Because Mars is farther from the Sun than the Earth, scien-

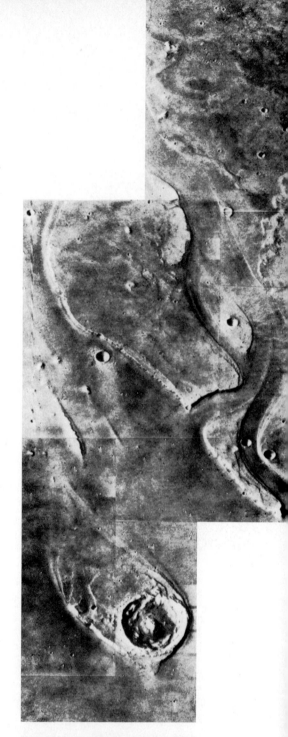

Teardrop-shaped islands within the meandering, intertwining channels on Mars's Chryse plain present strong evidence that water once flowed there.

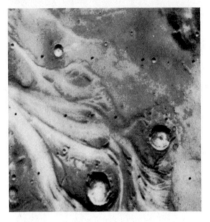

A dry, barren scene, *top,* changes character shortly after dawn. The surface warms slightly, driving off water vapor that condenses as fog in the cooler, low-lying areas such as craters and channel bottoms, *above.*

Yuty crater was probably formed when a meteorite collided with Mars. Layers of broken rocks, thrown out by the shock of impact, surround the crater.

tists expected that more of the volatile elements—those that condense at fairly low temperatures, such as the noble gases—would have been available when the planet formed. But again the Viking data presented contradictions. The low proportion of argon 36 and the enriched nitrogen 15-nitrogen 14 ratio hint at an early inventory of volatile gases like that associated with a class of meteorites called Type 1 carbonaceous chondrites that are believed to originate in the asteroid belt between Mars and Jupiter. On the other hand, Viking's instruments reported finding more krypton than xenon. This is just the reverse of what is believed to be true of carbonaceous chondrites.

Which brings us back to water. The Earth's atmosphere also has more krypton than xenon, which some researchers believe is because xenon clings more readily to the surfaces of shales and sediments. Is the lesser quantity of xenon relative to krypton in the Martian atmosphere a sign that past epochs of flowing water created fresh sediments capable of capturing the xenon?

One of the more intriguing speculations about Mars in recent years has been the possibility that its hypothetical early, more temperate climate may reappear in cycles, perhaps every billion years or so. Once again, Viking provided tantalizing support on both sides of the question. McElroy's calculations of the ratio of oxygen 18 to oxygen 16 suggest to him that, on a billion-year cycle, the surface may alternately release and recombine with vast quantities of oxygen, perhaps representing a major "pressure pulsation" of the atmosphere. Furthermore, Harold Masursky of the U.S. Geological Survey in Flagstaff, Ariz., and other geologists cited the differing numbers of craters in various channels as evidence that the seemingly water-caused features were of widely differing ages, ranging from hundreds of millions of years old down to as little as 100,000 years old.

But if there are climatic cycles on Mars, there must also be a reservoir of carbon dioxide somewhere on the planet from which the gas can be released to create a denser atmosphere. The polar caps were formerly considered the most likely site of that reservoir, but the discovery that the permanent polar caps are made of frozen water rather than frozen carbon dioxide argues against that possibility, according to planetary scientist Hugh H.

Kieffer of the University of California, Los Angeles. Kieffer heads the team whose study of the temperature measurements taken by the orbiter yielded the conclusion that the permanent polar caps are composed primarily of water ice.

There is certainly room for a lot of carbon dioxide to be mixed in with the soil, perhaps to a considerable depth, but that poses another problem. If the carbon dioxide were concentrated in one area, such as the polar caps, a little extra heat, perhaps from a change in Mars's tilt, could release enough of the gas to trap more heat, release more gas, and keep the process going long enough to build up a substantial atmosphere. But if the carbon dioxide reservoir is diffused throughout the planet, Kieffer said, the initial temperature rise would not release enough carbon dioxide in any one place to trap the additional heat necessary to keep the process going.

If Mars lost part of its atmosphere because temperatures fell low enough to freeze out the gases, what caused the cooling? One possibility, though it would seem to rule out a cyclic climate, is simply the natural cooling of the planet after it formed. Early in 1977, however, another possibility was raised by planetary scientists Joseph A. Burns of Cornell University in Ithaca, N.Y.; William R. Ward of the Center for Astrophysics in Cambridge, Mass.; and Brian Toon of the Ames Research Center in Mountain View, Calif.

Like other planets, Mars is not perfectly round; it is flattened at the poles because of its spin. Various crustal features and density irregularities put it further out of round. Burns, Ward, and Toon calculated that if Mars were as little as 7 per cent rounder, its axis of rotation would tilt considerably more than it does today, thereby exposing the polar regions to greater concentrations of the Sun's heat for longer periods. Temperatures there would then rise above the melting point

The carefully crafted Viking lander was put through detailed tests before it was launched on its Martian mission.

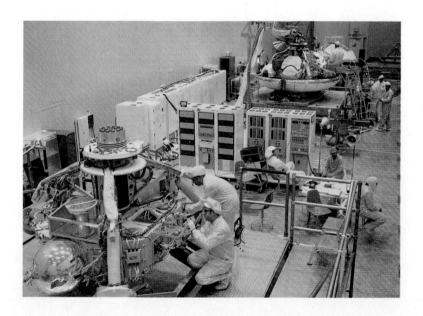

A Laboratory on Mars

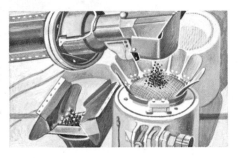

A long, mechanical arm on a Viking lander stretches out to scoop up a soil sample. The arm rolls back, swings around, and deposits the sample in chambers for the chemical and biological experiments.

of water, and far above that of carbon dioxide, for weeks at a time. But the significant point, according to the three researchers, is that 6.4 per cent of the planet's irregular shape is due to a single Martian feature—the massive bulge thousands of kilometers across known as Tharsis, on which the planet's four largest volcanoes sit. Another 3.6 per cent of the planet's shape is due to other crustal features and density variations. At least some of these are presumably related to the internal upheavals that formed Tharsis. In other words, climatic conditions could have been very different before the Tharsis bulge bulged.

And what if Tharsis were to collapse? Even before Viking data were available, it was suggested that the planetary crust in the Tharsis region may be too weak to support itself and that internal activity might be holding it up. If such activity were to die down and the crust collapsed, the planet's shape might be changed enough to tilt its polar regions further toward the Sun. Thus, even if Mars's climate is not cyclic, it might be reversible.

The key to determining Mars's internal activity is given by quakes. Each Viking lander carried a seismometer to detect quakes, signs of an internally active planet. Unfortunately, the Lander 1 seismometer stuck. This was a major blow to seismologists' plans because they need two instruments in different locations to pinpoint the position of any tremors that are detected.

However, the Lander 2 instrument worked well. On Nov. 4, 1976, it detected what Viking seismologists thought was the first Marsquake ever recorded. The team, headed by Don L. Anderson of the California Institute of Technology, later decided the instrument was simply recording the effect of a strong wind. But 20 days later, the instrument recorded a second event centered about 100 to 200 kilometers (60 to 120 miles) from the lander. From the seismic results, Anderson estimated that the crust at the Lander 2 site is about 15 kilometers (9.5 miles) in thickness.

The surface of Mars is remarkably diverse. There are heavily and lightly cratered regions, but there are also many signs of liquid activity, volcanism, erosion, fracturing, possible glaciation, and other phenomena. In the northern hemisphere, where much of the early Viking photography was concentrated, a great deal of the surface appears old and greatly altered by a variety of processes. Earlier observations had indicated that the Martian winds may be responsible for much of the erosion seen on the planet. In some areas, however, many small craters have survived, suggesting that wind erosion is not always as effective as had been supposed.

In the search for landing sites, the orbiter imaging team, headed by Michael H. Carr of the U.S. Geological Survey, discovered a number of features that were little known from previous missions. There are regions of collapsed terrain, where the surface is lower in elevation and more chaotic than the surrounding land. Perhaps this is caused by the sudden melting of subsurface permafrost. Other regions show the same pattern of collapse, but without the blocklike rubble, as though the collapse might have come more slowly.

The sharp orbital photographs also gave scientists a clear look at some of the Martian volcanoes. The lack of erosion suggests that some of them are quite young in geologic terms, perhaps only a few hundred thousand years old, according to some geologists. Distinct layers of ejected material on the volcanoes' slopes are allowing geologists to construct the chronological sequence of their formation. The location of the four largest volcanoes, including the towering Olympus Mons, on the Tharsis uplift could well represent a major irregularity in the planet's interior structure.

The landers, although restricted to their immediate viewing area, provided the most detail about the Martian surface. The pictures they sent back showed both fine- and coarse-grained rocks, together with signs of wind-shaping, layering, fracturing, and the popped gas-bubble remnants known as vesicles. The dominant color in the early photos was a brilliant orange, but continuing processing has refined this to a less-dramatic brown. Scientists believe the coloring is caused by an

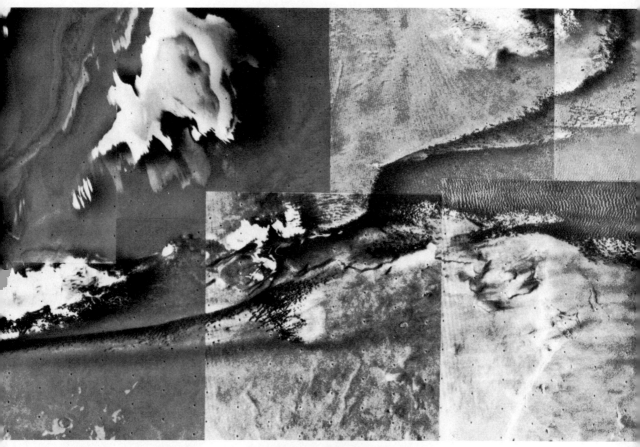

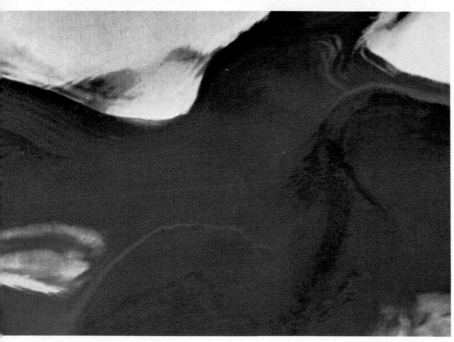

Patches of water ice lie
at the edge of Mars's
north polar cap, while
dark sand drifts along
a gently curving channel
to accumulate in the
rows of a broad dune
field, *above*. Melting
of the upper crust of
frozen carbon dioxide
in the summer reveals
layers of water ice and
soil, perhaps related to
climate changes, *left*.

iron oxide coating, although such a coating, if continuous, would have to be as thin as 0.25 micron (0.25-thousandths of a millimeter) to agree with the soil analyses at the landing sites.

Element for element, the fine-grained surface materials analyzed at the two sites were almost indistinguishable—rich in silicon and iron, with significant amounts of magnesium, aluminum, sulfur, calcium, and titanium. The material turned out to be quite tacky, sticking readily to the landers and their instruments. The experimenters also deduced that it must contain from 3 to 7 per cent of some mineral that can be magnetized—hematite, magnetite, maghemite, or goethite. All in all, Viking researchers concluded, the combination of various materials seemed unlike any simple major soil or rock known on either the Earth or the Moon.

Mineralogically, however, the fine grit turned out to have a significant link with Earth. When scientists tried to produce model soils in a laboratory to match the Viking data, they came up with a mixture that suggested the Martian soil could include as much as 80 per cent of iron-rich clays much like those produced on Earth by the weathering of volcanic basalts. Researchers believe that the soil similarity at the two Martian sites is caused by large-scale wind distribution of the fine soil around the planet.

The most spectacular wind-related feature on the planet is the vast belt of dunes circling the north polar cap. Discovered by Viking 1, the belt has been called "perhaps the largest dune field in the solar system." It includes regions where unbroken rows of dunes, hundreds of meters from crest to crest, march on for hundreds of kilometers.

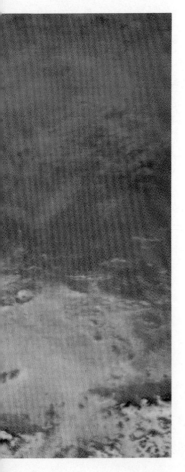

A turbulent dust storm forms in the Argyre basin with the approach of summer in Mars's southern hemisphere. Local dust storms may spread widely to cover vast areas of the planet.

The north polar cap itself is spectacular, scalloped with ice-draped cliffs, melt patterns, and other erosion features. Perhaps most interesting are the regions where parts of the icecap can be seen in cross section, even from orbit. Successive layers of ice and dust clearly show what is believed to be evidence of seasonal freezing and thawing—and perhaps of longer cycles.

Early in 1977, as winter was approaching in the Martian north and summer was coming to the south, photos from orbit showed water-ice hazes in the northern atmosphere and wide-ranging dust clouds in the south. The dust storms had only recently become conspicuous, but the ice hazes had been seen at varying altitudes through most of the mission. The subtle temperature balance between atmosphere and surface also produced early-morning fogs, wave clouds associated with surface features, and simple, individual clouds in the equatorial regions.

The most striking atmospheric phenomenon is the color of the Martian sky. Scientists initially thought it was blue, but careful color-balancing of the photographs showed it to be more of a salmon-pink. This is because the thick atmosphere responsible for the Earth's blue sky is all but missing on Mars and 99 per cent of the light is scattered by suspended dust particles blown up from the surface. And, in a nearly uncanny reversal of the terrestrial color scheme, the Martian sunset may be a pale, washed-out blue, because only at sunset do the Sun's rays pass through enough atmosphere for the blue light-scattering properties to overcome the pink tinge from the dust.

Each Viking lander's arsenal of scientific instruments included an experimental package designed to look for signs of life. A 3-meter (10-foot) arm with a scoop on the end collected soil samples which were distributed among chambers for three experiments. One experiment exposed the soil to an atmosphere similar to that found outside the lander, but with radioactive carbon dioxide and carbon monoxide added. A living organism would take in the radioactive carbon and convert it to organic material. After several days, the radioactive atmosphere was flushed out of the chamber, and the sample was then heated and vaporized. The resulting gases were tested to see if any of the radioactive material had been taken in.

In another experiment, the soil sample was moistened with a radioactive nutrient solution. After several days, the enclosed atmosphere was tested for radioactivity, which might indicate that the nutrients had somehow been metabolized. In the third test, the sample was soaked in a nonradioactive nutrient solution, while the atmosphere was monitored for signs of change that might indicate metabolism.

Despite months of study and attempted interpretation, the results of the experiments were inconclusive. The tests did show strong reactions from the samples, however, grounds for further study even if there is no life on Mars. "The soil didn't just lie there," one team member said. "It got up and worked."

Following the success of the initial phase of the mission, Viking was scheduled to continue sending back information until at least mid-1978. The early phase focused on biology, chemical analysis, and photography. As the operation continues, the seismology and meteorology investigations are becoming more prominent.

Centuries of fascination with the planet, augmented by the Viking results, have prompted increased scientific interest in future missions to Mars. An unmanned roving vehicle could conduct more elaborate versions of the Viking experiments at several different sites. Penetrators, dropped from orbit with instruments inside, could be a relatively inexpensive way of getting instruments such as seismometers into the ground at widely separated points on the planet. There is the possibility of sending a robot craft to bring a sample of the Martian surface back to Earth—an idea that raises objections from some scientists—both those who fear alien contamination and those who worry that a sample may not survive the trip back to Earth unchanged.

The most exotic possibility lies in the distant future. It may never happen, and no one is even seriously proposing such a plan as yet. But a year before the Vikings even reached Mars, the National Aeronautics and Space Administration sponsored a study on the possibility of "terraforming" the planet—using bacteria, climate modification, and other means to make Mars habitable for human beings.

Genes: Handle With Care

By Daniel L. Hartl

Scientists are assessing the risks and rewards of one of the greatest developments of the 20th Century — moving genes from one organism to another

Genes are hereditary voices from the past. These tiny bundles of chemicals, found in every cell of every living thing, tell each organism what it will be, based on what its ancestors have been. Different sets of genes instruct a bean seed to become a bean plant, a pea seed a pea plant; they tell one egg to make a sparrow, another to make an eagle. Although a species may change slowly over the long course of evolution for environmental or other reasons, it retains most of the characteristics unique to it.

Now, molecular biologists have a technique that permits them to put together genes from different organisms. The scientific potential for this technique is enormous in medicine, agriculture, and many other fields. But it has also aroused public concern that scientists may create new and unpredictable forms of life that cannot be controlled. The work is called recombinant DNA research because it involves breaking apart chains of deoxyribonucleic acid (DNA), the chemical that carries genetic data, and recombining them in various ways.

Genes have already been transplanted into bacteria from such diverse organisms as fruit flies, sea urchins, toads, and rabbits. Recombinant DNA research may someday lead to genetic engineering in which scientists will actually manipulate the genetic material of plants, ani-

In Michelangelo's painting, the hand of God gives life to Adam.
Scientists now have in their hands the ability to create new life forms.

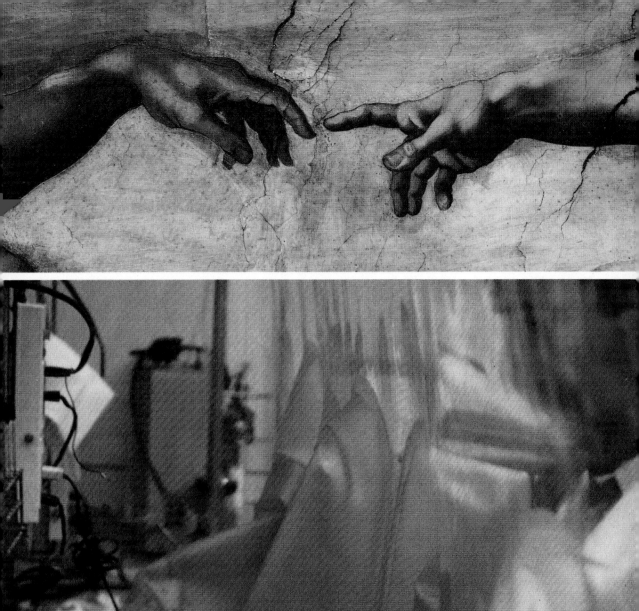
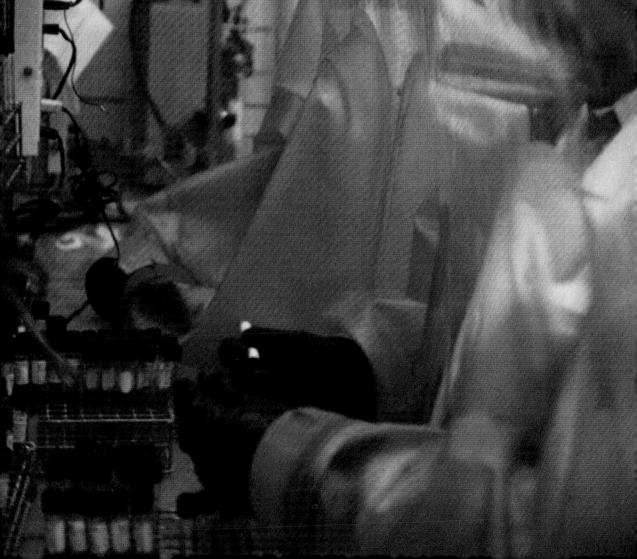

mals, and even human beings. The possibility that recombinant DNA research might create dangerous microorganisms has led some biologists, health officials, political leaders, and environmentalists to urge that work of this kind be stopped or severely restricted.

The molecular biologist's new technique, developed by biologist Stanley N. Cohen and his associates at Stanford University in California, resembles that of a jeweler expanding a ring. The jeweler cuts the metal band, inserts a new piece of metal in the gap, then welds the free ends together. In much the same way, the biologist can chemically insert a piece of DNA that carries genetic instructions from one organism into the DNA of another organism.

Using a laboratory centrifuge, scientists can separate ring-shaped molecules of DNA called plasmids from ground-up cells of certain strains of *Escherichia coli,* a useful and usually harmless bacteria found in the intestinal tracts of humans and many other animals. Plasmids appear to have important functions in the bacteria, such as carrying genes for resistance to antibiotics. They multiply and pass from generation to generation much like ordinary bacterial genes.

Like any other DNA molecule, plasmids are composed of two parallel strands, somewhat like two ticker tapes aligned face-to-face. Each strand has a "backbone" composed of deoxyribose sugar $(C_5H_{10}O_4)$ units joined end to end by atoms of phosphorus (P). Chemical groups called bases are attached to the sugars and jut off the backbone at regular intervals. DNA contains four bases—adenine (A), guanine (G), thymine (T), and cytosine (C)—and any sequence of these bases can occur along the backbone. The sequence determines the hereditary information in a gene.

The bases jutting from the parallel DNA backbones meet in the middle to form base pairs. Because of their chemical attractions, only certain complementary base pairs can be formed. If one strand has an A at a certain position, the other must have a T at the corresponding position; where one strand has a G, the other must carry a C.

Scientists snip open the plasmid rings by adding restriction enzymes to a solution of the plasmids. Molecules of these enzymes recognize certain base sequences of the DNA, attach to them, and cut them at specific points called restriction sites. For example, restriction enzyme Eco RI, which is widely used in recombinant DNA research, attaches to the base sequence $\frac{\text{-CTTAAG-}}{\text{-GAATTC-}}$, its restriction site, and then snips the backbones between the A and G bases on each side. This creates ends that terminate in two short single-stranded regions, $\frac{\text{-CTTAA}}{\text{-G}}$ and $\frac{\text{G-}}{\text{AATTC-}}$. These two ends may be thought of as being "sticky" because they have complementary base sequences and tend to pair with each other and with other pieces of DNA cut by the same enzyme.

The donor DNA to be inserted into the plasmid DNA can be obtained from ground-up cells taken from almost any organism—fruit flies, frogs, mice, even humans. The relatively small plasmid DNA that is generally used carries only one restriction site where Eco RI

The author:
Daniel L. Hartl is
a professor of biology
at Purdue University.
A geneticist, he also
writes the Genetics
article in *Science
Year's* Science File.

can cut, but donor DNA has many restriction sites. So Eco RI can split donor DNA into many fragments, and each will have a sticky, single-stranded region at both ends. When such fragments of donor DNA are mixed with opened plasmids in solution, the sticky ends of a donor fragment may pair by chance with the sticky ends of an opened plasmid ring. This association can break up easily unless it is made permanent by adding to the solution another enzyme, DNA ligase, which fuses DNA backbones together.

Of course, sticky ends of donor and plasmid DNA form various other combinations, too. Fragments may fuse with each other, or plasmids may re-form without an inserted piece. Sometimes, researchers will spin the solution in a centrifuge to separate DNA molecules according to size and remove those that are larger than the original plasmid. This group contains the so-called recombinant DNA molecules–plasmid molecules with a piece of donor DNA inserted.

When researchers mix recombinant DNA with a special strain of live, plasmid-free *E. coli*, some of the recombinant DNA infects, or enters, the cells. The plasmid used in much recombinant DNA research contains a gene that confers resistance to the antibiotic tetracycline, so adding tetracycline eliminates all but the infected cells. The surviving cells multiply, and each produces a clone, a group of genetically identical cells, for further laboratory research. Every cell in a clone carries one or more copies of the recombinant plasmid. The entire procedure, from extraction of the donor DNA and the creation of a genetically new organism to production of the clone, is relatively simple and can be completed in less than a week.

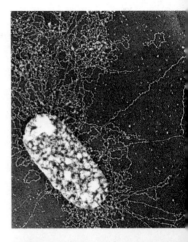

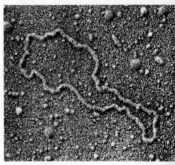

When the cell wall of *E. coli* is broken, the bacterial DNA streams out in long strands, *top*. The closed ringlike strand at the upper right is a plasmid that was in the bacteria. Another plasmid, *above*, was the gene carrier utilized in the first recombinant DNA tests.

Although research reported in February and March 1977 indicates that yeast and mold genes can function when transplanted into bacteria, genes from higher, multicellular organisms have not yet functioned when so transplanted. Gene *expression* (function), which is usually the production of a protein, normally occurs in two stages. In the first step, called transcription, the two strands of a DNA molecule separate, and one strand becomes a *template* (pattern) for the production of a ribonucleic acid (RNA) molecule that is complementary to it in base sequence. After this transcription, the DNA strand rejoins its companion strand. In the second step, called translation, the information encoded in the RNA molecule is used by the cell to manufacture a particular protein.

Many plasmids can be passed from one *E. coli* cell to another or to a related species naturally, so they are not used in recombinant DNA research. The scientists use only those plasmids that are not naturally infectious, to ensure the safety of the research.

Donor DNA can also be combined with vectors, or carriers, other than plasmids. One such vector is the DNA from the virus lambda, which infects only *E. coli*. Researchers are now developing methods for transferring genes into cells of higher organisms in the hope that some transplanted genes will function. Several other viruses, which are es-

Modifying the Chain of Life

Cutting the Chain

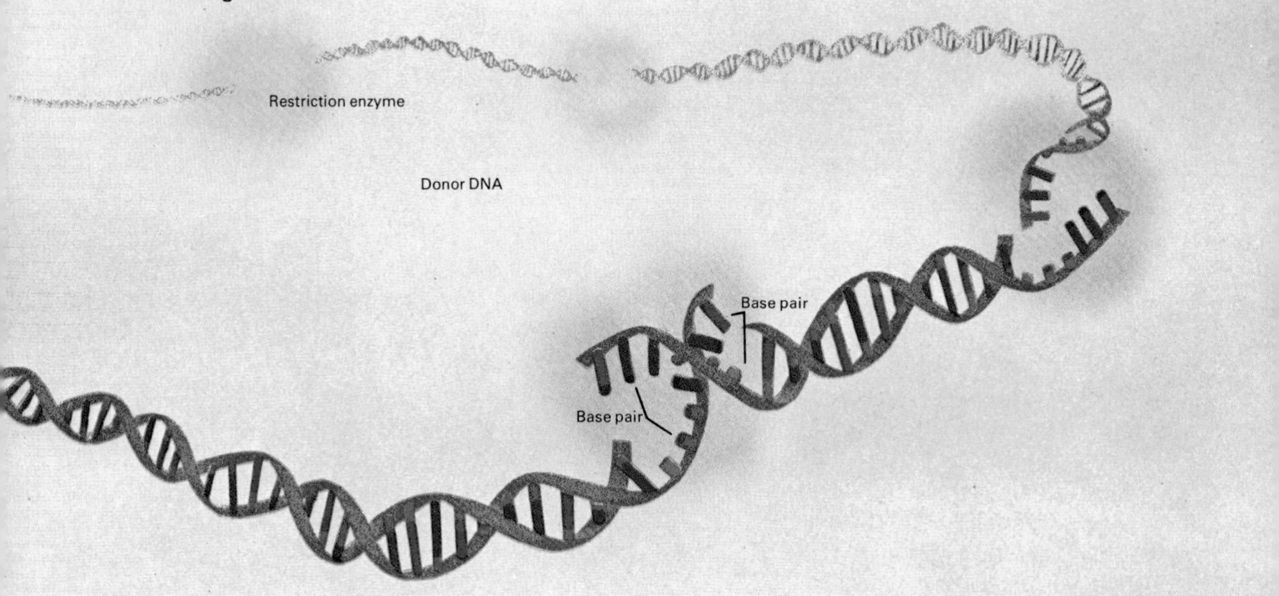

Restriction enzyme

Donor DNA

Base pair

Base pair

Welding a New Chain

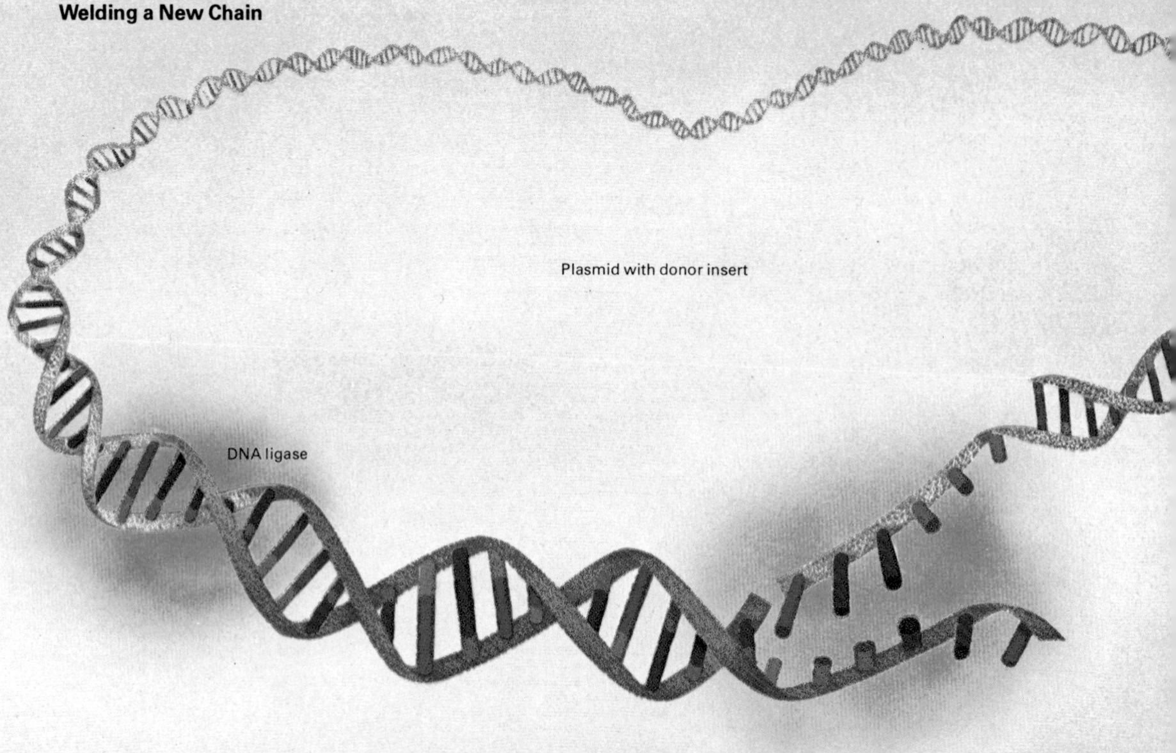

Plasmid with donor insert

DNA ligase

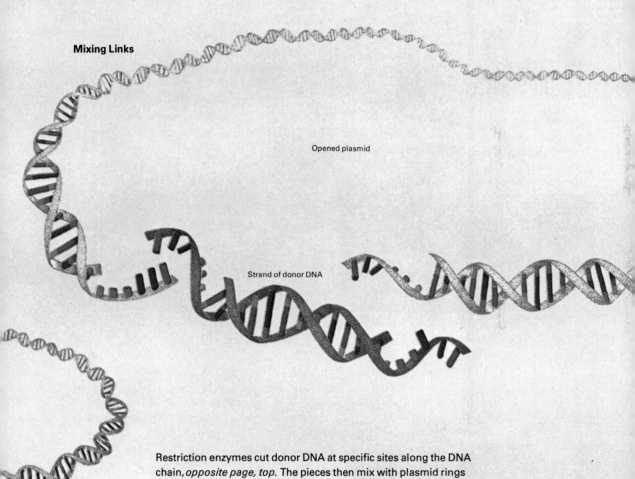

Mixing Links

Opened plasmid

Strand of donor DNA

Restriction enzymes cut donor DNA at specific sites along the DNA chain, *opposite page, top.* The pieces then mix with plasmid rings that have been opened by the same restriction enzyme, *above.* Because the same enzyme cuts both the donor DNA and the plasmid, each has short, single-strand ends that complement the other in base sequence. The chemical attraction between the complementary bases allows the donor and plasmid pieces to link up. The connection becomes permanent when another enzyme, DNA ligase, is added. This seals the base bond, opposite page, bottom, making a larger plasmid that contains genetic information from the donor organism as well as its own.

sentially little more than packets of DNA, are potential vectors. One is polyoma virus, which infects mouse and hamster cells; another is SV40 virus, which infects both monkeys and humans. The possible use of polyoma and SV40 as vectors has caused some concern, particularly because SV40 can infect human cells and evidence of SV40 infections has been found in some cases of neurological and malignant diseases in humans. But scientists have not yet found any connection between polyoma virus and human disease, nor demonstrated that SV40 can cause human disease.

In addition to the choice of vector, the method of recombination is important in weighing potential benefits and hazards. The most difficult types of recombinant DNA experiments to assess are "shotgun" experiments, in which all the donor DNA is fragmented and the pieces are allowed to fuse randomly with plasmids. Researchers have no control over which fragments might be cloned. Because many higher organisms contain DNA base sequences similar to those in cancer-causing viruses, the danger that these fragments might be transferred to *E. coli* has caused some concern. There is less potential hazard if the donor DNA used in experiments is partially purified. For example, fragments resembling DNA from cancer-causing viruses might be removed and destroyed. Even so, no experimenter can predict which of the remaining fragments will be cloned. And some of these might produce organisms with unknown and possibly harmful traits.

Pure DNA taken from *E. coli* cells drips from a glass rod, *below.* Bacterial DNA and plasmid DNA form two separate bands due to differences in density, *bottom,* after they are separated in a laboratory centrifuge.

The safest experiments are those performed with superpure DNA, the DNA of a single, known gene. Superpure DNA can be prepared using RNA transcribed from DNA in the course of gene expression. If RNA transcribed from a single gene is purified, a DNA strand complementary in base sequence to the RNA can be produced by adding to it the enzyme reverse transcriptase and the chemicals of which DNA is composed. Reverse transcriptase, as its name implies, makes the RNA reverse its normal activity in transcription and produce DNA instead of the protein coded for in the gene. From the one superpure DNA strand, a complementary strand can be produced by adding a different enzyme. Superpure DNA obtained in this manner has been used to clone the rabbit gene for hemoglobin, the major oxygen-carrying protein present in blood.

The potential benefits of recombinant DNA research are enormous. If genes from higher organisms could be induced to function in bacteria, scientists might be able to transfer such human genes as those for insulin, growth hormone, or blood-clotting factors to *E. coli.* Then, these scarce substances could be produced in enormous batches. Useful genes from agriculturally important bacteria, such as those that fix nitrogen from the air for use by legumes and other plants, could be transferred to other organisms and perhaps provide a new industrial source of nitrogen fertilizer. Researchers could use clones to obtain large quantities of certain types of DNA so that they could study the complicated base sequences of many genes. They could also use recombinant DNA to study cell differentiation and development, including such cellular abnormalities as cancer.

But there are also dangers in such work, even though most biological laboratories have strict safety rules. Disease-causing cells might accidentally escape from a laboratory and spread throughout the world. Certain naturally occurring mutant strains of *E. coli* are known to have originated in one place and spread very widely. Despite precautions, an escaped clone might transfer its plasmid to other bacterial species. Escaped polyoma or SV40 clones might be especially dangerous, even though the viral strains used in recombinant DNA research are genetically defective; they can multiply only with the aid of "helper" viruses not usually associated with them in nature.

Microbes can accidentally escape from laboratories in countless ways. Human error or carelessness is a prime cause, and they can escape in vented air or in liquid or solid wastes. Even ants, spiders, flies, or cockroaches may carry microbes out of the laboratory.

Critics suggest that natural barriers prevent a gene exchange between most species, perhaps for good biological reasons. Indeed, recombinant DNA research has already breached the greatest biological barrier of all—between the procaryotes, organisms such as bacteria and blue-green algae whose cells have no well-defined nucleus, and the eucaryotes, higher organisms whose cells have a well-defined nucleus. Toad and fruit fly DNA, for example, have already been fused with plasmid DNA and inserted into bacteria.

Some scientists feel that bridging this procaryote-eucaryote barrier, which has existed for several billion years, could have unpredictable consequences. Once created, certain self-reproducing procaryotic cells with functioning eucaryotic genes might escape, threatening animal and plant life with new forms of disease. Other scientists cite evidence that this barrier has already been breached in nature.

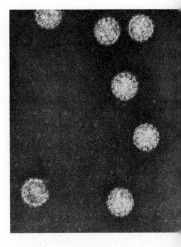

The virus SV40, magnified 200,000 times, could be a vector in recombinant DNA research. There are some scientists who are concerned because this virus can cause cancer in animals and can also infect human cells.

Nevertheless, enough concern existed so that a committee of scientists of the National Academy of Sciences took an unprecedented step. On July 26, 1974, 11 distinguished molecular biologists published a letter in the United States journal *Science* and the British journal *Nature* that called for a voluntary, worldwide ban on certain gene-transplant experiments until the "potential hazards" could be explored and safeguards could be developed. Among the signers were biologists Paul Berg and Stanley N. Cohen of Stanford University; James D. Watson of Harvard University, who shared the Nobel prize for physiology or medicine in 1962 for determining the molecular structure of DNA; and David Baltimore of the Massachusetts Institute of Technology. Baltimore shared a Nobel prize in physiology or medicine in 1975 for the discovery of reverse transcriptase.

The letter led to a four-day conference in February 1975 at the Asilomar Conference Center in Pacific Grove, Calif., at which 139 scientists from 16 countries drew up safety guidelines designed to reduce the risks of gene-transplant experiments. Some experiments, such

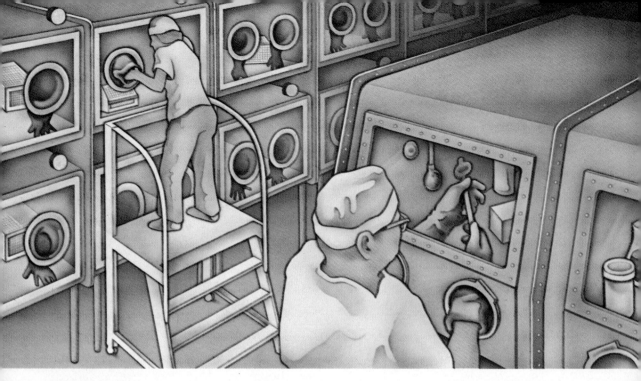

Reducing the Risks
Supersafe P4 "hot" labs
are designed for the
most dangerous DNA
work. Researchers use
glove boxes so they
never come in contact
with test organisms.
They shower and change
clothing when entering
or leaving the laboratory.
All materials and wastes
must be sterilized in the
sealed laboratory area.

as transplanting genes from cancer-causing viruses into bacteria, were judged too hazardous to be carried out at all.

The Asilomar guidelines were unofficial, but their publication in *Science* in June 1975 attracted public and governmental attention in many countries. The day after the Asilomar conference ended, the National Institutes of Health (NIH) began drafting its own official safety standards, which were published in July 1976. These guidelines provided the background for the public controversy over the research which continued into the summer of 1977. Scientists still could not agree on how serious the risks actually are.

Those who drew up the NIH research safety guidelines in 1976 were aiming at reducing the dangers of recombinant DNA research while permitting many types of experiments to continue. The NIH guidelines, which apply only to NIH-supported research, prohibit certain kinds of experiments and regulate others. Cell clones carrying recombinant DNA cannot intentionally be released into the environment, nor can they be introduced into humans or animals in clinical tests. The transfer of genes from disease-causing organisms is prohibited, as is the transfer of genes for deadly toxins, venoms, or poisons. And the NIH must review all proposed research to ensure that the projects comply fully with the standards.

The NIH safety standards established two forms of containment for any permissible gene-transplant experiments–physical containment, which refers to the security of laboratory facilities, and biological containment, which requires the use of bacteria-vector combinations that can survive only in the laboratory. The levels of containment are geared to the estimated degree of risk in each kind of experiment. In

general, the stricter the physical containment, the less strict is the biological containment.

The more complex the test organism—that is, the closer it is to humans on the evolutionary scale—the greater the presumed need for laboratory security. Thus, there are four levels of physical containment, graded from P1 to P4. P1 laboratories are those used for experiments involving the least hazardous microorganisms. Scientists can use P1 laboratories to transfer genes from species that normally exchange genes with *E. coli* and to study safe clones produced elsewhere. Recombinant DNA experiments involving multicelled plants or lower animals can be performed only in P2 laboratories. P2 labs are similar to P1 in design, but they must also have sterilizers available for decontamination of liquid and solid recombinant DNA wastes. Access to a P2 laboratory is limited to people aware of the hazards, and there are stricter rules for decontaminating materials and equipment.

Scientists can transfer genes from nonprimate mammals, birds, and animal viruses to *E. coli* only in P3 facilities. Such laboratories are designed and equipped for physical containment and are posted with biohazard warning symbols. All wastes must be decontaminated. Hazardous work must be performed in glove boxes, enclosed working spaces with rubber gloves attached which seal off the experiment so that workers never come in contact with it.

High-risk experiments with genes taken from primates, for example, can be performed only in P4 facilities, which must be completely isolated, have ventilation systems that filter exhaust air to remove dangerous material, and decontaminate wastes in special sterilizers. Those who work in P4 labs must shower and change their clothes before entering and leaving the laboratory. There are only a few P4 facilities located in the United States; one is located at Fort Detrick, Maryland, which was once the center of the United States Army's biological warfare research.

The biological containment of *E. coli* is graded in three levels, EK1 to EK3. EK1 experiments utilize the standard strain of *E. coli*, called K-12, which is used routinely in laboratories. Although this strain survives poorly in nature and normally cannot live in the intestines of humans, pigs, calves, and mice, it is possible that the organism could survive under certain conditions.

Systems certified as EK2 will be bacteria-vector systems that carry genetic defects permitting fewer than 1 cell in 100 million to survive in the natural environment. For example, some *E. coli* strains disintegrate without the help of other molecules. Only one system, X1776, was certified as an EK2 by the summer of 1977.

NIH scientists will test the genetic defects of EK2 strains as they are developed and certify as EK3 those they consider to be suitably safe. In some cases, the biological containment that is required for an experiment can be lowered one grade if the physical containment is increased one grade.

All government-supported research must adhere to the NIH guidelines, but private research and that supported by foundations or industry is currently unregulated. Pharmaceutical companies such as Upjohn Company, Eli Lilly and Company, Merck and Company, and Miles Laboratories, Incorporated, are all active in recombinant DNA research and object to an NIH guidelines requirement that clones be grown only in batches too small to be commercially useful. Other industrial companies are also interested in the research. For instance, General Electric Company is trying to develop a bacteria that will be able to digest oil slicks.

The United Nations World Health Organization, the European Molecular Biology Organization, and the International Council of Scientific Unions have each established committees to monitor safety standards for recombinant DNA research, but uniform international control is unlikely. For example, research involving mouse cells infected with cancer-causing viruses that could be done in the United States or Great Britain only with the strictest containment is currently underway in Switzerland under much more lax rules.

Critics, most of whom are scientists, argue that the guidelines are too lax, that too little is known about *E. coli*'s relationship to its environment to assess risks and that the guidelines are not universally enforceable. They say that breaking the procaryote-eucaryote barrier poses unknown hazards, and that–if experiments are to be performed at all–they should employ bacterial species that cannot infect humans. Many scientists agree, but others regard the risks as minor and worth taking, in view of the potential gains in scientific knowledge and practical benefits.

The Scientists Respond

In 1973, in a move unprecedented in the history of science, a group of microbiologists publicly called attention to the possible hazards in the work they were doing—research with recombinant DNA. In the years since then, these and other scientists, as well as some laymen, corresponded, held meetings, and drew up a set of guidelines that they hoped would at least reduce, if not remove, the hazards. The issue has continued to be highly controversial and, by the summer of 1977, it was far from resolved.

Following is a brief history of some of the significant events in the continuing debate, from the initial concern of the researchers about the recombinant technique to a discussion of the inevitability of federal legislation to control it. Each event is described by a scientist who has been closely involved.

The Steps
To Asilomar

Several of the first reports involving the use of recombinant DNA techniques were made at the Gordon Research Conference on nucleic acids held in June 1973 in New Hampton, N.H. The molecular biologists attending that conference realized that some of the recombinant DNA molecules might prove hazardous to laboratory workers and the public. As a result, co-chairmen Maxine Singer of the NIH and Dieter Soll of Yale University sent a letter outlining the scientists' concern to the presidents of the National Academy of Sciences (NAS) and the National Institute of Medicine. This letter was also published in *Science*.

I later learned from Singer that the NAS Assembly of Life Sciences favored the recommendations of the letter and that Paul Berg was going to be asked to act on them. I corresponded with Berg, and in March 1974 he replied that the NAS had contacted him and he wanted to convene a meeting.

The meeting was held on April 17, 1974, at M.I.T. It was clear that recombinant DNA techniques were becoming rather easy to use. Since they held such promise, it seemed likely that many scientists would soon be using them for recombining DNA's from a variety of sources. Since experience with the technique was still limited, it was difficult to be sure that a more pathogenic *E. coli* would not result.

We finally agreed on four recommendations which were contained in a letter published in *Science* and *Nature* in July 1974. We recommended a moratorium on two particular kinds of experiments. We also pointed out some possible dangers of shotgun experiments–breaking up whole DNA from an organism and later replicating various pieces of it. But we did not recommend deferring such experiments because the potential hazards were not clear enough. All we could agree on was to recommend that such plans be weighed carefully.

We also called for the creation of an NIH advisory committee to oversee policy development. And we asked that an international meeting be convened.

As soon as the letter was published, the planning for the Asilomar meeting began. Berg, David Baltimore, and I started to discuss what kinds of scientists should be invited. We wanted experts on the infectious properties of organisms like *E. coli*, as well as experts in animal viruses and bacterial plasmids. We developed a list of scientists who were experts in areas that we felt should be evaluated prior to the meeting.

We asked four people to head four task forces. Each selected his council of experts, held meetings, and produced reports that formed the basis of the deliberations at Asilomar.

The organizing committee consisted of Berg, Baltimore, Singer, Sidney Brenner of Great Britain, Niels Jerne of Switzerland, and myself. We met with the four task-force chairmen at M.I.T. in September 1974 to plan the program.

Meanwhile, Berg was coordinating the list of possible invitees. He asked Daniel Singer, a lawyer in Washington, D.C., to organize the nonscientist part of the program.

The Asilomar conference was held from Feb. 24 to 27, 1975. The program included results of recent research using recombinant DNA techniques, and reports from the task forces which had been evaluating the potential biohazards of different types of experiments. In addition, there was considerable discussion about biological containment.

Singer and three other lawyers discussed the social responsibilities of scientists and stressed the ethical aspects of weighing risks and benefits. They also talked about the possibilities of institutional and personal liability in the case of injuries or accidents involving the recombinant DNA technique.

The Asilomar organizing committee stayed up most of the night of February 26 writing a summary statement. A central feature of this statement was the principle that evaluation of the biohazard potential and adoption of appropriate biological and physical containment were integral parts of experiments with recombinant DNA molecules. The statement also set up a system which matched levels of presumed risk with levels of containment for a variety of recombinant DNA experiments. After considerable discussion the next day, the majority of scientists attending the meeting endorsed the statement. It subsequently became the interim policy for recombinant DNA research by vote of the NIH Program Advisory Committee.   [Richard O. Roblin III]

The author:
Richard O. Roblin III is Head, Molecular Biology of Tumor Cells Group, Basic Research Program, Frederick Cancer Research Center. He is one of the 11 scientists who signed the letters to *Science* and *Nature* recommending limits on recombinant DNA research.

Molecular biologists from around the world met in the informal and bucolic setting of the Asilomar Conference Center in Pacific Grove, Calif., in February 1975 to discuss the possible hazards of recombinant DNA research.

Shaping The NIH Guidelines

The author:
DeWitt Stetten, Jr., is Deputy Director for Science, National Institutes of Health. He was chairman of the Recombinant DNA Molecule Program Advisory Committee of the NIH that was responsible for writing the guidelines.

The day after the Asilomar conference ended, the Recombinant DNA Molecule Program Advisory Committee of the NIH, of which I was the nonvoting chairman, held its first meeting. The committee was made up predominantly of non-NIH molecular biologists and microbiologists.

We were charged by Robert S. Stone, then director of NIH, to explore the nature of the hazard, to consider what the NIH might do to control it, and to determine how the NIH would support research. The last of these was, in effect, the charge to write guidelines.

Some members of the committee were actually doing or about to do recombinant DNA experiments. One was a specialist on lambda bacterial virus, two specialized on plasmids, and one on genetics. We have since added a political scientist and a theologian-ethicist.

At our second meeting, in April 1975, we set up a subcommittee to consider experiments which the NIH itself might do to test the hazards of the technique. We decided on the general nature of the guidelines and appointed a sub-committee, chaired by David S. Hogness of Stanford University, to draw up a first draft.

We analyzed this draft at the next meeting in July at Woods Hole, Mass. The Hogness guidelines closely resembled those in the Asilomar document. They listed six or eight types of experiments not to be done under any circumstances. They also defined physical and biological containment.

We distributed the Woods Hole guidelines widely, and the resulting correspondence varied substantially. It was clear that we needed to have another look. So I appointed another sub-committee headed by Elizabeth Kutter of Evergreen State College in Olympia, Wash., to write another draft. Now we had three sets of guidelines.

Before the committee met again, we made a "variorum edition"—one in which all the known variants are recorded together. This showed us where the three sets of guidelines agreed and where they did not.

We met next in La Jolla, Calif., in December 1975 to resolve some 250 points of variation. We also sharpened up the rules for the NIH experiments.

In February 1976, Donald S. Fredrickson, the new director of NIH, convened an advisory committee which included a judge, lawyers, physicians, and a consumer expert. We presented the history of recombinant DNA research and our guidelines, and invited the committee's opinions.

By June 1976, the guidelines had cleared through the secretary of Health, Education, and Welfare (HEW). They were issued on June 23, 1976, and were published in the *Federal Register* on July 7, 1976. In the meantime, we were drafting an environmental impact statement which we published in the *Federal Register* on September 9, 1976.

NIH is processing grant applications according to the guidelines. Every applicant's institution must have a biohazards committee, which must review the facilities and the training of the applicant and certify them to us. The applicant must also file with us a Memorandum of Understanding and Agreement—a detailed statement of precisely what experiments he is going to do, under what levels of containment. He must guarantee that he has read and understands the guidelines.

In the autumn of 1976, an interagency committee was established to which many federal agencies sent representatives, with NIH as the lead agency. The committee was told to find ways to extend the NIH guidelines to cover all recombinant DNA research performed in the United States.

The interagency committee report in March 1977 recommended legislation to control, through the offices of HEW, the preparation and use of recombinant DNA molecules in the United States and listed the major elements such a law should have. HEW drew up a bill based on the recommendations and submitted it to Congress in April.

We will re-examine the NIH guidelines at least once a year. Obviously, as scientists get into their experiments, new factors will appear. Some inconsistencies in the present guidelines came to light only after scientists started doing research.   [DeWitt Stetten, Jr.]

Scientists and laymen both participated in discussions at the National Institutes of Health in Bethesda, Md., that led to the research guidelines.

Adapting the Guidelines

The president of Princeton University appointed a Committee on Bio-hazardous Research in April 1976 to suggest policy and develop guidelines for conducting research into bio-hazardous materials. As at other research institutions, scientists at Princeton have long worked with bacterial, fungal, parasitic, and viral organisms that could conceivably pose hazards. Some of this work is already regulated by various agencies, and some is not. The debate over the dangers of recombinant DNA research in a sense catalyzed the desire to be assured that Princeton researchers in the life sciences were exercising exemplary caution.

The committee was to assess the advisability and need for life scientists at Princeton to participate in such potentially biohazardous work. The possible dangers of this research had to be considered in terms of its effect on the campus and the community. Also, the professional need for Princeton scientists to work in this field was obvious.

The scientists on the 10-man committee included myself (an ecologist) as chairman, three other life scientists, a physicist, a chemist, and a geophysicist. The nonscientists included a political scientist, the director of University Health Services, and the director of the local Office of Occupational Health and Safety. It is interesting to note that only 3 of the 10 members have interests that involve biohazardous materials.

The committee sought information in two ways. First, everyone read much of the material in print up to that time, starting with the NIH guidelines and the environmental-impact statement, and including all the arguments pro and con. Second, we called for testimony from specific people and invited anyone else interested in presenting a viewpoint. Among these people were university scientists; a professor of philosophy and a professor of religion; a Princeton, N.J., resident; and a member of the Friends of the Earth. We also solicited ideas from Princeton University graduates and undergraduates.

Work with infectious organisms has always carried a risk, but it has gone on nevertheless because of the potential benefits to humanity. The committee recommended that recombinant DNA work at Princeton be considered in this same light. In addition to potential values that would accrue to scientific knowledge in general, there are two arguments specific to Princeton.

First, scientists who wish to work in this promising field would be unlikely to come to or remain at Princeton if such work were forbidden there.

Second, much recombinant work at the P3 level is already underway throughout the world. To lessen the risks already being taken requires a better understanding of the biological processes involved. Institutions and individuals of Princeton's caliber should be contributing to this understanding.

So the committee recommended that Princeton life scientists be permitted to engage in research on recombinant DNA, subject to the NIH and other guidelines. These will be modified in certain ways.

First, no P4 facility will be built at Princeton University. The amount of security necessary is not compatible with a university setting.

The P3 facilities at Princeton will be built to standards more stringent than the NIH definitions. Each will be subdivided into an inner and an outer area, with negative air pressure in both, and a pressure drop maintained between the two. Waste material cannot be drained directly to the outside. The laboratory will use autoclaves and bioguard hoods, and all outgoing air will be filtered. Access will be through double doors.

People using a P3 facility must know how to handle hazardous materials. Those who do not will be required to take a formal, noncredit, course.

The committee concluded that biological containment is more effective and predictable than physical containment. Therefore, although NIH guidelines permit, for example, a P2 facility to be exchanged for P3 when EK2 bacteria is used instead of EK1, at Princeton the higher biological containment must be maintained.

The committee recognized that public interest in recombinant DNA research and the hazards it may pose will continue to grow. We therefore recommended that Princeton University disseminate information in ways such as distributing the report of the committee and offering public lectures on the subject.   [Robert M. May]

The author:
Robert M. May is Class of 1877 Professor of Zoology at Princeton University. He served as chairman of the Princeton Committee on Bio-Hazardous Research that recommended procedures on DNA research.

43

The Public Gets Involved

The Forum of the National Academy of Sciences (NAS) was designed to deal with scientific issues with implications for society like recombinant DNA research. There are an increasing number of scientific discoveries that affect society enormously. In the past, scientists made discoveries and applied them. Later, they faced whether or not they were exposing the environment or society to danger. We now recognize that some of the applications of technology, and perhaps some of the discoveries themselves, affect society.

The layman really decides all science issues in the United States because most members of Congress are not scientists. The problem, however, is that Congress depends on the recommendations of scientists or engineers who testify on highly technical issues, and some of those issues are not clear-cut. If all scientists agree that something has a certain risk, then the layman-legislator can decide early. But on such issues as Red Dye No. 2, nuclear energy, or the technology of coal, scientists disagree on the risks.

Recombinant DNA is also an excellent example of this problem. We could either stop everything and never find out what the risks are, or proceed with some degree of caution. Exactly how much control is exerted on that caution is very important. The only way this can be decided is to get scientists and laymen together in a discussion such as the NAS Forum.

The forum has an advantage over a legislative hearing in that scientists with opposing viewpoints can make statements to and cross-examine one another. In a legislative committee hearing, the scientists appear sequentially, and cannot question one another. Furthermore, the congressmen cannot be sure that they are phrasing their questions correctly to get at the issues.

The Forum on Recombinant DNA was held from March 7 to 9, 1977, in Washington, D.C. We had Environmental Protection Agency (EPA) officials, a philosopher, and a labor leader who discussed the health hazards to laboratory workers.

One question dealt with a very technical matter—is *E. coli* the proper bacteria to use as the research vehicle? But there were other questions, such as the danger of terrorism and the role of international law, that could be discussed by laymen as well as by scientists.

The forum has two outstanding virtues. One is that people confront each other. A scientist can say to another scientist: "You have just made the statement that recombinant DNA research is a hazard to mankind, but define what you mean by 'a hazard.' Can you prove that it is a hazard?" The other virtue is that the forum produces a published record—the "book." All the testimony can be read later and a person can decide at leisure which particular lines of reasoning seem justified and which do not. And, knowing that a book will be published helps control the inclination to make exaggerated statements.

There are two major innovations to recent forums. One is the case histories. There are a number of DNA experiments that even the advocates for this work say we should not do. There are other experiments which almost everybody agrees are relatively safe. To deal with those experiments that no one is sure of, we analyzed some case histories. For example, we studied the research on exactly where specific genes are located on the chromosomes of mammals. Paul Berg outlined the potential benefits of this research based on experiments with other types of research, and Robert Sinsheimer discussed the potential risks. Another example was the synthesizing of insulin in *E. coli*. Cornelius Petting of Eli Lilly and Company talked about these benefits and Harvard biologist Ruth Gordon about the risks.

These discussions were designed to see which experiments have very little risk and large benefits, and therefore should be done; which experiments should wait until we understand more; and which experiments should not be done at all.

The second new approach is the workshop, in which people meet to confront specific problems. In one workshop at the Forum on Recombinant DNA, scientists argued about whether or not to use *E. coli*. At another, lawyers argued about how to draw up legislation. I think that those discussions allowed us to come to grips with some crucial problems that may mean the difference between having acceptable legislation and legislation that will hamper research in the future.   [Daniel E. Koshland, Jr.]

The author:
Daniel E. Koshland, Jr., is Chairman, Department of Biochemistry, University of California, Berkeley. He is also Chairman of the General Advisory Committee of the Forum of the National Academy of Sciences.

The National Academy of Sciences sponsored a Forum on Recombinant DNA in March 1977 in Washington, D.C. Included were technical explanations, *top,* panel discussions, *above left,* and statements by scientists, such as Paul Berg, *above,* who initiated the precautions. People opposing the research supplemented vocal disapproval with signs, *left.*

Some Reservations

One of my major concerns about recombinant DNA research is that we are intervening in the evolutionary process. We are making available the entire gene pool of the planet for this research. For example, in the so-called shotgun experiments we apply a restriction enzyme to an entire set of DNA from drosophila, the fruit fly. This breaks the DNA into 20,000 or 30,000 pieces, each containing some group of genes which are put into bacteria to multiply.

Some scientists want to do that with pieces of higher organisms – rats or mice or human beings. It is no harder to insert monkey genes than it is to insert fruit fly genes. According to the NIH guidelines, you just have to take more precautions. Conversely, bacterial genes can be and have been put into monkey cells in culture.

There are safer techniques that can, in part, replace the shotgun technique. For example, Har Gobind Khorana completed synthesizing a gene at M.I.T. in 1976 and inserted it into bacteria. He and his colleagues knew exactly what they put in. Another example is the experiment in which you start with the messenger RNA for a specific protein, let us say hemoglobin. You make a DNA copy of that RNA, then insert it in bacteria to multiply. The RNA is probably not pure and you may have a few things with it that you are not quite sure about. It is also tedious, time-consuming, and much more difficult. But it is a lot safer than a shotgun experiment.

Let us look at another potential danger in this work. The exchange of genes in nature is very limited, and primarily within species. If we consider just the microbial world, we do not know what kinds of new properties we will give to these organisms by introducing all kinds of unknown genes. We might convert a normally harmless organism into a toxic one. Even if the genes themselves are not initially hazardous, you do not know how they are going to evolve.

Genes are changed in mutations, of course, but mutations take place one step at a time. In recombinant DNA experiments, you are suddenly providing whole blocks of genes, a far greater jump. We do not know enough about evolution to know what could emerge.

People tend to forget that human beings have evolved into a certain eco-logical niche. No microbial or other enemies have prevented us from occupying that niche, or destroyed us on the way into it. And we depend completely on all kinds of biological equilibriums that we do not control. We depend on the bacteria that restore carbon and nitrogen and purify our wastes, and we depend on plants that supply the oxygen in the air. We may be sowing the seeds for future human enemies, pathogens, or we may be sowing the seeds that might destroy some of the biological balances on which we depend.

I have suggested that all work that involves the creation of genetic combinations between species that do not ordinarily combine be done in P-4 facilities. This precaution would provide the maximum degree of physical containment. It would also limit this work to a few places where it could be supervised as carefully as possible.

I would also like to find alternatives to *E. coli* that would have a much more restricted environment and could transfer DNA to a much more limited range of organisms. The use of *E. coli* can be made safer by steps such as those taken in developing the EK2, but even those organisms could be provided with more sophisticated safeguards.

There are people who point out that nature in all its evolutionary wisdom has still permitted smallpox and various other diseases. That is true enough. Furthermore, the pathogens that we know now are certainly not the only ones that could be created, and in time new ones undoubtedly will appear out of mutations. We know of viruses that can only attack humans and must have appeared since humans appeared. That is part of evolution, and it has an effect on the whole evolutionary pool, too.

I am not arguing that we have no right to intervene in evolution. We already have, from the beginning of agriculture. But we should not intervene in a blunderbuss way with unpredictable consequences.

Furthermore, we need a broad social view, not just a scientific view. I am bothered by the benefit-risk assumption in which we simply add up benefits and risks and do not worry about who is getting the benefits and who is taking the risks. They are not necessarily the same people. [Robert L. Sinsheimer]

The author:
Robert L. Sinsheimer is Chancellor, University of California, Santa Cruz, and former Chairman of the Biology Department at California Institute of Technology.

Assessments For the Future

Edward M. Kennedy, senior senator from Massachusetts, is one of several congressmen who has sought testimony in order to write legislation that will permit DNA work to continue with reasonable precautions.

The author:
Clifford Grobstein is Vice-Chancellor for University Relations, University of California, San Diego.

The NIH guidelines are a good beginning but are hardly the end of the recombinant DNA policy story. They represent a partial moratorium with a significant degree of regulation. But something more is needed.

In talking about recombinant DNA hazards or benefits, we are talking about probabilities and uncertainties. For example, Sinsheimer is concerned with unprecedented crossing of the barrier between procaryotes and eucaryotes. But Bernard D. Davis, of the Harvard Medical School, argues that this crossing probably has already taken place many times in nature. Davis' view is based on fragmentary data that should be expanded to give us greater certainty on this issue.

Instead of arguing from incomplete data we ought to be scientifically investigating such issues to help us make decisions. I call this issue-oriented science, as distinguished from applied and basic science. It is becoming more important in many technical-political issues.

Many people believe that much of what is covered by the guidelines will turn out to be without hazard at all. So they expect the regulatory mechanism to wither away. I am inclined to doubt that. Even if recombinant DNA technology itself turns out to be relatively safe, the issues that are being raised under the broad heading of genetic engineering are not going to go away.

I think there is an even deeper issue—the increasing power we have to accomplish our own purposes. When these purposes are backed by the ability to modify our "inner selves" to "improve" the species, people feel threatened.

We need a comprehensive assessment of the recombinant DNA issue, preferably under the auspices of the federal government. We definitely need monitoring to find out where the research is going. The assessment should be made by people who are knowledgeable not only technically, but also about the structure and values of our society. In addition to scientists, the evaluators should include economists, philosophers, political scientists and others, each looking at this research from their own particular expertise. They should become familiar with the science and the people who do it, and try to project consequences, both good and bad. They should also try to establish priorities for practical applications.

This assessment might result in a guiding program or policy that would be used by organizations such as federal agencies, and for discussion with people of other countries.

I see the need for a comprehensive assessment lasting, perhaps, for two years. I expect that out of it might come a permanent commission. It might be called the Commission on Genetic Resource Management.

Private industry poses one of the chief questions that such a committee would have to consider. It may be desirable to declare a moratorium on private industrial application of recombinant DNA techniques for a time. Perhaps we should simply say that there will be no commercial application of this research for a certain number of years, long enough to discourage any industrial investment until we have reduced uncertainty.

It appears that no existing laws not specifically aimed at recombinant DNA carry the authority to deal fully with it. Accordingly, new legislation is being drafted that will extend regulation to all recombinant DNA research and use in both public and private sectors.

I would like to see the legislation take advantage of a number of new elements on the policymaking scene in Washington, D.C. The Office of Technology Assessment, for example, could act as a center to advise on legislative activity on recombinant DNA research. The re-established Office of Science and Technology Policy in the White House could also be used. And legislative committees and executive agencies such as the NIH or the National Science Foundation certainly ought to be involved.

At whatever level the assessment is made, it should result in federal legislation that covers everybody. It will have to be written to cut off the variety of local and state controls that have begun to appear, if we are to avoid complication and confusion.

There will undoubtedly be a procedure for licensing or registering each research project. The agency chosen to administrate the legislation will have to have the power to monitor all research and to revoke a license if the guidelines are violated or the work poses an unforeseen hazard.   [Clifford Grobstein]

47

The Two Faces Of Diabetes

By Thomas H. Maugh II

Having learned that diabetes is two diseases posing as one, scientists are seeking better ways to control the deadly sugar imbalances common to both forms

Margaret was 60 years old and 112 pounds overweight. She was also a diabetic who had been taking large doses of insulin for 18 years in a fruitless attempt to control the high blood-sugar levels that are characteristic of diabetes. Her physician referred her to John K. Davidson, an internist at Emory University School of Medicine's Grady Memorial Hospital in Atlanta, Ga.

Davidson admitted Margaret to the hospital, put her on a fast, and cut her insulin dose in half. Within three days, her blood sugar fell to an abnormally low level, and Davidson took her off insulin completely. After two weeks of fasting, Margaret was released from the hospital and put on an 800-calorie-per-day diet. She lost the 112 pounds in less than a year and has not used insulin since. That was in 1972. In 1977, she had no symptoms of diabetes.

In 1972, Davidson also treated a different kind of diabetic. Sue Ellen, an Atlanta high school junior, lost 15 pounds over a two-week period, even though she was eating and drinking much more than usual. Then she collapsed one evening while attending a football game and was rushed to Grady Hospital. Davidson determined almost immediately that Sue Ellen was in a diabetic coma. He quickly gave her an injection of insulin and intravenous fluids.

Within hours, Sue Ellen's condition stabilized and she regained consciousness. Davidson, a nurse, and a dietitian worked out a diet designed to maintain Sue Ellen's proper body weight. They also taught her how to inject herself with insulin. Sue Ellen later married and had a child. Unlike Margaret, Sue Ellen requires insulin injections twice a day. Otherwise, she is healthy.

Both women suffer from diabetes mellitus, a disorder of the process in which the hormones insulin and glucagon regulate sugar concentrations in the blood. Insulin is secreted by beta cells, specialized cells in the pancreas. The hormone regulates the transfer of energy-producing sugars from the blood to insulin-sensitive target tissues, such as liver, fat, and muscle. Insulin also helps regulate the synthesis of proteins and fats that are essential to the normal growth and repair of tissues. Glucagon, secreted by specialized pancreas cells that are called alpha cells, acts in the opposite way. Along with other hormones, glucagon regulates the liver's conversion of proteins and fats to sugars that are then released into the blood.

If the body fails to produce enough insulin or if insulin-sensitive tissues do not respond properly, sugars are removed from the blood more slowly than normal. As a result, high concentrations of sugar build up in the blood after eating. This condition is called hyperglycemia. And when insulin is not being used properly by the body, the concentration of glucagon in the blood increases. Scientists do not understand why this occurs, but the increased concentration of glucagon stimulates the liver to burn up protein and fats from tissues throughout the body, releasing sugars into the blood and producing persistent hyperglycemia.

The increased breakdown of fats and proteins stimulates the liver to produce biochemicals called ketone bodies. An abnormally high concentration of ketone bodies in the blood produces ketoacidosis, a condition marked principally by increased acidity of the blood. This leads to diabetic coma and—if the victim does not receive medical attention rather quickly—to death.

The case histories of Sue Ellen and Margaret illustrate some of the major differences between what scientists now recognize as two diseases that once masqueraded as one. Juvenile-onset and maturity-onset diabetes strike at different stages of life, have different causes, present different symptoms, and require different forms of therapy.

Sue Ellen has juvenile-onset diabetes, which is usually found in people under age 20. Margaret has maturity-onset diabetes, which most often affects people over 35. There are more than 1 million juvenile-onset diabetics in the United States and at least 9 million maturity-onset diabetics.

The National Commission on Diabetes, a panel appointed by Congress to study the problem, reported in 1975 that diabetes and its complications cause more than 300,000 deaths in the United States each year, making it the third leading cause of death, behind heart disease and cancer. The number of diabetics in the United States is doubling every 15 years. A newborn child now faces a 1-in-5 chance of

The author:
Thomas H. Maugh II is a staff writer for *Science* magazine.

developing diabetes. Ironically, this increase is a direct result of improvements in the treatment of the disease. Diabetics who once might have died young or during periods of stress, such as giving birth, are now living relatively normal lives. They are bearing children who are more likely to develop diabetes.

The commission's figures indicate that diabetes does not strike equally at everyone. Women are 50 per cent more likely than men to develop diabetes; nonwhites, 20 per cent more likely than whites; and the poor, three times more likely than the affluent. Studies of population groups having a high incidence of diabetes—such as the Pima Indians in Arizona—suggest the possibility of hereditary susceptibility, but there is no firm evidence to support this.

Researchers have been making great progress in learning about both forms of the disease. They now believe that juvenile-onset diabetes is probably caused by a viral infection. Epidemiologists have frequently shown that there is a close correlation between viral infections and the beginning of juvenile-onset diabetes. For example, Harry A. Sultz of the State University of New York at Buffalo studied the post-World War II medical records of people in Erie County, New York. In 1974, he reported that outbreaks of mumps in the county were followed about 3.8 years later by a high incidence of juvenile-onset diabetes. Other investigators have found similar correlations with measles, hepatitis, and upper respiratory infections.

Researchers have proved that viruses can cause diabetes in experimental animals by destroying pancreas beta cells. Around 1970, virologists John E. Craighead of the University of Vermont and Abner E. Notkins of the National Institute of Dental Research demonstrated that certain animal viruses can produce diabetes in rodents. Perhaps even more important, English virologists D. Robert Gamble and K. W. Taylor demonstrated in 1973 that human Coxsackie viruses, which produce upper respiratory infections, can also produce diabetes in susceptible strains of mice. United States virologists reported in January 1976 that they had confirmed this finding.

Gamble and Taylor also collected a great deal of evidence indicating that there is a close relationship between Coxsackie infections and juvenile-onset diabetes in humans. Gamble argues that Coxsackie viruses are probably the major cause of this disease. But he says it is also possible that Coxsackie along with several other viruses, such as mumps, could damage a victim's beta cells, and juvenile-onset diabetes might occur after a series of such damaging infections.

Obviously, not everyone infected by Coxsackie and other viruses develops diabetes. So it seems likely that a genetic factor is also involved. Some researchers believe that a genetic defect prevents the immune system of some individuals from effectively fighting off viral infections before they do the damage that triggers diabetes. Other investigators argue that a genetic defect may make the immune system respond too aggressively to a viral infection, so that the white blood

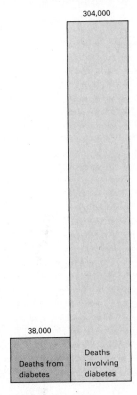

Diabetes' Hidden Toll

304,000

38,000

Deaths from diabetes

Deaths involving diabetes

Diabetes ranks officially as the fifth leading cause of death in the U.S., claiming 38,000 lives annually. However, the National Commission on Diabetes estimates that when the disease is counted as a major underlying factor in other fatal conditions, such as cardiovascular deaths, the toll reaches 304,000 deaths, making diabetes the third leading killer.

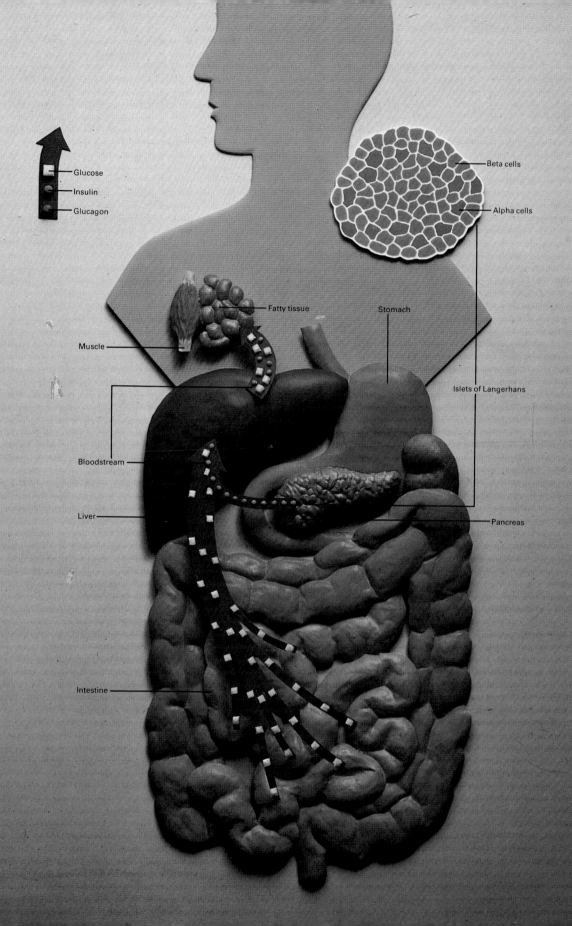

Glucose

Insulin

Glucagon

Beta cells

Alpha cells

Fatty tissue

Stomach

Muscle

Islets of Langerhans

Bloodstream

Liver

Pancreas

Intestine

Sugar on the Scales of Life

In normal individuals, food digested in the intestine releases sugar in the form of glucose into the bloodstream, *left.* This rise in blood-sugar levels causes beta cells in the pancreas to release insulin, which helps transport sugar from the blood to such tissues as fat, muscle, and the liver, where it is stored. During a period of not eating—for example, between dinner and breakfast—alpha cells in the pancreas secrete glucagon into the blood, *below.* Glucagon regulates the liver's conversion of proteins and stored sugar into energy-producing sugar that the blood carries to such vital organs as the brain.

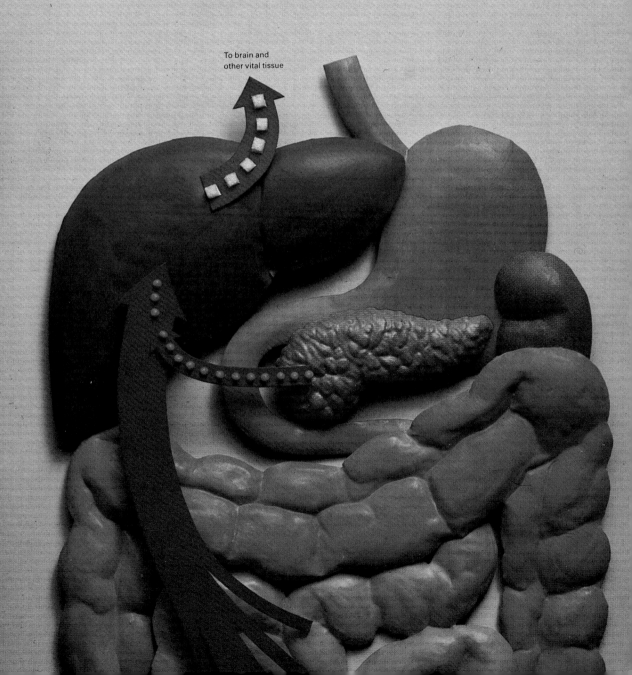

To brain and other vital tissue

cells that would normally attack only the virus also attack beta cells. There is not yet enough evidence to indicate which, if either, of these theories is correct, but the results are the same in either case. The body can no longer produce its own insulin, and the delicate balance between storage and release of sugars is severely disrupted. An individual can go from apparent health to a diabetic coma in a few days. Such sudden change is usually triggered by stress.

Juvenile-onset diabetes is treated with injections of insulin derived from the pancreases of cows and pigs. The patient's diet must also be controlled so that the minimum amount of insulin is required and so that concentrations of sugar in the blood do not fluctuate too widely.

With careful medical treatment, juvenile-onset diabetics can live relatively normal lives for many years. However, nearly all suffer eventually from complications such as impaired vision, deadening of nerves in the arms and legs, and kidney failure. Most researchers attribute these complications to persistently high concentrations of sugar—and perhaps of insulin received during the treatment—in the patient's blood. Therefore, some scientists are looking for more precise ways of controlling blood sugar and insulin levels.

Treatment for juvenile-onset diabetics may be improved someday by the use of the hormone somatostatin, discovered in 1972 by physiologist Roger C. L. Guillemin and his associates at the Salk Institute in San Diego. The main function of this hormone is to control growth. However, the researchers discovered that it also slows the secretion of both insulin and glucagon. Patients will not develop hyperglycemia from insufficient quantities of insulin if their release of glucagon is somehow inhibited. In tests on diabetics, researchers have administered somatostatin which reduced the high concentrations of glucagon. Perhaps even more important, endocrinologist John E. Gerich and his colleagues at the University of California Medical School in San Francisco have demonstrated that somatostatin is much more effective than insulin in treating the coma-producing ketoacidosis, because it inhibits the formation of ketone bodies almost immediately.

There are several problems involved with the use of somatostatin, however. It breaks down in the bloodstream very rapidly and therefore must be administered continuously by intravenous means. Also, because its main function is to inhibit the release of growth hormone along with other hormones, its effects must be carefully monitored in young people who are still growing. Finally, relatively limited amounts of somatostatin are available. Researchers are trying to develop synthetic forms of somatostatin that can be made in large quantities, will last longer in the body, and will not affect the growth hormone. Thus, it may be many years before there is any significant therapeutic use of somatostatin.

Another possible way to control juvenile-onset diabetes is the pancreas transplant. Surgeon Richard C. Lillehei of the University of Minnesota Medical School performed the first pancreas transplant in

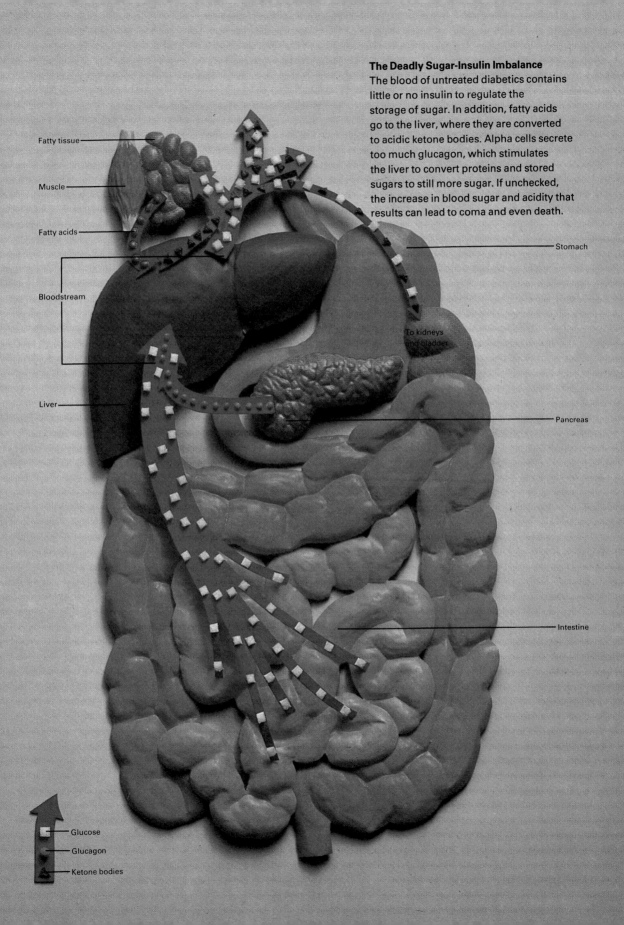

The Deadly Sugar-Insulin Imbalance
The blood of untreated diabetics contains little or no insulin to regulate the storage of sugar. In addition, fatty acids go to the liver, where they are converted to acidic ketone bodies. Alpha cells secrete too much glucagon, which stimulates the liver to convert proteins and stored sugars to still more sugar. If unchecked, the increase in blood sugar and acidity that results can lead to coma and even death.

Fatty tissue

Muscle

Fatty acids

Bloodstream

Liver

Stomach

To kidneys and bladder

Pancreas

Intestine

Glucose

Glucagon

Ketone bodies

December 1966. By the end of 1976, surgeons had performed 51 pancreas transplants on 49 patients in the United States and five other countries. Several patients lived for a year or more with the transplant, and one survived nearly five years. All the other patients died within a much shorter period, however. Some rejected the transplant. But most of the patients, already weakened from the complications of diabetes, could not withstand the surgery and the powerful drugs used to prevent their immune systems from rejecting the foreign pancreas.

Another problem is that certain tissues composing the pancreas, called exocrine tissues, secrete extremely corrosive digestive enzymes that can harm healthy tissues. These enzymes must be disposed of in a transplant operation. The most effective method seems to be by constructing an artificial passageway to shunt the enzymes from the new pancreas into the intestine, through which they can be eliminated from the body. However, surgeons must construct the passageway in advance and allow it to heal before the pancreas is transplanted.

The alpha and beta cells essential to blood-sugar control are located in small areas scattered throughout the pancreas called the Islets of Langerhans. Many researchers now believe it is better to transplant only the islets. Some, such as surgeon John S. Najarian of the University of Minnesota Medical School and pathologist Paul E. Lacy of the Washington University School of Medicine in St. Louis, have used rodents to show that islets can function normally when transplanted into genetically similar animals. The researchers injected fluid containing the islets into a major blood vessel, and the islets then found their way to the liver, where they functioned as though they were in the pancreas. In the process of separation, many islets are lost or destroyed, so pancreases from several donors are needed for each transplant. In addition, transplanted islets provoke a strong immune response that requires large amounts of immunosuppressive drugs.

Internist Josiah Brown of the University of California Medical School at Los Angeles and anatomist Orion D. Hegre of the University of Minnesota School of Medicine think that these problems can be solved by using pancreases from fetuses. The fetal pancreas contains less exocrine tissue in proportion to islets than the adult pancreas and can be grown in culture to produce a substantial increase in the number of islets. Both Brown and Hegre brought blood-sugar levels to normal in diabetic adult rats by giving them islet transplants from genetically similar rat fetuses.

In 1976, Brown transplanted one whole fetal pancreas that had been temporarily implanted in a healthy adult rat for about three weeks. The transplant to a diabetic adult rat reversed the disease. However, the rejection problem is still very severe when the fetuses used are not genetically similar to the recipient.

Lacy and Najarian have isolated significant quantities of human islets, and Najarian has tried to transplant them into humans. But the problems associated with immune suppression are so severe that pan-

receptors makes the cells more resistant to the effects of insulin and prevents the cells from being overwhelmed by high levels of the hormone in the bloodstream.

This adaptive mechanism is not perfect, however. When a new equilibrium between high insulin levels and fewer receptors is established, the metabolism—the process of turning food into energy—of the target cells becomes less efficient. The results are similar to those that develop from the decreased fuel efficiency of an automobile engine when the car is traveling faster than its optimum cruising speed. Just as the reduced efficiency of the automobile engine often exposes mechanical defects or tuning problems, the cell's decreased efficiency exposes latent metabolic defects. In people who are prone to diabetes, the high concentration of circulating insulin that accompanies obesity makes latent diabetes become overt.

Generally, physicians have been giving obese maturity-onset diabetics much the same treatment as juvenile-onset diabetics. The patients frequently receive insulin or oral antidiabetic agents, some of which are thought to stimulate production of insulin by the pancreas. The drugs may overcome insulin resistance in some patients for a time, but eventually this treatment makes the cells more insulin resistant. Also, the drugs increase patients' appetites so that they gain more weight, thereby making the problem even worse.

Roth and other researchers discovered this cycle can be stopped by restricting food consumption. Dieting causes both the number of receptors on the target cells and the concentration of insulin in the blood to return to normal. And the insulin-sugar balance stays normal as long as maturity-onset diabetics restrict their caloric intake or keep their weight at the proper level. Exercise also reduces insulin resistance, but scientists do not yet know why.

Davidson and his colleagues at Emory University have stopped insulin therapy in more than 1,000 maturity-onset diabetics since 1971, using a diet instead. Up to 95 per cent of the patients stick to the regimen and report that they feel better than ever. Doctors at Emory University now generally reserve insulin therapy for juvenile-onset diabetics, any diabetic in a coma, or such special maturity-onset cases as diabetic pregnant women.

The results of this work with diet indicate the importance of detecting maturity-onset diabetes as early as possible. If individuals knew they were susceptible to maturity-onset diabetes, they could adjust their diet, maintain the proper body weight, and probably never develop the full-blown disease. But there is still no generally accepted way to detect a genetic tendency toward diabetes. Biochemist Melvin Blecher and his associates at the Georgetown University Medical Center reported in 1975 that certain apparently healthy children of diabetics have a defect in the binding of insulin to white blood cells. Blecher believes this may indicate that these young people will develop maturity-onset diabetes. If he is right, the binding defect may someday be used as a predictor of the disease. For now, however, the only guideline for predicting who might develop the disease is that children of maturity-onset diabetics are much more likely to develop diabetes than are children of nondiabetics.

Maturity-onset diabetics who successfully lose weight and juvenile-onset diabetics who eventually will receive pancreas transplants or artificial sugar-regulating devices will be far less likely to suffer from such complications of diabetes as blindness, nerve damage, or atherosclerosis (hardening of the arteries). A growing body of evidence suggests that most of these complications may result from uncontrolled hyperglycemia, from continually high concentrations of insulin, or from a combination of both.

Davidson believes that sustained hyperglycemia may cause nerve damage. Many of the obese maturity-onset diabetics he has treated suffered severe pain in their legs because of nerve damage. He found that a reduction in their weight and the accompanying return of blood sugar to normal levels frequently brought a remarkable improvement in their condition after 12 to 18 months. Such improvement has never been consistently demonstrated with other forms of therapy.

Dealing with Diabetes

Mary Tyler Moore

Ron Santo

Thousands of diabetics have overcome the physical and psychological problems of their illness to live active, and even extraordinary, lives. They range from homemakers to executives to show-business personalities and even athletes.

The television actress Mary Tyler Moore, for example, learned she was diabetic in 1969, at the age of 32. She had just suffered a miscarriage, and the doctors treating her discovered that the concentration of sugar in her blood was far above normal.

At first, Moore found her illness difficult to accept, so she sought professional counseling to help her cope emotionally. She knew nothing about diabetes and feared that she would end up as a hopeless invalid. However, she soon discovered that this was not true. Moore exercised regularly, went on a strict diet, and learned to inject herself with insulin. Like other diabetics, she found that the disease is fairly easy to live with.

Such outstanding athletes as baseball player Ron Santo, tennis champion Billy Talbert, and hockey player Bobby Clarke became sports heroes in spite of diabetes. Santo, now an executive for a Midwestern oil company, learned he was diabetic when he was 18, right after signing his first professional baseball contract. For years, he kept his illness secret. "I didn't want people to think I was an invalid, to feel sorry for me," he says. "I wanted to prove myself before I told anyone." After gaining baseball stardom, Santo told the public about his disease and became active in efforts to help diabetic children.

In contrast, Clarke made no secret of the fact that he had been diabetic since the age of 15. When he became eligible in 1969 for the National Hockey League (NHL) draft, most teams rejected him because of his condition. Finally, he was drafted by the Philadelphia Flyers. Clarke went on to become captain of the team and twice won the NHL's Most Valuable Player award.

Such famous diabetics serve as examples of how the handicaps of the disease can be overcome. "When I talk to youngsters with the disease, I tell them we diabetics are really very lucky," says Santo. "There are lots of diseases you can't control, but this is one you can." [Darlene R. Stille]

High concentrations of insulin may be a factor in the development of atherosclerosis in diabetics. Studies have shown that diabetics who receive the largest concentrations of insulin also have the highest incidence of atherosclerosis. A similar correlation is also observed in obese nondiabetics who, because of their overweight condition, have high insulin levels. Also, internists Edwin L. Bierman of the University of Washington Medical School and Robert W. Stout of the Queen's University of Belfast in Northern Ireland found that smooth muscle cells from arteries grown in culture proliferate when they are exposed to insulin. Such proliferation of cells is believed to be the first step in the development of atherosclerosis.

Impaired vision caused by diabetic retinopathy afflicts more than 95 per cent of all diabetics who have had the disease for at least 25 years. In retinopathy, new blood vessels form over the retina, rupture, and leak blood into the normally clear vitreous humor, a jellylike substance that fills the eyeball. Scar tissue can also form when the ruptured blood vessels heal, pulling on the retina and eventually detaching it from the back of the eye.

Ophthalmologist Alan L. Shabo of the University of California School of Medicine in Los Angeles suggests that an immune response to insulin may cause these effects. Shabo repeatedly injected rhesus monkeys with beef insulin to make them unusually sensitive to the hormone. He then injected small quantities of insulin into the vitreous humor of the animals' eyes. Their immune responses to the insulin, he found, produced most of the symptoms of the worst form of retinopathy. He observed no symptoms, however, when he injected insulin into the eyes of animals that had not been given repeated injections of insulin. Shabo believes that insulin in the blood of insulin-dependent diabetics could produce the same effects he observed in the monkeys. The greater the amount of insulin in the blood, the greater would be the damage to the eye.

Despite all these discoveries, there are still many questions about diabetes to be answered. For instance, are the Coxsackie viruses a primary cause of juvenile-onset diabetes? If so, would immunization against the viruses reduce the incidence of diabetes? What precise role does the immune system play in the development of juvenile-onset diabetes, and what is the nature of the genetic defect in maturity-onset diabetes? What can be done to make pancreas and islet transplants more successful?

To find the answers to such difficult questions, the National Commission on Diabetes recommended that funding for diabetes research be increased from the $43 million spent in 1975 to $142 million in 1980. The commission also recommended that all new funding go to basic research on preventing and curing diabetes. With this increased funding and greater emphasis on basic research, some of these questions about diabetes could be answered much sooner than anyone would have predicted even five years ago.

Harnessing Earth's Fountains of Fire

By Thomas R. McGetchin

**In the search for ways to predict volcanic eruptions,
scientists are learning how to control lava flows
and tap the huge quantities of energy volcanoes contain**

The ground around La Soufrière volcano on Guadeloupe island in the French West Indies began to rumble and shake in July 1976, and small eruptions that produced ash and blasts of steam gurgled within its crater. The smoldering, trembling mountain concerned many of the more than 70,000 city dwellers and farmers who live in Basse-Terre at the base of the volcano and on its slopes. They recalled stories of the violent 1902 explosion of Mont Pelée on Martinique, two small islands away. That holocaust destroyed the coastal town of St. Pierre and killed all but one of its 38,000 inhabitants within minutes.

The Basse-Terre officials ordered immediate evacuation. Soon the roads were clogged with trucks and cars taking people and their most precious possessions to other parts of the island or to boats that would take them to other islands in the French West Indies.

But moving against the flow of inhabitants was a small band of foreigners intent on reaching La Soufrière as quickly as possible. They had come from all over the world, carrying heavily loaded backpacks and instrument cases containing magnetometers, gravimeters, tiltmeters, and seismographs. They were earth scientists assembling to collect and analyze samples from the volcano and to listen with their sophisticated, sensitive instruments to the variety of gurgles, groans,

A brilliant display from the volcanic eruption of the Icelandic island
of Heimaey lights the night sky around the village of Vestmannaeyjar.

Steam and ash poured from fissures at the summit of La Soufrière volcano on Guadeloupe island in 1976. The awakening volcano drew scientists from around the world to study it.

The author:
Thomas R. McGetchin is a geologist with the Invar Science Institute in Houston. He specializes in the study of volcanoes.

and pops emanating from within. A volcano's activity can provide a wide variety of information from which scientists can determine the composition, temperature, and gas content of the earth's interior.

Earth scientists hope to learn how to predict violent eruptions and thus reduce the toll in lives and property. They are also studying the possibilities of tapping the enormous amount of heat expended in eruptions and the even larger amount stored under and near volcanoes. Experimental drilling is now underway in the United States and Russia to determine if we can turn this heat into useful energy. In addition, many of the world's important ore deposits lie in or near volcanoes, because volcanic activity draws pure ores to the earth's surface from deep in its interior. Models of how volcanoes work and evolve have thus become important tools in mineral exploration.

La Soufrière and other volcanoes of its type have episodic cycles of activity; in some cases, ash eruptions grow in intensity until they culminate in a blast of fragments and searing gas that can destroy everything within a range of 10 to 20 kilometers (6 to 12 miles). The violent expansion of gas-charged molten rock sometimes tears a mountain apart, and the cloud of hot dust produced in such an eruption can drift great distances and kill many living things before it cools and dissipates. Even then, part of the cloud will remain in the atmosphere, drifting around the world for years. Climatologists and atmospheric physicists use high-altitude airplanes to collect samples of this dust for laboratory analysis. The composition and the amount of it in the atmosphere provide valuable data for inputs to computers that model the patterns of air movement and circulation of high-altitude dust particles in the upper atmosphere.

During the last 500 years, volcanoes have killed an average of 360 persons per year. Some of the more dramatic volcanic explosions produced catastrophes at Vesuvius, Italy, in A.D. 79; at Krakatoa, Indonesia, in 1883; and at Mont Pelée in 1902.

Volcanoes vary in size from small bulges with craters a few tens of meters across to the gigantic Mons Olympus on Mars. The tallest known volcano in the solar system, Mons Olympus is about three times as tall as the highest mountain on earth and as big as the state of New Mexico. The differences in size and shape of volcanoes are due to the total volume, consistency, and flow rate of the lava; the force with which material comes out of the volcano; and the amount of ash present. Hawaiian lava, for example, which hardens to black basaltic rock, is so fluid it looks almost like syrup as it flows down the sides of the Mauna Loa and Kilauea craters. This kind of lava builds up the crater walls as it cools, forming shallow-sloped mountains, shaped like nearly flat shields. In an eruption, Mauna Loa can add up to 15 meters (50 feet) to the height of its shield. Other lavas, such as the dark gray andesites of Santiaguito Peak in Guatemala, have a much thicker consistency and form bulbous domes. Santiaguito Peak is growing at a rate of only a few meters a year. Such lavas also once came from the volcanoes in the Cascade Range in Washington.

Volcanoes and earthquakes often occur together. Both result from the movement of the giant plates that make up the rigid outer shell of the earth. The continents lie embedded in these plates, which make up the earth's crust and, below, part of its mantle. The plates move about slowly in relation to each other, at a rate of only about 1.3 to 10 centimeters (½ to 4 inches) per year. Most earthquakes and volcanic eruptions occur at the edges of these giant plates where they push

Almost all of the most active volcanoes are located near the edges of the 20 giant plates that form the earth's crust. The rest well up through the center of the plates, forming volcanic islands such as those in Hawaii.

Major Active Volcanoes and Plate Boundaries

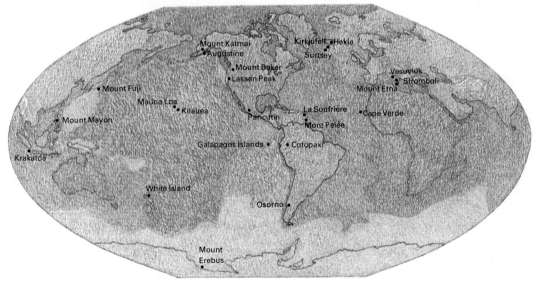

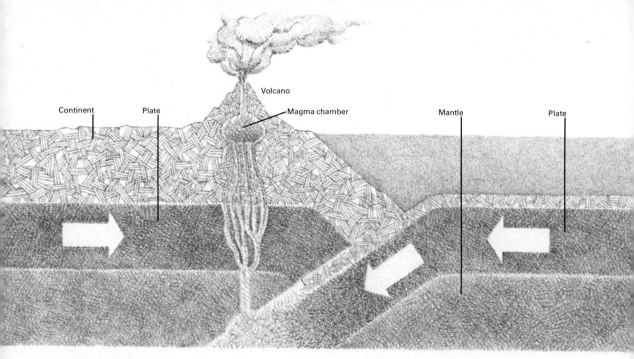

Continent Plate Volcano Magma chamber Mantle Plate

The Birth of Volcanoes

Volcanoes found near the edges of the giant plates are formed when a submerging plate's edge melts and the magma rises to the surface. Volcanoes in the middle of a plate are formed by deep, giant hot spots that heat large quantities of rock, creating magma that works its way through the earth's crust. This molten rock also oozes through the cracks between the separating plates at the mid-ocean ridges.

against each other or are pulled apart. Geologists describe the motion of the plates and the consequences of such motion as plate tectonics.

About 400 of the world's 600 to 700 active volcanoes are found near plate edges. They form there because the edge of one plate is heated as geologic forces push it down and under another plate. The melted rock of the severely strained and submerged plate edge then rises to the earth's surface to form a volcano.

Where two plates slowly pull apart, vast amounts of volcanic material ooze up to fill the growing crack and harden, adding to the plate edges. The folds of oozing lava build ridges at the plate edges which form a huge mountain system on the ocean floor. The top of this mountain system can be seen in Iceland where the Mid-Atlantic Ridge is exposed on dry land.

The present Mid-Atlantic Ridge is some 16,000 kilometers (10,000 miles) long. In a real sense, this is one continuous volcano erupting along a line instead of at one point. This underwater volcano continuously pumps out staggering amounts of energy and raw materials from the earth's interior. Though this underwater crack constitutes only a tiny part of the earth's total surface area, about 20 per cent of the earth's heat escapes from it.

Some volcanoes are also found far from plate edges. Here, for reasons that are as yet unknown, a channelway from the interior of the earth penetrates the plate and reaches the surface. The Hawaiian volcanoes are examples of midplate craters.

A rare but interesting type of volcano that also occurs far from the plate edges is the diamond-bearing kimberlite pipe. A diamond is a form of very hard carbon produced under great pressure deep in the earth. A kimberlite volcano has a single episode consisting of at most a

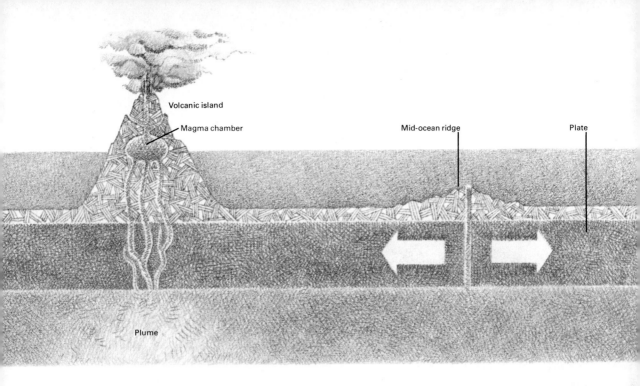

Volcanic island

Magma chamber

Mid-ocean ridge

Plate

Plume

few eruptions, bringing diamonds to the surface in violent gas-driven blasts that originate in a high-pressure gas pocket at depths of from 200 to 300 kilometers (120 to 180 miles). Although no one has seen a kimberlite pipe erupt, the violence of the eruptions can be reconstructed from the jumble of broken rocks that clog the volcano's vent, or crater. Diamond mines are man-made excavations, as deep as 1,050 meters (3,500 feet), into these volcanic vents.

With my colleague Wayne Ullrich, I spent several years calculating how kimberlite volcanoes operate. We believe that material from deep in the earth flows up through the vents in much the same manner as fuel travels through a rocket nozzle, except that there are many more solids than gas, and the temperatures are cooler. The gas, mainly water vapor and carbon dioxide confined at high pressures by the earth's crust, forces its way to the surface in a violent eruption. The vents flare outward like the bell on a trumpet as they near the surface. The explosion cracks the earth's surface and scatters ash and huge chunks of rock for many kilometers around the vent.

Kimberlite pipes contain an intriguing variety of rocks. Many are *xenoliths* (foreign rocks) that are either torn from the vent walls by the explosion or formed deep in the earth where the diamonds are formed. Experts estimate that these fragments are forced to the surface from 80 to 160 kilometers (50 to 100 miles) below in a quick trip that takes only about two hours.

Geologists have spent a great deal of time investigating kimberlite pipes because these volcanoes contain a variety of fragments from the earth's lower crust and upper mantle that can be examined and studied. Such types of rock are found only in and around these pipes, and are one of the best sources of direct data about the earth's interior.

A Kimberlite Pipe Blows
Unique volcanoes called kimberlite pipes are essentially vents in the earth's crust. They are drilled by gas-driven debris, including diamonds, from deep in the earth's interior.

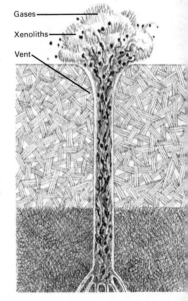

Gases

Xenoliths

Vent

67

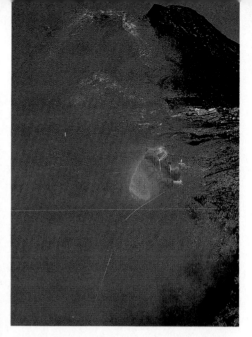

Hawaiian Volcano Observatory geologists use
a seismograph to measure movements of the
earth, *left;* take gas samples for analysis, *above;*
and insert a thermocouple into the center of
an active lava pool to measure its heat, *below.*

Pressures equal to those deep in the earth have been duplicated in laboratories by shock waves produced by explosives and high-velocity impacts. Studies of rocks and minerals under such laboratory conditions have revealed much about the behavior of materials deep in the earth. But our understanding of volcanoes is still sketchy and speculative. Our ability to predict eruptions is based largely on observations of their periodic activity rather than on any fundamental understanding of their physical and chemical processes or on exact mathematical models of their behavior.

Volcanic eruptions in isolated areas, where most take place, are little more than curiosities to most people; they present little danger to life or property. But we need to determine exactly when volcanoes that lie near cities and towns, such as La Soufrière, may erupt. And we also need to monitor the temperature and other conditions of extinct volcanoes near population centers, such as Mount Baker in Washington. Mount Baker periodically heats up internally. It may not become active again in our lifetime, but such heating can cause massive mudflows of ancient volcanic debris and ice that lie on the mountainsides. Flows of this sort, called lahars, have long been a recognized part of the geologic history of the Pacific Northwest. Tacoma, Wash., is built on a lahar that flowed from Mount Rainier in prehistoric times.

Should a lahar occur on Mount Baker, it would suddenly slide down on two hydroelectric dams and the small village of Concrete at the mouth of a narrow canyon below Mount Baker's east flank. Water levels were lowered in the dams for eight months in 1975 and 1976 and access to camping areas below the glaciers was restricted because the mountain's temperature rose as indicated by steam activity at the

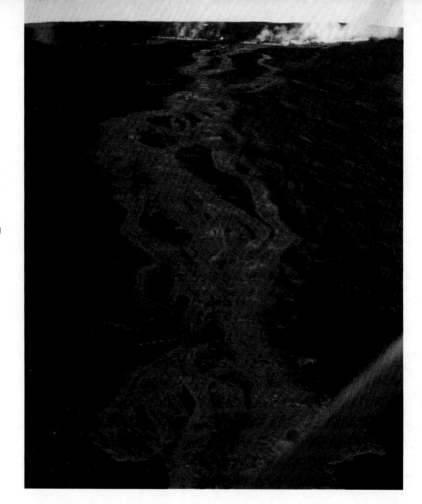

Red-hot lava flows like syrup down the side of the Mauna Loa volcano, burning and smothering everything in its path.

summit vent and an increase in the production of melted water in streams beneath the glacier at the summit. Because the activity subsided toward the end of 1976, it is likely that the heat supply has been dissipated and the danger is probably over.

Hawaiian volcanoes are among the most thoroughly studied and systematically monitored active craters in the world. Unlike the potentially violent but rare eruptions on Guadeloupe and Martinique, the Hawaiian volcanoes erupt so frequently that the pre-eruption signals are well recognized. Typically, deep earthquakes are registered on the tiltmeters and seismographs of the Hawaiian Volcano Observatory on Kilauea and Mauna Loa about three weeks before the eruptions occur. In time, the quakes work their way toward the surface, heralding the movement of molten rock in the crater vent. Later, the mountain swells enough to be measured by tiltmeters. Just before it erupts, the mountaintop deflates and another type of earthquake, called a harmonic tremor, occurs. Scientists believe a harmonic tremor is caused by the lava near the top of the volcano being forced through an underground horizontal dike system that branches out from the main crater vent. Deflation, following a long period of mountain swelling and coupled with the onset of a harmonic tremor, has

Hot wax poured on a scale model of Mauna Loa helps researchers at the Los Alamos Scientific Laboratory to study the path of lava flow and ways the lava might be diverted or otherwise controlled.

become a reliable indicator that a Hawaiian eruption is imminent. The same pre-eruption signals do not necessarily apply to any other volcanoes, because different volcanoes will have a great many different kinds of underground plumbing.

Mauna Loa awoke from a 25-year silence in July 1975, filling its summit-pit crater with lava. Some of the melt spilled over the rim and ran down the north slope. Since then, small earthquakes and swelling of the mountain indicate that Mauna Loa's plumbing system is again active. Studies indicate that the next eruption of volcanic lava from Mauna Loa may occur high on its east flank, where outbreaks occurred in 1880 and 1942. Hilo, a city of 26,000 and the island of Hawaii's principal port, lies at the bottom of a broad valley between Mauna Loa and Mauna Kea. Lava erupted from Mauna Loa's east flank would flow down into this valley toward Hilo. The 1880 flow reached the outskirts of Hilo, and part of the city is now built on top of that hardened flow. The 1880 flow, larger of the two, was about average as Mauna Loa eruptions go; it contained some 300 million cubic meters (354 million cubic yards) of lava. In 1950, Mauna Loa erupted at least 600 million cubic meters (708 million cubic yards) of lava, which flowed westward into the Pacific. Geologists of the U.S. Geological

Steam rises as hoses spray cold water on lava in an effort to cool it, *left,* and stop its flow toward Vestmannaeyjar, Iceland. Bulldozers also pile up barriers of hard lava, *above,* to block the flow.

Survey and Hawaiian Volcano Observatory anticipate another eruption within about two years.

Kilauea, Mauna Loa's nearby little sister, has erupted nearly every year for several decades. But its eruptions are so mild that people run toward it with cameras rather than fleeing for their lives.

Neither volcano is as dangerous and unpredictable as Guadeloupe's La Soufrière. La Soufrière did not erupt as expected in 1976. But because of the research work, the French government convened a commission in November. The commission concluded that a dangerous eruption at La Soufrière could be detected in advance with the improved monitoring program that geologists have established. As a result of this decision, evacuees returned to their homes on December 1.

Scientists made dramatic advances in predicting eruptions in 1976 at two isolated volcanoes in the Arctic. Augustine volcano, on an island in Cook Inlet, Alaska, staged a spectacular series of moderate-sized ash eruptions in January and February that culminated on February 6 with a vigorous blast of lava and ash. The ash flows destroyed a crude scientific station on the island, and the cloud was easily visible 125 kilometers (75 miles) away in Anchorage. For about a year prior to this eruption, a University of Alaska research team had been studying the chemical evolution of Augustine's rocks, quakes in the area, and the geology of previous lava and ash deposits.

Jurgen Kienle, a young Swiss-American geophysicist, noticed an odd pattern in the earthquake frequency data for Alaska that correlated with the development of Augustine's growing eruption episodes. It related to ocean tide cycles in a fairly simple way. The amplitude of the tides varies during the month; it is largest at new and full moon. Kienle found that the eruptions coincided with the minimum high tides. Apparently, the lunar tides helped to trigger the volcanic eruptions in some unknown way. Perhaps the most appealing of Kienle's several theories to explain this is that the tides are pumping the molten rock back and forth in an unstable, gas-charged chamber in the volcano, causing it to fracture. The pumping is most vigorous at maximum high tide and may cause minor fractures through which some gas in the underground chamber can escape. But the pumping is less efficient at the minimum high tide, and fractures do not develop. Then the gas pressure builds up until the volcano erupts.

Across the Bering Strait, a team of Russian volcanologists from the Institute of Volcanology, led by S. A. Fedotov and seismologist Pavel Takarev, watched intently through June and early July 1976 as earthquake frequency grew along a volcanically active section of the Kamchatka Peninsula. After a thorough study of the earthquake data, the volcanologists predicted both the time and place of an eruption. Geologic teams and camera crews assembled, waited, swatted black flies and cursed the mud and makeshift roads, but they were not disappointed. When basaltic lava burst spectacularly from a fissure 30 kilometers (19 miles) long on July 6, the camera crews captured some of the most dramatic volcanic-eruption films obtained anywhere. The prediction, made on the basis of earth tremors, was a great triumph for the Russian scientists.

Besides making advances in prediction, geologists have experimented with ways to control and divert the flow of lava from an erupting volcano so that it does no damage. During the Heimaey island eruption in 1973 in Iceland, large naval fire pumps were used to pour seawater on the advancing lava flow. The water chilled and hardened the lava, retarding its flow toward the island's harbor. The economically crucial fishing port of Vestmannaeyjar was saved in this way, though its residents were temporarily evacuated. Man-made barriers have diverted lava flows at Mount Etna in Sicily and at the Aso volcano in Japan. The U.S. Army Corps of Engineers also is designing a system of rock and mud lava-diversion barriers on Mauna Loa's east flank to protect Hilo.

Another control method is the use of high explosives to collapse the lava-tube system that commonly forms in highly fluid Hawaiian lava flows. These tubes, many as large as subway tunnels, develop as complex branching systems under the cooled solid crust of a lava flow and are filled with large quantities of flowing lava when a volcano is erupting. Scientists assume that bombing would collapse the roofs of these tunnels and the lava would cool and harden when exposed to the air.

Although an enormous amount of volcanic material can pour from a crater, only a small fraction of the molten material in the volcano actually reaches the surface. Much more remains inside to cool slowly. This cooling and hardening mass contains sulfur, water, and a number of gases, as well as various metals including copper, gold, molybdenum, tin, tungsten, and zinc.

Natural geothermal wells are used to produce electrical power in Wairakei, New Zealand, *above.* In experimental design for a heat well, *right,* cold water will be heated in deep holes near a young, dormant volcano in New Mexico.

Tapping the Earth's Heat

Power station

Hot water   Cold water

Hot material containing all these elements continues to rise from deep in the earth long after the volcano stops its surface eruptions and is considered extinct. The new material mixes with the older magma and ground water to form a rich broth of metals and gases. The continuing pressures of the rising material squeeze the broth out into the cracks and veins around the volcano's base, where it cools and hardens into rich mineralized zones of copper, lead, zinc, and other metals. The rich copper mines of the American Southwest and the Peruvian Andes were formed in this way, and recently discovered metal deposits in New Guinea were found by drilling into extinct volcanoes. Gold, mercury, and sulfur, which tend to stay in liquid form longer than most other metals, are forced into the cracks nearest the earth's surface where they can be mined relatively easily.

Geologists hope eventually to use the heat energy in volcanoes to produce electrical power. The Energy Research and Development Administration (ERDA) laboratory at Sandia, N. Mex., has been experimenting with ideas and methods for drilling into the molten rock chambers of active volcanoes and inserting a heat exchanger. There is enough heat in a relatively small volcano to generate energy for a city as big as San Francisco for about 60 years, even if no new hot rock rises into the chamber. Needless to say, there are enormous engineering problems to solve in converting the heat into useful energy. The Sandia laboratory, in cooperation with the Hawaiian Volcano Observatory and a number of university scientists, has recently conducted a series of small-scale drilling and geophysical experiments in the lava lakes at Kilauea. These scientists also are doing experimental work in large laboratory tanks, producing rock melts and synthetic magmas with which to study the properties of such molten material under controlled conditions.

Another idea for using volcanic geothermal heat is soon to be demonstrated at the ERDA laboratory in Los Alamos, N. Mex. The Los Alamos scientists have drilled twin holes 3 kilometers (1.9 miles) deep into granite in northern New Mexico near the edge of the Valles Caldera, a large but dormant volcano similar to those found in Yellowstone National Park. They have created a large fracture at the bottom of one of the drill holes by pumping fluid into it at high pressure—a technique called hydraulic fracturing that is commonly used in oil fields. In theory, when cool water is forced down the first hole, it will flow over the hot surface of the fracture, become heated, and will then flow to the surface through the second hole. The hot water will then go through a heat exchanger to produce energy. The scheme is particularly appealing because it is a closed-loop system in which water is recycled; it does not require natural hydrothermal water in the ground, as do the geothermal heating systems used to heat homes in Reykjavík, Iceland.

A test of the Valles Caldera system appears to be only two or three years away. Estimates of the power output of a single operating loop are about 100 megawatts of electricity, so that a group of 10 or 12 twin-hole loops would produce about 1,000 megawatts, the average consumption of a large metropolitan city such as San Francisco. There are from 15 to 25 sites in the Western United States that geologists consider suitable for the development of these systems.

The materials dissolved from the rocks by the circulating hot water might also be used. These include ores of copper, uranium, silver, gold, molybdenum, zinc, and lead. They will cake on the walls of the heat exchanger, from which they could be removed easily during periodic shut-downs.

Volcanoes, then, provide us with both challenges and opportunities. We may soon be able to predict eruptions of most known active or temporarily dormant volcanoes, saving lives and property. And, in the process of studying volcanoes, we may be able to harness the energy and mine the wealth of minerals that these awesome natural wonders bring to the surface from deep within the earth.

Signals from Inner Space

By Barbara B. Brown

Using sophisticated biofeedback devices, we may learn to control the innermost workings of our bodies to cure or prevent a wide range of maladies

Imagine that you are a bright young student who has developed a problem about taking exams. Although you know the material, you get the jitters when exam time approaches. Your neck becomes painfully tense, your head throbs, and your memory often blanks out during the test. These are symptoms of a fairly common student ailment, exam anxiety. To ease these symptoms, your school psychologist or physician might suggest that you should try a revolutionary type of treatment, biofeedback.

This is a method of giving people information about what is going on inside the body, and it may be one of the most dramatic developments in the history of Western medicine. Using biofeedback, you can learn how to regulate internal bodily activities—heart rate, blood pressure, muscle tension, sweating, brain waves, perhaps every function

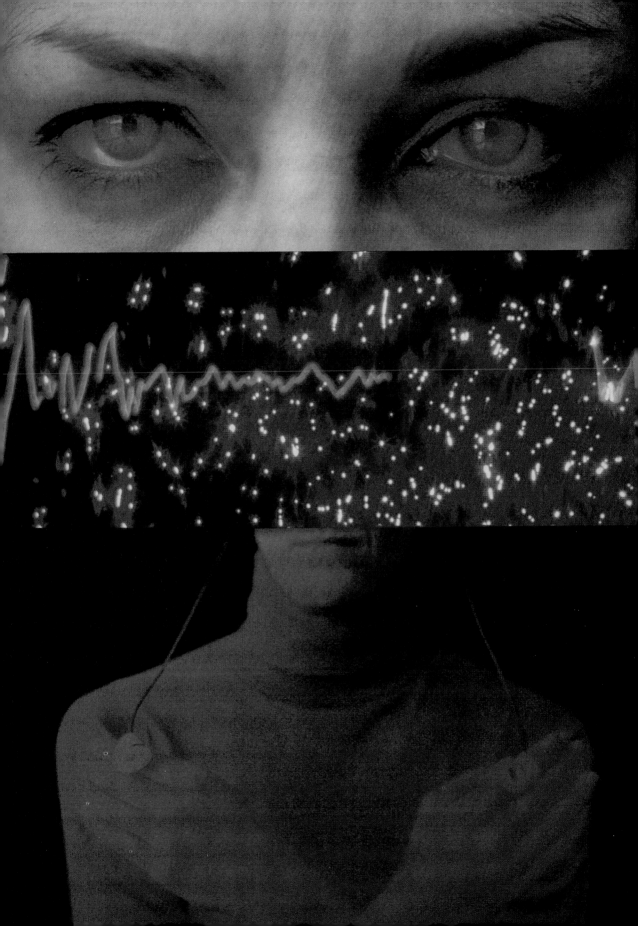

The pervasive effects of anxiety on the body, caused by the stress of modern life, may account for from 60 to 80 per cent of all ills.

down to those within the cells themselves. This seemingly mysterious ability is achieved through a combination of applied physiology, psychology, and sophisticated electronic equipment that can pick up signals within the body and feed them back to you.

At your first biofeedback session, the technician introduces you to an instrument that records muscle tension, a subtle tautness of which you are unaware. The technician pastes sensors, small metal disks, over certain muscles, perhaps on your head or neck. Wires relay electrical activity detected in your muscles by the sensors to a biofeedback instrument that amplifies the activity and converts it to a biofeedback signal. The signal can be a tone of changing pitch, a light that changes in brightness, or a meter whose needle swings back and forth as your muscles tense or relax.

In this case, the biofeedback therapist hooks you up to a meter on which you can read numbers from 1 to 20. The needle swings between 10 and 15, indicating that your muscles are very tense. The therapist then tells you to use some mental process to bring the needle down.

You probably have no idea how to do this. So the therapist suggests some techniques for using your mind—which we can define as conscious or subconscious awareness—to affect your bodily processes. In one technique, called Progressive Relaxation, you alternately tense and relax groups of muscles to become aware of degrees of muscle tension and of the sensation of relaxation. In another, called Autogenic Training, you suggest to yourself that various parts of your body feel heavy, warm, relaxed, and calm. You might also try to calm your thoughts by visualizing relaxing scenes or by meditation.

These mental techniques bear certain similarities to yoga and self-hypnosis. However, there are two important differences. First, an instrument constantly monitors the effectiveness of the biofeedback technique. When yoga or self-hypnosis are used alone, there is no immediate way to determine how effective they are. Second, the biofeedback signal itself is a powerful aid in learning to direct bodily functions. When the signal tells the brain what is going on inside the body, some unconscious mental process acts to alter internal functions. In this process, your brain apparently sorts out and evaluates the biofeedback information, then directs messages along nerves to change the level of muscle tension.

The author:
Barbara B. Brown, a pharmacologist and biofeedback researcher, has written and edited several books about biofeedback techniques.

Learning to manipulate internal bodily functions is quite a different experience for most people. So, in addition to teaching you mental techniques, the therapist will help you by giving you psychological support—encouragement and praise when you make the needle move.

By practicing with the machine and doing relaxation exercises at home, you learn how to regulate your unfelt tension, just as you would regulate any familiar muscle activity, such as moving your legs to walk. After perhaps 5 to 10 training sessions, you develop an awareness of the previously unfelt muscle tension and are able to relax at will, without the machine.

The therapist might then advise you to practice regulating muscle tension while imagining that you are taking an exam. Eventually, you learn how to keep your muscle tension at the proper level under all conditions. You prevent it from giving you headaches and causing your mind to go blank. But you also keep a little tension during a test because it is useful in staying alert and productive.

Even though you now know how to control an unfelt bodily function, you do not know exactly how you do it. The mental processes that take over automatic functions usually controlled by the autonomic nervous system are all subconscious. I call this a "biological awareness" that knows the state of the body at all times. This awareness can trigger some mechanism to alter bodily functions after a conscious decision is made to do so. The decision, however, must be based on information consciously obtained through one of the senses.

The technique of biofeedback came to light only in the late 1960s, and researchers have been busy ever since exploring its possible applications in the treatment and prevention of disease. Although biofeedback itself is new, experiments with relaxation treatment have been going on since the 1920s. Physiologist Edmund Jacobson of Harvard University, for example, theorized that tension was an underlying cause of many illnesses. He also realized that body muscles could remain tense even when a person was resting. He believed that learning deep relaxation of the muscles could be helpful in treating illness. In the 1920s, he developed the Progressive Relaxation technique. Also in the 1920s, German neurologist and psychiatrist Johann H. Schultz developed Autogenic Training.

Now, however, the idea of controlling life-sustaining activities within the body has captured the fancy of scientists and nonscientists alike. Eastern mystics have long claimed that they can stop or slow such vital organs as their hearts. But Western medicine had maintained that the activities of the internal organs, automatically regulated by primitive lower-brain areas and the autonomic nervous system, were beyond voluntary control.

The simplest example of control over an automatic function is blinking and winking. Blinking is an automatic reflex activity that distributes moisture over the eyeball. Winking is conscious control of this up-and-down movement of the eyelid. Children learn to wink fairly easily. And they can use a simple form of biofeedback—looking in a mirror—to help them.

When you feel your own pulse or take your own temperature or blood pressure, you are receiving information directly about parts of your body that you do not normally control. For example, your pulse provides important information about your heartbeat through the sense of touch as your fingers feel the rhythmic throbbing in your wrist. But most internal bodily functions cannot be detected directly through the senses, so electronic devices are used to transmit the signals in a form that one of the senses can recognize.

One such device is the instrument you used to relieve exam anxiety. This biofeedback machine is similar to an electromyogram (EMG), which records muscle activity. Special biofeedback instruments have been developed to provide continuous information about other bodily functions. For example, blood-flow meters help people with blood-circulation problems learn to control the diameter of blood vessels so that more blood can flow to their hands and feet. Electronic thermo-meters measure skin temperature, reflecting both blood flow and mus-cle tension. An electroencephalograph (EEG) is used to pick up brain waves. Researchers have found that persons suffering from anxiety generate few of the long, slow brain waves called alpha and theta. But these waves are abundant in persons experiencing the tranquillity of certain forms of meditation. In an experimental form of biofeedback, ulcer patients swallow sensors that measure and give them information about the gastric acid level in the stomach.

Biofeedback techniques grew out of the work of researchers in sev-eral disciplines—psychology, neurophysiology, muscle rehabilitation, and cybernetics, the comparative study of animal and mechanical systems and how they are controlled automatically. In the late 1960s, these researchers—most of them unknown to each other—were con-ducting experiments leading to the same conclusion: Some mental process can control internal functions if the person is somehow made aware of these functions.

Experimental psychologists claimed that animals could be trained to develop control over such automatic functions as heart rate and blood pressure. Some physicians experimented with showing partially paralyzed patients that remnants of their muscles were undamaged. The patients watched the electrical activity of their muscle cells on an oscilloscope, a device with a televisionlike screen, thereby learning which groups of cells to exercise and strengthen. But the muscle-reha-bilitation researchers did not develop a biofeedback technique.

Biofeedback work began when researchers let patients see their bod-ily activities being recorded by electronic medical instruments and asked them to change the recorded signal. These researchers found that all the average person needs in order to learn how to control an internal function is reasonably accurate information about the func-tion and a request or instruction to change it.

In my own case, I was conducting experiments on the effect of color on brain waves. In my laboratory at the Veterans Administration Hospital in Sepulveda, Calif., I attached electrodes to the scalps of volunteers and made EEG recordings of changes in their brain wave patterns when they were shown lights in a variety of colors. I then began to wonder if the experiment would work in reverse. Could the volunteers' various brain waves be used to turn on different colored lights? I began with a single blue light that glowed when the person hooked up to the EEG was producing alpha waves. But it soon became apparent that something else was going on. The wiggly line produced

on the EEG graph told me that the volunteer was producing more and more alpha waves. One person after another produced the same result. The conclusion about why this happened became obvious–the volunteers were receiving information about their own brain waves from watching the dimming and brightening of the blue light and using this information to control the amount of alpha waves their brain produced. Then I began to hear that researchers in other fields had discovered similar effects.

The scientists investigating this phenomenon, like me, were relatively isolated. Psychologists communicated with other psychologists but not with neurophysiologists, and vice versa. In September 1969, I determined to bring these biofeedback researchers together and organized the first meeting, in Los Angeles, of what became the Biofeedback Research Society. By 1977, there were branches of this organization in almost every state.

At the meeting, we all became aware of the great extent of research on this phenomenon. For example, anatomist John V. Basmajian of Emory University Medical School in Atlanta, Ga., reported that it was possible to control the activity of a single nerve cell. Basmajian's work centered around nerve cells in the spinal cord. A single such nerve cell controls from three to several hundred muscle cells connected to the nerve cell by nerve fibers, which carry the nerve cell's commands. Basmajian knew that the electrical activity of each nerve cell and its connected muscle cells produces a unique electronic signature that he could monitor with sensitive electrodes attached to the skin above the muscle cells.

So that he would be monitoring as few cell groups as possible, Basmajian used tiny electrodes 25 micromillimeters in diameter. He wired these electrodes to an instrument that converted the electronic signatures to rhythmic "bips." Each cell group had a rhythm of its own. After listening for a few minutes, a volunteer could pick out individual rhythms from the jumble of bips and learn to manipulate the sounds. In effect, the volunteer was turning the nerve cells that activated the muscle cells on and off at will. None of the volunteers knew how they did this. They experienced no sensations and had no clues about what they were doing, except the sound of the rhythmic bips.

Beginning in 1969, I took time out from my own experiments to review reported research on biofeedback and examine the results. Eventually, I brought together much of the work that had been scattered throughout various scientific disciplines in two books, *New Mind, New Body* (1974) and *Stress and the Art of Biofeedback* (1977).

From the clinical and experimental findings of many scientists, it appears that biofeedback is a valuable technique for treating a wide range of diseases, particularly stress-related illness, whether simple tension headache or complex emotional illnesses, such as depression. It

Imagining peaceful scenes while listening to a biofeedback signal can aid in learning to control bodily functions that are usually automatic.

shows promise as a treatment for drug abuse and other sociological problems, and for psychosomatic disorders—mental disturbances that result in physical symptoms, such as high blood pressure or ulcers. All of these problems have symptoms in common—an anxiety underlying physical or emotional tension, or both.

Many scientists believe that the pervasive effects of anxiety on the body account for many of today's illnesses. Various medical experts have estimated that from 60 to 80 per cent of all illness today can be traced to the stress of modern life. There are four major categories of illnesses—emotional, psychosomatic, physical, and those of unknown cause. Only physical disorders, such as infections or injuries, are not triggered by stress. Illnesses of unknown cause may or may not be. Regardless of origin, all illnesses and disorders are aggravated and magnified by stress. In fact, just the stress that stems from being sick can severely impede recovery.

In the early 1970s, psychologist and electrical engineer Thomas Budzynski and psychophysiologist Johann Stoyva of the University of Colorado Medical Center in Denver did experiments to determine whether deep relaxation combined with biofeedback would be beneficial in treating stress-related illness. For their study, they specifically chose people suffering from tension headache, because muscle tension was definitely the cause of the pain, and anxiety and stress were very likely the underlying factors.

Making EMG recordings of volunteers suffering from tension headaches, Budzynski and Stoyva found that tension was particularly high in their forehead muscles. The researchers then divided the volunteers into three groups. One group received biofeedback training in learning to relax the forehead muscle. They listened through earphones to a tone that indicated when muscle tension there increased or decreased. Another group listened to random low tones that were not biofeedback, and the third group received no treatment. The three groups kept detailed records of their headaches during the study. After four weeks, the volunteers treated with biofeedback experienced greatly reduced muscle tension and a 75 per cent drop in tension headaches. However, there was very little change in the tension levels of those in the other two groups.

Although subsequent research has shown that biofeedback does indeed relieve tension headache, this same research has raised questions about the role of the forehead muscle in bringing about this relief. It could be that biofeedback from this muscle draws attention to the general tension level of all the head muscles and that the brain uses this to relax those muscles.

Beginning with Jacobson, many researchers have come to believe that if muscles are deeply relaxed, anxiety cannot exist. Following through on this concept, a number of experimenters and therapists in

External electronic sensors from biofeedback machines pick up internal information, such as skin temperature, muscle tension, and brain waves.

the 1970s successfully combined EMG biofeedback with relaxation techniques to reduce anxiety in some patients.

How does reducing tension in the skeletal muscles relieve anxiety? Physiologically, it is a complicated story. Muscle tension is regulated automatically because of certain internal control systems. These systems use information derived from special sensory nerve cells in the muscles and tendons. Nerve sensors detect levels of muscle tension and relay this information along sensory nerves to muscle-control areas in the midbrain and lower brain. These brain centers evaluate the information to determine the appropriate action. If adjustments in muscle tension are required, the control areas send impulses down the motor nerves telling the muscles what changes to make. All this activity is unseen and unfelt. However, when thoughts or emotions change muscle tension, higher, more complex, areas of the brain take over and impose their decisions on the automatic regulating systems.

Anxiety does this. In states of anxiety, the higher brain areas tell the automatic control centers in the lower brain and autonomic nervous system to stay alert and keep the muscles prepared for action. But at the same time, sensory nerves are signaling the automatic control centers that the muscles are already tense, and this heightens the sensation of anxiety. The higher brain areas are so preoccupied with the situation causing the anxiety that they have little or no conscious awareness of the body's tense condition.

Anxiety is an extension of the normal alertness that occurs when we are uncertain about the outcome of a situation. There are many degrees of anxiety, conscious or subconscious, but they all develop from fear of the unknown. And this fear alerts the body to danger, preparing it to take some action for survival. This is called the fight-or-flight response. The body first tenses its muscles in a way identical to the "get ready, get set" state before the "go" signal. The body's internal systems come to the aid of the muscles. The heart and lungs work harder to supply more oxygen through the bloodstream. Other functions, such as digestion, stop or slow down drastically.

These are the kinds of bodily changes that cause butterflies in the stomach and dryness in the mouth. This reaction is clearly felt and quite understandable if we step off a curb and see a truck bearing down at high speed. But psychologists claim that we can become conditioned to reacting in a similar way to relatively minor threats, such as taking exams. These reactions to minor worries can snowball into serious anxieties and severe disturbances of bodily functions. Persons suffering from anxiety cannot fully relax even in sleep. They may think they are resting, but their muscles remain tense.

A number of biofeedback researchers theorize that in learning to control muscle tension using biofeedback, the mind receives information about the unfelt tension from the biofeedback signal and uses this

After biofeedback training, a person is aware of previously unfelt muscle tension and can relax at will without the aid of a machine.

information to relieve the tension. At the same time, watching or listening to the biofeedback signal diverts a person's attention from the problem causing the anxiety, and this reduces the need to keep muscles tense and alert. Once the higher brain becomes aware of excessive tension, it corrects the situation, freeing the midbrain and lower brain to better regulate other automatic functions. The higher brain can then turn its attention to coping with the cause of the anxiety.

In addition to causing skeletal muscle tension, anxiety and stress cause tension in the muscles of internal organs and blood vessels. Therefore, the stress of modern living may play a large role in high blood pressure, for example, or stomach ulcers.

The common denominator in all stress illnesses, and in many other disorders as well, is a fundamental disturbance of nerve activity. In stress illnesses, the nerves that connect to the muscles, blood vessels, internal organs, and brain nerve tissue are overactive. Some researchers assume that a similar nerve overactivity occurs in migraine headache and epilepsy. In biofeedback training, patients with these disorders can learn to reduce the overactivity of nerves. On the other hand, some biofeedback is directed toward increasing nerve activity. For example, a patient whose nerves have been damaged by a paralyzing stroke might regain the use of muscles by retraining the nerves.

Stroke victims often have difficulty walking because of paralysis in muscles around the ankle, a condition called foot drop. Many must use canes and leg braces for support. Basmajian reported in February 1977 that he and his colleagues at Emory University's Regional Rehabilitation Research and Training Center had successfully used biofeedback to treat patients with this disability.

Basmajian's team used a biofeedback device as small as a cigarette pack that the patients can place in their pockets or hang on their clothing. Two electrodes leading from the device are attached to the patient's lower leg over the muscle that bends the foot up at the ankle. Each time a patient activates the weakened leg muscle, the device gives off a buzzing sound. Basmajian can adjust the biofeedback instrument so that the patient has to produce more and more muscle movement to make the buzzer sound.

Out of 39 patients, 28 showed marked improvement. And of 25 patients who had to wear leg braces, 16 were able to walk without them after from 3 to 25 half-hour biofeedback sessions. Stroke patients also use the device to help them learn to walk with less of a limp. After about five weeks, Basmajian begins to wean them away from the biofeedback buzzer. He reports that some can even walk without canes.

Biofeedback may work because of our ability to direct the activity of nerve impulses. Training nerves to make the body perform is a fundamental ability of human beings. This is demonstrated by every movement we master—learning to walk, throw a ball, drive a car, or play the piano. An extension of this nerve training is learning to control automatic functions.

Biofeedback therapy is an entirely different way to treat illness. The patient, rather than the doctor, has most of the responsibility for the cure. Most therapists do not believe, however, that the patient can become his own therapist or doctor. Not only does the patient lack critical information for proper learning, but there may also be unseen medical and psychological risks. While biofeedback has no known serious side effects, difficulties can occur when the technique is used by patients with endocrine disorders, such as diabetes or hyperactive thyroid. Such patients are already being treated with drugs, and biofeedback may affect their endocrine glands, causing what was an acceptable drug dosage to become excessive.

It is also essential that the proper information be fed back. For example, the placement of the electrodes in EEG biofeedback is critical. A man once came to my laboratory to check his progress in trying to learn by himself how to produce more alpha brain waves. I checked the electrodes after he placed them on his head and found that, in their position, the electrodes were picking up eye muscle movement. He had been training an eye muscle, not his brain.

Nearly all the research and clinical evidence I have reviewed points to the importance of the mind in biofeedback learning. This is a very new aspect in Western medicine. Generally, biological science has dealt only with things that could be measured in physical terms; but thoughts, concepts, or judgments cannot be put in a test tube. Now biofeedback is allowing us a glimpse of the enormously complex tasks our brains are capable of performing. And it shows that human beings suffering stress illnesses may be the only ones who can cure themselves.

A great deal of research on biofeedback remains to be done. Often therapists have used biofeedback to treat an illness and reported success, but how this works has not yet been explained. More sophisticated biofeedback instruments must also be developed. For example, it is difficult to measure blood pressure accurately, and constantly inflating the blood-pressure cuff is uncomfortable for patients. Investigators are trying to develop biofeedback techniques that the majority of people suffering from high blood pressure can use more easily.

As of the summer of 1977, about 2,000 researchers and therapists were conducting experiments or treating patients with biofeedback. We can anticipate that people will become increasingly educated about their internal bodily functions, biofeedback devices, and the concept that the cure for stress illnesses lies in the mind. When this comes about, there may be fewer stress illnesses.

Further, because it seems that we can use biofeedback information to learn control over the activity of a single cell, it may eventually be possible to regulate damaged cells and speed the recovery of tissues. The ultimate use of biofeedback might well be in teaching people to become so aware of every cell, tissue, organ, and system that the body would operate at peak efficiency and therefore become extraordinarily resistant to the great majority of diseases.

Engines for The Eighties

By Frederick R. Riddell and Donald M. Dix

**The twin problems of dwindling energy supplies
and air pollution have spurred engineers
to look for better motors for the family car**

The sleek, streamlined automobiles rolling off the world's assembly lines are a result of more than 100 years of engineering. They fit well with modern-day jet planes, microwave ovens, color television sets, and pocket computers. They have automatic transmissions, hydraulic brakes, and power steering. Even their rubber tires have been reinforced with nylon, fiberglass, and steel so that they wear well. Yet their engines are essentially the same as those that powered Henry Ford's Model T more than 50 years ago—four-cycle Otto engines.

The first of these engines was designed in 1866 by Nikolaus August Otto and Eugen Langen of Germany. Its descendants have survived challenges from many competitors. Some of the 4,000 vehicles built in 1900 had electric motors, others were driven by steam. Only a few had the Otto-cycle, or "explosion," engines. None of these cars traveled more than a few miles per hour, and they were used only for short trips. But there was much discussion in the growing automotive industry about which type of engine—electric, steam, or explosion—would eventually predominate.

As demands for speed and range increased, the electric car became the first casualty. It was highly popular before 1920, but the weight of its batteries was a problem. Later versions of the electric car weighed 2,268 kilograms (5,000 pounds), and the batteries accounted for 60 per cent of that weight.

Early steamcars were lighter, but they had to stop every few miles to take on water. The addition of a steam condenser overcame that problem, but it made the steam engine too bulky and too heavy. One of the last "steamers" on the market, the 1930 Doble, could travel at speeds of up to 122 kilometers (75 miles) per hour and go as far as other cars before refueling. But it weighed even more than the heavy electric car and cost over $10,000.

By 1920, most engineers agreed that the Otto was the best engine for the family car. Because it showed greater potential for economic fuel consumption and weighed less, it would best meet demands for increased speed and range as they developed. Since then, numerous improvements that have provided more and more power for the given weight have consolidated the Otto's position.

However, when the automobile industry had to confront the dual problems of environmental pollution and dwindling oil reserves in the early 1970s, the Otto engine suddenly came up wanting. Research had shown years before that unburned hydrocarbons and nitrogen oxides in automobile exhausts react in sunlight to form photochemical oxidants, or smog. Subsequent research found that these two pollutants plus a third closely identified with automobile exhausts, carbon monoxide, were dangerous to human health even in relatively small amounts. They interfere with respiration and the supply of oxygen in the blood, aggravating a variety of respiratory and circulatory ailments. Automobiles may account for as much as 80 per cent of all carbon monoxide emissions, 66 per cent of all hydrocarbon emissions, and 50 per cent of all nitrogen oxide emissions in the atmosphere in metropolitan areas. These emissions may cause as many as 4,000 deaths each year in the United States, about 8.5 per cent as many as traffic accidents cause.

The authors:
Frederick R. Riddell and Donald M. Dix, engineers at the Institute for Defense Analyses in Arlington, Va., specialize in assessing potential advances in technology.

In December 1970, the U.S. Congress imposed strong controls on exhaust emissions. The Federal Clean Air Amendments Act of 1970 required that by 1976 carbon monoxide and unburned hydrocarbon emissions be reduced by a factor of about 40 from their previous levels and nitrogen oxides by a factor of about 15. Automakers and others protested that these standards were too costly and unrealistic to implement. As a consequence, the government moved back the deadline to 1978, and it may be pushed back again, to 1982.

The petroleum shortage became acute in 1973. The Arab oil embargo that year shocked Americans into the realization that even a temporary petroleum shortage can disrupt our lives. Automobiles consume about 416 billion liters (110 billion gallons) of fuel each year in the United States, so they became prime targets for conservation

Progress Towards Federal Requirements

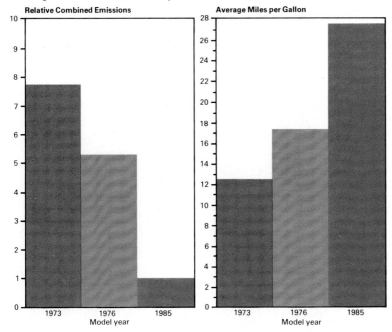

Relative Combined Emissions | **Average Miles per Gallon**

Model year: 1973, 1976, 1985

Federal standards to cut exhaust emissions and improve gas mileage on American automobiles by 1985 (blue bars) are forcing manufacturers to make changes in car models that each year come nearer to these standards, as shown by the averages for 1973 and 1976 model years.

measures. Government regulations issued on Dec. 17, 1975, require that the average fuel consumption efficiency of automobiles must be increased to 11.9 kilometers per liter of gasoline (27.5 miles per gallon) by 1985. In 1973, the average American car got 5.3 kilometers per liter (12.5 miles per gallon).

By 1976, the average had improved to 7.4 kilometers per liter (17.5 miles per gallon), but it will be difficult to get better mileage with the conventional Otto engine.

The Otto is an internal-combustion engine in which power is produced by burning the fuel, mixed with air in a carburetor, inside its cylinders. The hot gases from the exploding mixture expand rapidly, pushing pistons up and down within the cylinders. This motion is converted to rotary motion by the crankshaft and sent through the transmission to the drive shaft and rear axle to the wheels to propel the automobile. During the up-and-down movements, or strokes, of the pistons in the cylinders, the fuel mixture enters the chamber, is compressed, ignited, and finally, the burned and unburned gases are expelled into the exhaust system.

The Otto engine is surprisingly inefficient in converting the energy in its fuel into useful power—an ultimate determinant of gas mileage. Its efficiency is limited by six factors. The first limitation is the maximum compression ratio (the ratio of the volume of fuel mixture in a cylinder at the beginning of the compression stroke to the volume at the end) attainable without knocking or pre-ignition. Knocking occurs when the temperature of the compressing gases in the cylinder rises

Inside an Otto Engine

Intake stroke

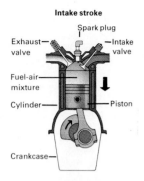

Spark plug

Exhaust valve — — Intake valve

Fuel-air mixture —

Cylinder — — Piston

Crankcase —

Compression stroke

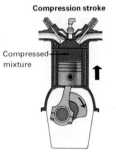

Compressed mixture

Power stroke

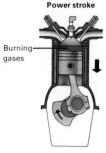

Burning gases

Exhaust stroke

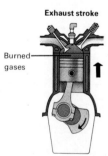

Burned gases

The four-stroke Otto cycle begins as the cylinder draws in its fuel-air mixture. This is compressed, burned, and wastes are expelled.

enough to ignite the gasoline before the piston is in the proper position. This wastes power, and about 40 per cent of the fuel energy is lost in avoiding pre-ignition. Another 15 per cent of the energy is lost because the chemical composition of the air in the cylinder changes as the burn starts, making the remainder of the burn less efficient and producing less power during the piston's downward, or expansion, stroke. About 10 per cent is lost because of the time required for combustion near the top of the compression, or upward, stroke. During that time, energy that might otherwise be used to expand the gases and drive the piston down in the cylinder merely heats the gases or escapes through the cylinder walls.

Friction between the pistons and cylinder walls and in the various bearings wastes another 5 per cent. In addition, most automobile driving utilizes only a small fraction of the engine's maximum power. At reduced power levels, less fuel is supplied, and, if the air flow is not reduced, the carbureted mixture will not contain enough fuel to sustain combustion in the cylinders. The air flow is reduced in an Otto engine by throttling, or "choking." However, because the flow is reduced, throttling forces the engine to work harder to supply the air-fuel mixture. This, combined with the fact that a little more fuel is supplied than would otherwise be necessary, accounts for about a 10 per cent loss of energy. Finally, about 5 per cent of the fuel energy is lost because the engine cannot operate at peak efficiency under most conditions. To do so would require that the engine have an infinite number of gears in its transmission.

The remaining 15 per cent is all the fuel energy that is delivered to the transmission to power the car. This level of efficiency in the Otto is attained at an engine weight of about 1.8 kilograms (4 pounds) per horsepower. If its efficiency could be increased to, say, 20 per cent, automobiles in the United States would consume about 104 billion liters (28 billion gallons) less gasoline each year.

The pollution problems of the Otto engine are equally complex. The combustion of gasoline in air should ideally produce only carbon dioxide and water vapor. However, carbon monoxide is produced when there is insufficient oxygen in a cylinder to burn all the fuel (a so-called fuel-rich mixture), or when the fuel is not in the cylinder long enough to burn completely. Similarly, unburned hydrocarbons are produced in fuel-rich conditions or when the mixture is not hot enough to burn completely. Nitrogen oxides are produced at high temperatures in either somewhat fuel-rich or fuel-lean (more oxygen than is needed) mixtures. Otto engines produce carbon monoxide mostly while operating at low-power levels, or slow speeds, when the engine operates somewhat fuel-rich. Unburned hydrocarbons are produced at all power levels in the relatively cooler gases adjacent to the cylinder walls. Nitrogen oxides are produced just after the fuel mixture ignites, in the hot gases at the beginning of the expansion stroke. These pollutants are a relatively small part of an automobile's

Propulsion with Electricity or Steam

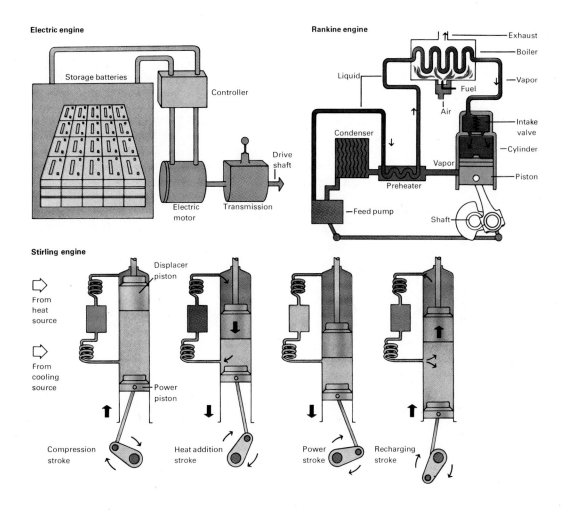

exhaust. In present automobiles, typical emission rates are in the range of 60 grams per kilometer (3.6 ounces per mile) for carbon monoxide, 12 grams per kilometer (0.7 ounce per mile) for unburned hydrocarbons, and 3 grams per kilometer (0.1 ounce per mile) for nitrogen oxides. Still, controlling them by simply modifying the Otto engine will not be easy.

Reducing carbon monoxide emissions by operating at fuel-lean levels increases the output of nitrogen oxides. Unburned hydrocarbons can also be reduced by operating at fuel-lean levels and by heating the exhaust gases so the hydrocarbons will burn. But to do this, the igniting of the fuel must be delayed at the top of the piston's expansion stroke, which reduces power and efficiency. Nitrogen oxide emissions can be reduced by lowering the maximum temperatures at the top of the stroke. But if this is done by operating at fuel-rich levels, more carbon monoxide and less efficiency result. And if this is accomplished

Electric engines use power that is produced elsewhere, stored in, and released from a battery. The external-combustion Rankine and Stirling engines allow greater control of how fuel burns because it burns outside the cylinders. Neither the electric nor the external-combustion engines produce the exhaust pollution that is typical of the Otto.

by delaying ignition, the engine is less efficient and produces less power. The same is true if incompletely burned exhaust gases are recirculated to the cylinder during the intake stroke. These conflicting factors were perhaps best exemplified in the engines used in most 1973 and 1974 automobiles. They were designed to reduce emissions, with much-delayed ignition at the top of the compression stroke, and much of the exhaust gas was recirculated. The result was sluggish and considerably less efficient performance.

Yet automobile engineers have made some improvements in the Otto engine. The 1977 models incorporate refinements in control of the fuel-air mixture ratio. A small amount of exhaust gas is recirculated. And a catalytic reactor has been inserted in the exhaust stream that permits excess carbon monoxide and hydrocarbons to be burned at relatively low temperatures. These improvements have lowered emission rates to about 2.5 grams per kilometer for carbon monoxide, 0.25 gram per kilometer for unburned hydrocarbons, and 1.2 gram per kilometer for nitrogen oxides. Unfortunately, they have added about $200 to the price of an automobile and have raised the engine weight to approximately 2 kilograms (4.5 pounds) per horsepower. Furthermore, the catalytic reactors must be replaced frequently, cannot tolerate leaded gasoline, and must operate at such high temperatures that they can sometimes be a fire hazard. Because of the added weight, they have also reduced the car's gas mileage.

To satisfy the now-delayed 1978 emission standards, nitrogen-oxide emissions must still be reduced to 0.25 gram per kilometer. The only immediate solution seems to be to add an improved catalytic reactor in the automobile's exhaust system that would reduce the nitrogen oxides as well as the carbon monoxide and hydrocarbons. But reactors for nitrogen oxides cannot operate in fuel-lean conditions, which precludes much further improvement in engine efficiency. Hence, new and more efficient types of engines should be considered.

Possible alternatives to the Otto engine include three other internal-combustion engines—the diesel, gas turbine, and Wankel. Other alternatives include three external-combustion engines still in various stages of experimental design, and modernized electric cars and modernized steam engines.

The electric car's major attractions are that it uses no petroleum fuel and emits no pollutants. It also is quiet, reliable, and easy to operate. However, it still requires very heavy batteries. And even the best batteries available today will provide electric cars with only 20 per cent of the power and 20 per cent of the range that the Otto engine produces. Other energy-storage devices, such as flywheels, fuel cells, and thermal systems, have been tested from time to time, but none has appreciably better prospects for powering the electric car than does the traditional heavy battery.

External-combustion engines, such as the early steam engine, burn their fuel outside the engine. The energy from the burned gases is then

A Brayton engine uses high-speed rotating turbine instead of pistons to compress its working fluid. A diesel uses the heat of compression to ignite the fuel-air mixture. The Wankel uses spark-plug ignition, but substitutes rotary motion for piston motion to compress and expand its working fluid.

transferred to a working fluid (steam in the case of a steam engine) that powers the engine. External-combustion engines can meet the proposed emission standards because their combustion occurs under well-controlled, fuel-lean conditions. But the external-combustion engine must have a heat exchanger similar in principle to the conventional automobile radiator to transfer the energy from the burned gases to the working fluid. The heat exchanger makes the engine bigger, heavier, and somewhat more expensive.

The Rankine, Stirling, and Closed Brayton are the three types of external-combustion engines that could possibly challenge the Otto. The Rankine steam engine compresses water in a pump and then passes it through a boiler that is heated by external-combustion gases to produce steam at relatively high pressure. The steam goes into a cylinder and drives a piston, then enters a radiator, recondenses into water, and begins the sequence again.

The Rankine's primary advantage is that it takes much less power to compress a liquid than it does to compress a gas. But this advantage is more than offset because making steam from water requires some fuel energy that contributes nothing to the power produced. Estimates indicate that the Rankine's maximum efficiency is from 10 to 20 per cent less than that of the Otto, and it weighs and costs substantially more.

The Stirling engine uses the basic piston-cylinder form of the Otto, but it operates quite differently. As the working fluid, hydrogen gas, is cycled in the engine, it is alternately heated and cooled, which varies the pressure of the gas. This alternating pressure drives the pistons. Estimates indicate that the Stirling engine potentially may be as much as 40 per cent more efficient than the conventional Otto. But no Stirling engine has yet been built that approaches this potential. Also, the Stirling must have many heat exchangers to operate properly, and they tend to make it big, heavy, and expensive.

Propulsion with Gas or Petroleum

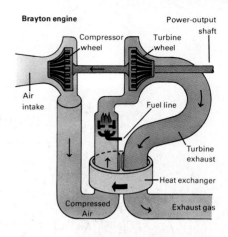

Brayton engine

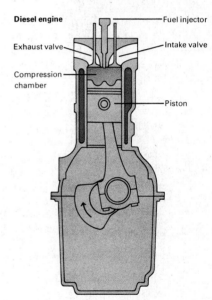

Diesel engine

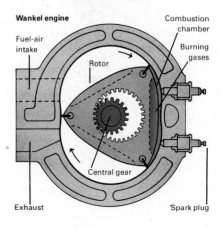

Wankel engine

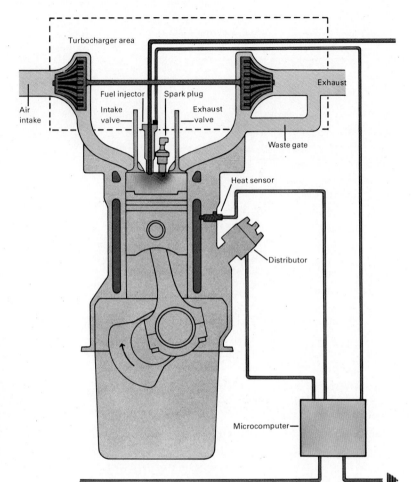

A charge-stratification engine has electronic feedback circuits in a miniature computer that control the flow of fuel and air according to cylinder temperatures. This increases engine efficiency and reduces the amount of unburned hydrocarbons because an optimum fuel-air mixture is maintained throughout the cylinder. The turbocharger makes use of exhaust gases to increase efficiency of the engine system.

The Closed Brayton engine uses a turbine, essentially a vaned wheel that is turned by the flow of gas. The Brayton compresses and expands the working fluid, again hydrogen. The gas is compressed in a high-speed compressor and heated to drive a high-speed turbine. It then cools in a radiator and is ready to repeat the sequence. Unfortunately, the Closed Brayton engine is no more efficient than an improved Otto. And, like the Stirling, it is larger, heavier, and more expensive.

Alternative engines more closely related to the Otto—internal-combustion engines—are the gas turbine, diesel, and Wankel. The gas turbine is a Brayton-cycle engine that burns its fuel directly in the high-pressure air between the compressor and the turbine, and all excess heat goes out through the exhaust. This eliminates the heat exchangers needed by the Closed Brayton. As a result, the engine is lighter, more compact, and less expensive, but produces the same power. Because combustion occurs in a flow of excess air, the gas turbine produces no carbon monoxide. And unburned hydrocarbons are not a problem because combustion can be kept away from the combustion chamber walls where low temperatures occur. On the other hand, the gas turbine produces more nitrogen oxide, because the gases stay hot longer than they do in the Otto.

In theory, the gas turbine is more efficient than the conventional Otto engine, but its potential efficiency has never been reached in test automobiles. And although it is smaller and weighs less than the Otto, so far this advantage has been more than offset by an estimated 50 per cent higher manufacturing cost.

The diesel engine offers greater efficiency than the conventional Otto by allowing greater compression ratios. This is done by compressing the air first and then spraying the fuel into the combustion chamber. The diesel uses a high-pressure fuel nozzle in each cylinder instead of the Otto engine's carburetor. The penalty the diesel pays for this advantage is in additional weight; it weighs from 20 to 30 per cent more than an equivalent Otto engine. This weight problem in diesel-powered automobiles cannot be expected to change.

The Wankel, or rotary engine, uses rotary motion for compression and expansion rather than the reciprocating motion of a piston engine. Apart from this, its operating processes are identical with those of the Otto—a mixture of gasoline and air enters the combustion chamber where it is compressed, burned, and expanded to produce power, and then exhausted to complete the cycle. So the Wankel has the same practical limits on compression ratio, and hence on efficiency, as does the Otto. Its major advantage is that it weighs as much as 30 per cent less than a conventional Otto engine of equivalent power. But Wankels now being used have not yet matched the efficiency of the Otto.

So we are back to the Otto engine which, for all its drawbacks, is still the most likely candidate to power the 1985 family car. Engineers are considering three modifications that can improve it: lean-burning, turbocharging, and charge stratification.

In lean-burning, an engine constantly operates at fuel-lean conditions by precisely controlling the fuel-air mixture. This is accomplished by measuring the fuel-air ratio and controlling the fuel flow by means of the electronic feedback circuits in a miniature computer. Such control provides a way to reduce the throttling loss at low power levels because more air can be handled with the same fuel flow. It also reduces the loss associated with the chemical composition of the burned gas and increases engine efficiency an estimated 10 per cent (increasing the overall efficiency from 15 per cent to 16.5 per cent). An exhaust reactor would still be required to reduce unburned hydrocarbons, and nitrogen oxides would still be a problem. Such miniature computers have already been installed in some models of General Motors and Chrysler automobiles.

Turbocharging consists of using a high-speed compressor to pressurize the intake air before it enters the carburetor. The compressor is

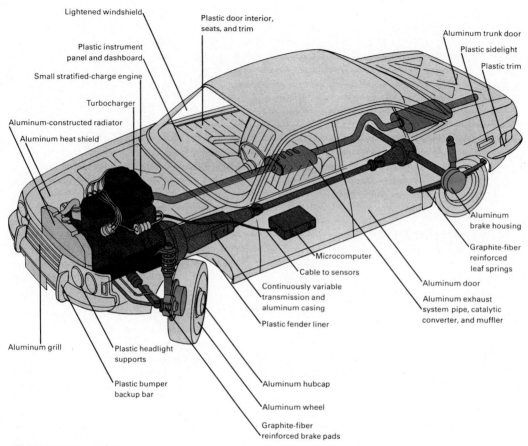

Lightened windshield

Plastic door interior, seats, and trim

Plastic instrument panel and dashboard

Small stratified-charge engine

Turbocharger

Aluminum-constructed radiator

Aluminum heat shield

Aluminum trunk door

Plastic sidelight

Plastic trim

Aluminum brake housing

Graphite-fiber reinforced leaf springs

Microcomputer

Cable to sensors

Continuously variable transmission and aluminum casing

Aluminum door

Aluminum exhaust system pipe, catalytic converter, and muffler

Plastic fender liner

Aluminum grill

Plastic headlight supports

Plastic bumper backup bar

Aluminum hubcap

Aluminum wheel

Graphite-fiber reinforced brake pads

Microcomputer engine controls and improved transmissions will help improve gas mileage. Aluminum and plastic body parts will result in cars that are lighter in weight and thus use less energy.

driven by a turbine that is powered by the flow of exhaust gases from the engine. With turbocharging, a smaller engine can produce the same power as the conventional Otto, and there is no need for throttling at low power levels. However, the engine's compression ratio must be reduced to avoid knocking, which partially offsets other gains in efficiency. Estimates indicate that perhaps a 10 per cent improvement in efficiency could be gained (from 15 per cent to 16.5 per cent), with a 25 per cent reduction in engine weight. Exhaust emissions could be controlled with catalytic reactors, as in the Otto.

There are two types of charge-stratification, carbureted and direct-injection. In both, fuel in the cylinders starts to burn in a small fuel-rich zone and burning then spreads through the rest of the mixture, which is fuel-lean. The overall fuel-lean operation tends to control carbon monoxide, and it produces fewer unburned hydrocarbons because there is little fuel near the cylinder walls. There are also fewer nitrogen oxides because only the fuel-rich part of the mixture burns at the beginning of the stroke.

In the carbureted versions, such as the Honda CVCC (controlled vortex combustion cylinders), each cylinder has a separate small chamber surrounding the spark plug that is supplied with a fuel-rich carbureted mixture. The main cylinder chamber receives a fuel-lean mixture from another carburetor. Because this type of engine operates somewhat fuel-lean overall, carbon monoxide emissions are controlled. Nitrogen oxides are suppressed by reducing temperatures and burning fuel-rich near the beginning of the stroke. At the same time, there is less fuel close to the cylinder walls, so fewer hydrocarbons are produced. The engine meets current emission standards with no loss in efficiency and virtually no treatment of the exhaust. But it does not meet the emission standards scheduled to go into effect in 1978, and it is no more efficient than the conventional Otto.

In the direct-injection charge-stratification engines, also called fuel-injection, the fuel is injected directly into the cylinders. No carburetors are needed. For greater efficiency, the compression ratio can be increased because the nonuniform mixture has less tendency to produce knock. And, because a locally fuel-rich mixture can be maintained at overall fuel-lean conditions, the engine can operate without throttling. Estimates indicate that such engines would be perhaps 20 per cent more efficient (from 15 per cent to 18 per cent), while satisfying current emission standards. At present, however, there is no known way to maintain this substantial efficiency improvement and simultaneously satisfy 1978 nitrogen oxide standards.

Finally, the efficiency of any Otto engine can be increased by developing better transmissions. A continuously variable transmission, which has smooth, mating tapered disks that rely on friction to transmit forces, could provide as much as 20 per cent more efficiency than the conventional transmission, which has a limited number of separate gears. A direct-injection, stratified-charge engine with such an improved transmission could provide perhaps 30 per cent more efficiency (from 15 per cent to 19.5 per cent), with a somewhat increased cost and weight. If a way to reduce nitrogen oxides to 1978 standards is discovered, or if the standard is relaxed, this combination could raise the average gasoline mileage from the current 7.4 kilometers per liter to about 9.6 kilometers per liter, with no other changes.

Further developments in gas-turbine or Stirling engines may provide additional benefits after the 1980s. But the Otto will probably power most cars during the 1980s. Its efficiency, weight, and cost, although affected by government regulations on fuel economy and pollution control, still make the Otto best for the family car. But if the automobile industry is to meet the standards set by the government now, most of us must be willing to trade in our large automobiles for compact, lightweight models at prices that reflect the engineering efforts behind making them both cleaner and more efficient.

The Wolf's Last Stand

By Mark Perlberg

The slavering beast of myth and legend haunts his descendants who still are the objects of hatred and fear

John Harris tours parts of the United States and Canada in a small van with two wolves as traveling companions. He exhibits the wolves at schools, trying to show that these animals do not deserve their man-killing reputation. On the other hand, Harris explains that wolves are wild animals whose behavior in some circumstances can be dangerous, and they should never be considered pets.

This is only one man's campaign to try to combat the fear and resulting hatred of wolves—a state of mind that has involved human beings in a centuries-old war against these animals. This war has led to the extermination of the wolf from 99 per cent of its original range in the United States outside Alaska. And as a result, the U.S. Department of the Interior in 1967 listed the wolf as one of the endangered species in the continental 48 states.

Wolf hatred is an emotional garment woven of fear and myth, spun out of the fables of Aesop, fairy tales such as "Little Red Riding Hood," and Russian stories of hungry wolves racing over the snow after a sleigh filled with terrified passengers. European settlers in America brought their fear of the wolf with them, along with their furniture and dishes. The residents of the Massachusetts Bay Colony in the 1600s fixed a one-penny bounty on wolves. By 1900, these animals had been exterminated from the Eastern United States.

Wolves freely roamed the Western prairies until the last of the bison, their chief prey, were killed in the late 1800s by hunters, settlers, and ranchers who had taken over much of the range. Deprived of their main food source, some wolves preyed on sheep and cattle. In 1915, the final campaign to destroy the wolf in the United States was launched. Congress authorized the control of animals on federal lands, and during the next 15 years, federal agents poisoned, trapped, or shot more than 24,000 wolves.

Despite the fact that human beings have long pictured the wolf as a cunning personal enemy, all research shows that wolves, more than most other animals, are anxious to avoid humans. It is difficult to prove that a healthy, nonrabid wolf has ever harmed a human being in North America. The Sault *Daily Star* of Sault Sainte Marie, Canada, for years offered a reward to anyone who could verify that they were attacked by a wolf. Each of 80 to 90 claims were investigated, and all proved groundless.

Attacks have been reported in southern and central Europe, where the wolf has lived much closer to people than in North America. But careful examination indicates that rabid wolves or wolf dogs, not normal wolves, were the culprits.

Wolves once inhabited an estimated 18 million square kilometers (7 million square miles) in North America, from central Mexico to the Arctic–the greatest range of any large North American land mammal. They lived everywhere except in rain forests.

Today in North America large populations of wolves live only in Canada and Alaska. Canada has about 40,000 wolves; Alaska, perhaps 10,000. An estimated 1,000 to 1,200 live in northern Minnesota, including Superior National Forest, and some 40 roam Isle Royale National Park, an island in the ice-blue waters of Lake Superior. A few wolves live in Michigan's Upper Peninsula, and some may exist in Louisiana and eastern Texas. Outside of the United States and Canada, only the mountains and forests of Siberia, eastern Europe, Greece, Spain, northern Italy, and the Balkans harbor wolves.

What we commonly refer to as tundra, timber, or gray wolves are known by the scientific name *Canis lupus*. They are related to such animals as dogs, coyotes, and jackals. In fact, most authorities believe dogs evolved from wolves domesticated by primitive man.

The author:
Mark Perlberg is a Chicago-based editor and free-lance writer.

With the exception of some large domestic dogs, wolves are the largest members of the genus *Canis*. Adult males measure from 1.5 to 2 meters (5 to 6.5 feet) from nose to tail tip and usually weigh from 43 to 45 kilograms (95 to 99.2 pounds). Females are about 30 centimeters (12 inches) shorter and weigh from 4.5 to 7 kilograms (10 to 15 pounds) less. Wolves have long slender legs, narrow chests, and flowing bushy tails. Their coloring varies from silvery-white through all shades of gray and brown-gray to black.

Wolves mature sexually at about 2 years of age. Depending on how far north they live, they mate between late January and early March and the female gives birth about 63 days later. Wolf pups are usually born in spring. Litter sizes vary from 1 to 13 pups, but the average is between 4 and 6. The pups are born and spend their first few weeks in a den. Sometimes the mother may use a hollow log or a cave as her nursery den, or she may dig one in sandy earth or take over and enlarge another animal's den, such as that of a fox.

The pups live on their mother's milk for about three weeks, but then they begin to eat food regurgitated by the adults. When they are 2 months old, the young wolves leave the den and live in the open with the adults. By late autumn, each pup has grown its first adult coat, an outer layer of long hairs covering a thick, soft underfur. This natural garment enables a wolf to curl up on snow and sleep even if the temperature falls to −46°C (−50.8°F.).

By examining wolf droppings and the stomach contents of dead wolves, scientists have determined that wolves living in the wild eat animals ranging in size from birds, beaver, and rabbits to caribou, deer, elk, and moose. In addition, wolves sometimes prey on such domestic animals as sheep, cattle, horses, dogs, and cats.

But the wolf's ecological niche–its place in nature–is that of the chief predator of large mammals in the Northern Hemisphere. Wild-life biologist L. David Mech observed this relationship between wolves and moose on Isle Royale from 1959 to 1961 as part of his graduate studies at Purdue University. During winter, Mech flew over the island in a light plane, tracking the wolves and counting both wolves and moose. During summer, he used a powerboat to move around the island's shoreline, and he also traveled 2,250 kilometers (1,398 miles) on foot. Among other things, he learned how fast wolves travel. They can cover as much as 72 kilometers (44.7 miles) in a day.

Wolves have long been portrayed as cruel killers in fairy tales such as "Little Red Riding Hood."

Mech examined 51 moose killed by wolves and found that most of the dead animals were between 8 and 15 years of age, while 18 were calves. He also found that 39 per cent of the dead moose had been ill.

From such data, Mech concluded that wolves help keep the moose herd healthy. By killing the sick, the old, and some calves, they help prevent the moose population from growing so big that it literally eats up its environment and dies of starvation.

This actually happened to the Isle Royale moose in the 1930s and 1940s. There were no wolves on the island then to prey on the moose, so the herd increased until there was not enough vegetation to feed all of them. As a result, the herd suffered from malnutrition and disease. Scientists speculate that during the cold winter of 1949 a wolf pack crossed the ice that formed between Isle Royale and the Canadian mainland. After the wolves took up residence on the island, they started to kill the old and sickly moose, the size of the herd dropped, and vegetation grew back. Eventually, the balance of nature was restored, and the herd became healthy again.

A wolf pup views the
world from the safety
of his den entrance,
above, a few weeks after
birth. Venturing forth,
a family of pups test
the waters of a stream,
while another playfully
taxes the patience of
an ever-watchful parent.

Wolves are highly social animals, living and hunting together in packs. Swiss biologist Rudolph Schenkel in 1947 became the first scientist to describe in detail the social organization of wolves. By carefully observing wolves that were kept in a large room at the Basel Zoo, he learned that each pack has two dominance orders, one for the males and one for the females. The chief male, or alpha wolf, is the pack leader, and all pack members defer to him. The chief female, or alpha female, dominates the pack's females, and she may dominate most of the subordinate males as well.

Schenkel described and sketched more than 50 postures and gestures of dominance and submission among the wolves he observed. Most dominant wolves have bright, alert expressions. They carry their ears cocked and their tails held high or in a comfortable drooping position. To show their authority, they will raise the tail or, on rare occasions, stand at right angles to a subordinate animal and put their forepaws across its shoulders.

There are two principal ways in which a subordinate wolf shows submission, depending on its rank within the pack. When challenged by one of higher rank, a relatively high-ranking wolf lowers its ears, places its tail between its legs, and pulls its lips back. It may wag its tail from side to side and sniff and lick the muzzle of the dominant wolf in a friendly manner. Low-ranking wolves are much less sure of themselves. Sometimes they snap and bark defensively and wag their tails. Often they roll over on their backs at the feet of the dominant animal.

Because of the dominance order, serious fights rarely break out among wolves in the same pack. When trouble occurs, higher-ranking wolves rush over and assume threatening postures.

Dominant and submissive behavior play an important role in a wolf ritual that scientists call greeting. After even a brief separation, two wolves from the same pack will greet each other, the lower-ranking wolf sniffing or licking the muzzle of the higher-ranking one. The entire pack greets its leader by nuzzling him when he returns from even a short jaunt by himself.

For years after Schenkel's pioneering work, zoologists believed that only the alpha male mated and that he had his pick of the females in the pack. But research by George B. Rabb, director of the Brookfield Zoo near Chicago, and his wife, Mary, has shown that this is far from what actually happens. Since 1959, the Rabbs have been studying a pack of wolves that lives in a 0.3-hectare (0.7-acre) enclosure at the zoo. During that time, the pack has been led by three different males. The Rabbs found that in the Brookfield pack the lower-ranking males mate more frequently than the alpha wolves. "It seems that the alpha male's responsibilities, because they take up so much of his time, to some extent exclude him from mating activities," says George Rabb.

So dominance has responsibility as well as privilege. The Brookfield alpha wolf performs such duties as patrolling the fence and breaking up fights. In the wild, the alpha wolf has first pick of meat from a kill and pick of the favored resting sites. But he also leads the hunt, which often puts him in a highly vulnerable position. For example, he may be kicked to death by a struggling moose.

The Rabbs, who name the Brookfield wolves by their birth order, call the current pack leader Male 6. Although he has been the leader since 1970, Male 6 has never mated because he suffers unrequited love. He prefers Female 4, the pack's oldest female, but she will have nothing to do with him. To make matters even more complicated, Females 6 and 8 are deeply attracted to Male 6, but he will have nothing to do with them because he is stuck on Female 4. "It's a real soap opera," says Mary Rabb.

By observing these relationships between the wolves, the Rabbs have proved that wolves form long-lasting bonds with their mates. Because of this, along with the dominant and submissive social structure, there are definite constraints within a pack on breeding. Their conclusions have been supported by other researchers. The Rabbs

From her small booth at the edge of Brookfield Zoo's wolf enclosure, Mary Rabb observes as two of the pack's wolves greet each other, *above right.* Nearby, another wolf crouches as though it were stalking some prey, *below right.*

have also observed that the female controls mating. If she doesn't want to mate, mating will not occur.

By studying wolves in captivity, the Rabbs can observe fine details of behavior that are probably impossible to view in the wild. On the other hand, situations develop at the zoo that wild wolf packs might never encounter. For example, wolves in wilderness areas where they encounter no humans live an estimated 8 to 10 years. In captivity, they live 15 to 16 years. Brookfield Female 4 is 16 years old and arthritic. Yet the Rabbs, who do not interfere with the pack in any way, do not want to remove her.

"We're watching to see how the wolves will resolve their complex psychological situation," says Mary Rabb. "When the old female dies, will Male 6 choose a mate and how will he do so?"

In addition to the rigid social control within wolf packs, the size of the pack appears also to be under some sort of control. Erik Zimen, a West German wolf biologist, in 1976 developed a computer model to

The Brookfield wolves exhibit various levels of dominant and submissive behavior. One cowers and snaps, *above left.* A lower-ranking wolf rolls over on the ground at the feet of the pack's leader, *left.*

simulate how the size of a wolf pack might be regulated. Zimen knew that the size of the average pack stays relatively constant, rarely growing to more than 13 wolves. Most packs have from two to eight wolves. Like other biologists, Zimen believes that the exact size of each pack is determined by the number and size of prey animals available in the environment as well as by the social habits of the wolves. For example, there appear to be limits to how many attachments a wolf can form and how much competition it can tolerate from other wolves.

Using data from ongoing research in Minnesota and from his own observations of wolves living in large enclosures in the Bavarian National Forest, Zimen's model suggests that several factors tend to keep the size of wolf packs stable. For example, wolf pups have a high death rate; only alpha females tend to mate and reproduce; and pack wolves keep strange wolves from joining the pack.

Zimen also theorizes that if the ecological balance between the number of wolves and the food supply is disturbed, certain mechanisms are set in motion to correct the balance. For example, if the death rate among pups drops, or if the food supply diminishes, the wolves become very aggressive about getting meat from the animals they kill. At this point, Zimen believes high-ranking wolves force some lower-ranking animals to leave and forage on their own. He speculates that this might partly account for lone wolves.

The lone wolf, who ekes out a precarious existence along the territorial boundaries of existing packs, is an intriguing problem. Working in northeastern Minnesota as a research biologist with the U.S. Fish and Wildlife Service, Mech has been studying the lone wolf since 1968. He captured a number of lone wolves in Superior National Forest, put collars containing radio transmitters on them, and then released them. By monitoring signals from the radio collars, he could follow the movements of each animal.

"There's been a lot of speculation," says Mech, "that lone wolves are old animals who've been kicked out of the pack. Although a very small percentage probably do fit that category, the largest number by far are very young animals, ranging from less than 1 year to 3 years of age. We're not sure how or why, but during certain periods of the year these animals tend to leave their packs and strike out on their own."

Mech has found two types of lone wolves. One wanders in a vast ring around the pack's territory. The other type may travel more than 209 kilometers (130 miles) away. Both types wander until they find their own hunting area—if one is available—and a wolf of the opposite sex. "These are the wolves that start new packs," says Mech, "and inhabit areas not occupied by a pack."

But if the lone wolf fails to find a suitable area, it will not mate. It will live out its life on the fringe of packs, and it probably will not live as long as pack wolves.

The size of a wolf pack's territory varies enormously depending on the season, location, and amount of prey available. Biologists have

A moose heads into the trees on snow-covered Isle Royale, trying to escape the relentless pursuit of a wolf pack led by the alpha male.

A scientist fastens a radio transmitter on a tranquilized wolf to track its movements in hunting prey, *right,* while another scientist tries to estimate the age of a common prey, a moose that was killed and eaten by wolves.

reported winter territories ranging from 93 square kilometers (36 square miles) in Minnesota to 12,950 square kilometers (5,000 square miles) in Alaska. Scientists have long believed that wolves, like dogs, stake out their territories through a process called scent marking, primarily by urination. Working with Mech, psychologist Roger P. Peters of Fort Lewis College in Durango, Colo., used the radio collar technique to find out if this is true. Mech put radio collars on 96 wolves in Superior National Forest. Peters tracked the collared wolves during the winters of 1971 through 1974 and studied scent marks made by urinating, defecating, or scraping in the snow. Recording when, where, and how often the marks appeared, Peters and Mech learned that the alpha male and female mark prominent objects, such as trees and rocks, on the edge of the pack's territory. In this way, they create a kind of invisible fence that allows pack wolves to know when they are within their own territory and tells alien wolves when they are entering the pack's turf.

Scent marking and various postures provide the wolf with an effective system of communication. But doubtless its best-known, yet least understood, means of communication is the famous howl. It seems clear that each animal has its own voice, just as humans do, and scientists have observed that some wolves howl when they are by themselves while others do not. But in a pack, all the wolves join in the thrilling chorus howl, and this seems to be the case both in the wild and among captured wolves. When one animal begins to howl, the others cluster around and join the chorus with much tail wagging and an air of general good fellowship.

David Mech returns from an airplane tour of wolf country carrying a dead radio-collared animal, whose part in a research project on wolves was cut short by a hunter.

Interestingly, when wolves howl in chorus, they do not howl in unison. Each wolf keeps its tone separate. If one animal's voice approaches the pitch of another, one of them adjusts to a different pitch. Ethologist Erich Klinghammer of Purdue University is studying howling behavior in two packs of captive wolves at the North American Wildlife Park Foundation in Battle Ground, Ind. He theorizes that the chorus howl may give each wolf the opportunity to learn the voices of other pack members so they can recognize each other if separated.

Another of Klinghammer's theories is that a type of wolf behavior called scent rolling is a form of communication. The animals roll in substances giving off odors to pick up the scent. Klinghammer believes that in the wild a wolf might roll in evidence of something it has found and bring back the scent to let other pack members in on the discovery, perhaps something like a nice tasty moose.

Clearly, wolves are fascinating creatures, vastly different from the way they have been pictured in song and story. Yet, whenever farmers and ranchers have pushed close to wolf country, the result has been disaster for the wolves.

In Alaska, one of the wolf's last strongholds, conservationists are doing battle with the Alaska Department of Fish and Game (ADFG). In February 1977, a federal court halted an ADFG program to elimi-

Silhouetted against the setting sun, two wolves raise their voices in the plaintive howl that haunts listeners in the few areas that are still the wolf's domain.

nate large numbers of wolves after several conservation groups and individuals filed a lawsuit against the state. However, Alaska filed a countersuit, and observers believe the legal wrangling may continue for some time. Alaska's controversial program authorized hunters to shoot wolves from small airplanes.

Bob Rausch, director of the Game Division of the ADFG, explains that a drastic drop in the number of caribou was the reason for authorizing aerial shooting. Between 1971 and 1976, the number of caribou fell from 242,000 to between 50,000 and 60,000. Rausch blames the drop on three factors. First, and most important, 25,000 to 30,000 caribou are killed each year by Indians and Eskimos, who use the animal skins for clothing. Rausch also admits that these hunters shoot many more caribou than they use; hunting is easier now because they use snowmobiles to reach the caribou herds. Second, wolves kill 10,000 to 15,000 caribou each year, and finally, some recent severe winters have been hard on the herd.

"We restricted hunters to a total of 3,000 bull caribou a year as of September 1976," Rausch says. "Killing males won't affect the rate of the herd's recovery. The only other factor we can control is wolf predation." So ADFG announced that it intended to shoot 80 per cent of all wolves in the Brooks Range. "Wolves became the innocent victims in a situation where humans are still partially dependent on wild animals," Rausch concludes.

This policy has wolf experts and conservation groups up in arms and led to the lawsuit. But some scientists, looking at the problem from an ecosystem view, believe that both hunting restrictions and some kind of short-term wolf-control program may be necessary. Ecologists Gordon C. Haber and Carl J. Walters of the University of British Columbia have made computer models, based on field data, of the seven major caribou herds in Alaska and the Yukon Territory in northwestern Canada. They found that in the past, each herd reached periodic levels of high population, producing more animals than the grazing area could support. Large animals then left the herd, and wolves reduced the remaining caribou to a low population level. After some years, animals from another, high-density herd would move into the area and bring the first herd up to high levels again. Then the process would start over.

Unfortunately, since the 1950s, all seven of the caribou herds have been at moderate to low population levels because of overhunting. Haber and Walters believe that unless hunting by humans and wolves is restricted to one or two herds, none of the rest will be able to build up to high population levels.

Minnesota has a wolf situation that is just as complex, but apparently much more emotional. Wildlife experts know that the deer population in northeastern Minnesota has declined because of a shortage of food. In the past, the deer ate bushes and small trees after loggers cut down forests. But now the forests are growing back, not enough big trees are being cut down, and the small plants that the deer fed on are being crowded out. Hunters blame wolves for the decline in the number of deer. Meanwhile, wolves sometimes move out of the wilderness area and stake out new territory on farmland. Consequently, farmers are demanding that wolves be removed from the endangered species list so they can be hunted.

Animal preservationists say that farmers and hunters are exaggerating the situation. They also speculate that many reported "wolf" sightings are actually coyotes or large dogs.

To find a solution acceptable to both sides, a group called the Eastern Timber Wolf Recovery Team has been formed by Mech and seven other biologists, representing three states—Michigan, Minnesota, and Wisconsin; and three federal services—fish and wildlife, forest, and national parks. These scientists have developed a plan that they hope, if it is adopted, will make Minnesota safe for its wolves.

The plan calls for a program to educate citizens about the true nature of wolves and to award full protection to wolves in a 25,900-square-kilometer (10,000-square-mile) wilderness area. But it also calls for changing the animal's status in Minnesota from "endangered" to "threatened." This would allow the killing of wolves that attack livestock where human settlement edges on the wilderness.

"Some wolves will have to be sacrificed," Mech admits, "because when wolves conflict directly with people, it fosters an outright hatred for the animals." And so, in a sense we are back to where it all began, with the age-old human hatred for the wolf.

Cores in Collision

By J. Rayford Nix

In search of new matter and knowledge, researchers use a worldwide arsenal of accelerators to smash atomic nuclei

Bang! Bursts of projectiles, pushed by invisible electromagnetic forces, race toward a target at the end of a long tunnel. Most of the projectiles pass through the target and continue straight ahead. Others emerge slightly off course. Still others strike the target, then almost instantaneously spin out of control and break apart.

Is this ray-gun target practice in the year 2000? No, the tiny projectiles are atomic nuclei, the dense cores of atoms. The targets are other nuclei that researchers working in laboratories throughout the world are trying to learn more about. The scientists want to know if nuclei are fluid like water, or viscous like honey. Can new nuclei be made? Do new forms of matter appear when nuclei smash together and become compressed to more than their normal density?

Because nuclei are too small to be seen, scientists must study them indirectly. They do so by shooting beams of nuclear projectiles at nuclear targets and measuring the charges, masses, and energies of the emerging nuclei, as well as the angles at which they emerge. British physicist Sir Ernest Rutherford used just such an experiment to discover nuclei in 1911, disproving the early theory that atoms were uniform spheres of positive charge peppered with negatively charged electrons, like plums in a pudding. Rutherford's associates bombarded a thin foil of gold with alpha particles that are spontaneously ejected at high speeds from radium and other radioactive elements. Some of the alpha particles emerged going in a much different direction. A few even ricocheted back toward the source. Rutherford was surprised. "It

Drift tubes in a section of the heavy-ion accelerator in Darmstadt, West Germany, guide projectile nuclei toward stationary target nuclei.

was almost as incredible as if you had fired a 15-inch shell at a piece of tissue paper and it came back and hit you," he recalled later. From this experiment he reasoned that the positive charge and most of the mass of the atom was concentrated at its center.

Within a few years, scientists had worked out the present view of the atom. The positively charged protons in the atom's nucleus are about 2,000 times as heavy as the negatively charged electrons that move in orbits around the nucleus. In fact, an atom is mostly empty space—if an atom were the size of a house, its nucleus would be only about the size of a pinhead. Overall, atoms are electrically neutral because the electrical charges of the protons and those of the equal number of electrons cancel one another.

Scientists learned in 1932 that, in addition to protons, nuclei contain uncharged neutrons that are slightly more massive than the protons. An element usually has an equal or larger number of neutrons than protons. For example, the lightest element, hydrogen (1 proton, 0 neutrons), can also exist as deuterium (1 proton, 1 neutron) or as tritium (1 proton, 2 neutrons). These different forms are isotopes.

Protons and neutrons, which are both called nucleons, are held together by strong forces that act like a nuclear glue over very short distances. Packed closely together, like marbles in a bag, the nucleons orbit around the center of the nucleus within imaginary shells of increasing radius, like the layers of an onion. Nucleons in the outer shells are more energetic than those in the inner shells. A proton shell can hold only a definite number of protons. Likewise, each neutron shell can hold only a fixed number of neutrons. When a shell is filled, remaining nucleons go into the shell of next larger radius and higher energy. Such nuclei as oxygen, calcium, tin, and lead contain completely filled outer proton shells. They are said to be magic because they are relatively more stable, or resistant to change, than nuclei with either a larger or smaller number of protons. Certain isotopes of these elements are said to be doubly magic because they also contain filled neutron shells. The magic numbers turn out to be 2, 8, 20, 28, 50, 82, 126, and 184 neutrons and 2, 8, 20, 28, 50, 82, and 114 protons. There are fewer protons than neutrons in high-energy shells because of the strong electrical repulsion between protons.

Rutherford had to depend on naturally radioactive elements for nuclear projectiles, but today nuclear scientists can draw on an arsenal of modern accelerators in France, Russia, West Germany, and the United States to hurl nuclei at one another. The accelerators strip atoms of their outer few electrons to yield positively charged ions that electromagnetic fields accelerate to great speeds. With their higher energies, these nuclear projectiles can overcome the electrical repulsion of the protons in even the largest target nucleus, penetrate it, and combine with it to form an entirely new, but unstable, nucleus. This compound nucleus decays in order to get rid of the excess energy from the collision. For example, it can emit neutrons, alpha particles (he-

The author:
J. Rayford Nix is a theoretical physicist at the Los Alamos Scientific Laboratory in New Mexico.

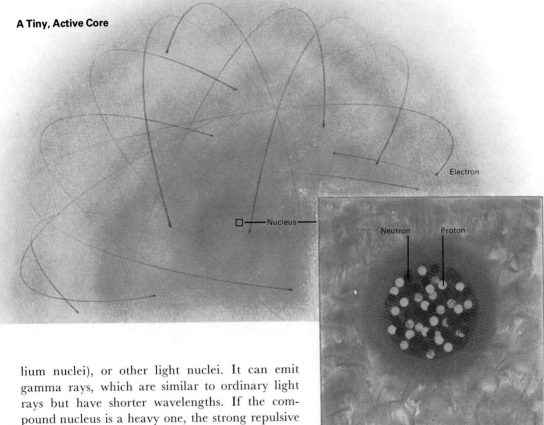

Electron

Nucleus

Neutron   Proton

lium nuclei), or other light nuclei. It can emit gamma rays, which are similar to ordinary light rays but have shorter wavelengths. If the compound nucleus is a heavy one, the strong repulsive electrical forces between protons can split it into two fragments of roughly equal size—the process of nuclear fission.

These stabilizing changes also occur in nuclei of natural radioactive elements. The relative stability of an isotope is determined by its half-life, the time required for half of any quantity of its nuclei to decay. Isotopes whose half-lives are comparable to the age of the universe—about 10 billion years—are considered stable for all practical purposes. About 300 such stable isotopes, produced when the universe was created, still exist in nature. These range from hydrogen, which contains a single proton, to plutonium 244, which contains 94 protons and 150 neutrons. Some 1,200 additional isotopes have been produced artificially by the reactions that occur when projectiles strike targets. Their half-lives range from a fraction of a second to years.

Nuclear scientists have found it useful to draw a kind of map of nuclei in order to compare them. Each nucleus is placed on the map by first moving northward by the number of protons it has, then continuing eastward by the number of its neutrons

One ten-thousandth an atom's size, a nucleus, *above,* is made of protons and neutrons that circle in imaginary shells of increasing radius. Calcium 40, *below,* has completely filled shells. Such nuclei are "magic" because they are more stable than slightly larger or smaller nuclei.

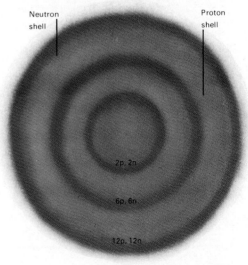

Neutron shell

Proton shell

2p, 2n

6p, 6n

12p, 12n

to reach its position. The plot of the 1,500 isotopes forms a narrow thumb, or peninsula, that extends diagonally from the southwest to the northeast, with the heaviest nuclei closest to the tip. A sea of instability, representing nuclei that have not yet been observed, surrounds the peninsula. Like a real peninsula, this one is higher inland than along its shore because scientists use height above sea level to represent stability. Stable nuclei tend to fall in the middle of the peninsula. Magic nuclei such as lead or tin tower above the others.

When nuclei along the coasts of the peninsula decay, they move toward more stable nuclei in the peninsula's interior. Nuclei along the southeastern coast decay by emitting an electron, which converts a neutron into a proton and moves the nucleus to the northwest. In contrast, nuclei along the northwestern coast decay by capturing an electron, which converts a proton into a neutron and moves the nucleus to the southeast. Nuclei along the northeastern tip decay either by emitting an alpha particle, which consists of two protons and two neutrons, or by nuclear fission. Both alpha decay and nuclear fission are caused by electrical forces, which grow stronger as the number of protons in the nucleus increases.

Until about 1940, researchers thought that the nuclear peninsula ended at uranium 238, which contains 92 protons and 146 neutrons. However, the peninsula has been extended again and again since then by bombarding known nuclei with neutrons and with heavier projectiles. Many of these new elements have been made in recent years by Albert Ghiorso, Glenn T. Seaborg, and their colleagues at the Lawrence Berkeley Laboratory in Berkeley, Calif. The search for everheavier nuclei has also been carried out by Georgii N. Flerov and his colleagues at the Joint Institute for Nuclear Research in Dubna, Russia. The heaviest nucleus produced to date contains 106 protons and 157 neutrons, a total of 263 nucleons.

About 1965, theorists realized that an island of relatively stable superheavy nuclei may exist beyond the tip of the peninsula in the vicinity of 114 protons and 184 neutrons. These magic numbers correspond to the next completely filled proton and neutron shells after those of the heaviest known filled-shell nucleus, lead 208. The nucleus with 114 protons and 184 neutrons should be the most stable against spontaneous fission, but it should be less stable against alpha decay than nuclei that contain fewer protons. When all possible modes of decay are taken into account, the nucleus with 110 protons and 184 neutrons should be the most stable. Calculations indicate that it should live on the average for more than 1 billion years.

How can we journey from the peninsula to this predicted island of stability? As is true for any island, there are two general ways to go—by sea and by air. By sea, imagine that you can sail to the southeastern, or neutron-rich, shore of the island through the multiple capture of neutrons. In this process, a nucleus first moves east by capturing one or more neutrons, then moves northwest by converting a neutron into

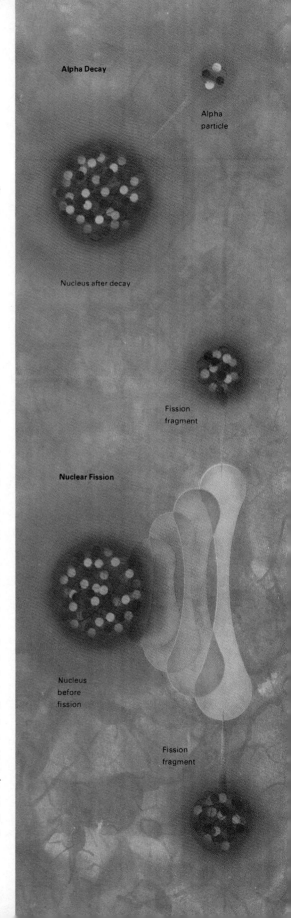

Alpha Decay

Alpha particle

Nucleus after decay

Fission fragment

Nuclear Fission

Nucleus before fission

Fission fragment

Stabilizing Nuclei

Unstable heavy nuclei can rid themselves of excess energy either by emitting an alpha particle (helium nucleus) or by splitting into two lighter fragments in the process of spontaneous nuclear fission.

a proton. Many naturally occurring nuclei were made in this way from the enormous quantities of neutrons that were released in supernovas, stars that explode suddenly.

The belief that multiple neutron capture may also have produced some superheavy elements has caused many scientists to search for them in nature during the past 10 years. The quest has led researchers to some unusual spots—from a 14th-century Russian Orthodox Church to the floor of the South Pacific Ocean off the Fiji islands, from California platinum and gold mines to meteorites and moonrocks. The Russian scientists examined the church's old stained-glass windows, which are held together by strips of lead. They hoped to find element 114 along with the lead because the chemical properties of these two elements should be similar. If element 114 were present in the lead strips and had decayed by spontaneous fission during the past 600 years, the fission fragments would have left tracks in the glass that would be visible under a microscope. The ocean search involved examining manganese nodules that form around objects such as sharks' teeth. Many rare elements are found in greatly concentrated form in these nodules, and researchers hoped that superheavy elements would be among them. Scientists studied platinum and gold because their chemical properties should be similar to those of elements 110 and 111, respectively. They examined samples from outer space because these could have been formed under conditions more favorable to the production of superheavy elements than those on earth.

The most recent search for superheavy elements in nature has been in monazite minerals from Madagascar, an island off the southeastern coast of Africa. Monazite is a phosphate mineral containing primarily rare-earth elements but also some thorium and uranium. Many of the small grains of monazite found embedded in mica are surrounded by spheres of radiation damage that look like halos when examined under a microscope. Most of the

Journey to the Island of Superheavy Nuclei

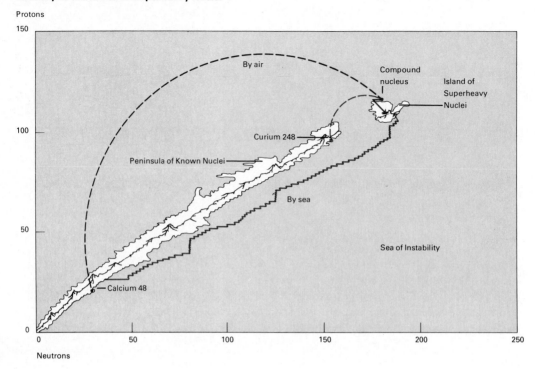

Placed on a map according to their number of protons and neutrons, the 1,500 nuclei found in nature or made in accelerators form a peninsula. Beyond its tip may lie an island of relatively stable superheavy nuclei. But all attempts to reach the island by air (nuclear collisions) or by sea (neutron capture) have failed.

halos are caused by the alpha decay of thorium and uranium in the monazite, but the origin of some larger so-called giant halos is not completely clear. In a 1976 experiment, monazite grains inside giant halos were bombarded with protons from a Van de Graaff accelerator. Some of the X rays produced by the collisions had the same energies as those expected from element 126 and other superheavy elements. However, later experiments showed that these radiations came from nuclear reactions involving lighter nuclei rather than superheavy elements. In all, about 15 claims for the discovery of superheavy elements in nature have been made since 1965, but none have been substantiated by further investigations.

The failure to find superheavy elements in nature probably indicates that they cannot be made by multiple neutron capture. Several calculations suggest that the heavy neutron-rich nuclei formed in the sea between the peninsula and the island of superheavy nuclei undergo fission and thereby "sink the ship."

To journey to the island by air, imagine that you can fly to its northwestern, or proton-rich, side by means of a collision between two heavy nuclei. There are two promising ways to do this. In the first approach, researchers fire a projectile such as calcium 48 at a target such as curium 248. The "landing strip," or site, for this collision is located at 116 protons and 180 neutrons, somewhat northwest of the island's center. In most cases, the colliding nuclei break apart into smaller nuclei instead of merging to form a compound nucleus. When a compound nucleus does form, its nucleons are moving so wildly that it undergoes nuclear fission. Occasionally, however, it decays to a more stable nucleus by emitting neutrons, alpha particles, or other light nuclei, which leads scientists to hope that some superheavy nuclei that live for years may result. In the second approach, the nuclear chemist fires a somewhat heavier projectile, part of which may combine with the target nucleus to form a superheavy nucleus while the remaining portion flies on past the target.

During the past 10 years, many nuclear scientists have taken off for the island of stability. None have yet reached the island, but with the development of new accelerators, many future flights will be attempted. Meanwhile, the attempts already made have led to several important discoveries, including a new type of reaction between nuclei. This highly inelastic scattering reaction was discovered in the early 1970s by scientists at the Joint Institute for Nuclear Research in Dubna and at the Institute of Nuclear Physics in Orsay, France. Scientists working at the Lawrence Berkeley Laboratory have also added to our knowledge of this new reaction.

The importance of this reaction can best be described by considering what happens in a typical experiment in which a beam of krypton 84 nuclei strikes a thin foil of bismuth 209. The results depend in part on how close to the center of a target nucleus a projectile strikes. Because the bismuth atoms are mostly empty space, most of the krypton projectiles miss the target bismuth nuclei by a wide margin.

Many of the krypton nuclei pass close enough to the positively charged target nuclei for the target and projectile to repel each other just as in Rutherford's classic experiment with alpha particles and gold foil. A projectile nucleus can come so close that the nuclear surfaces almost touch and the strong force that acts like a nuclear glue attracts nucleons in one nucleus to those in the other. In this case, the krypton nucleus emerges from the target at a smaller angle than it would have in the presence of the repulsive electrical force alone. The collision process is called elastic scattering because the internal structures of the two nuclei do not change.

Some projectiles just graze their targets, so that the nuclear surfaces barely overlap. This produces inelastic scattering, in which some of the energy from the projectile's original motion may be converted into random motion of the nucleons inside the target or projectile or both. Sometimes, nucleons in one nucleus move to the other nucleus.

Dubna, Orsay, and Berkeley experimenters found that something radically new happens if the projectile strikes closer to the center of the target nucleus. The two nuclei flow together, forming an unbalanced dumbbell. Most of the projectile's energy of motion is converted into energy of motion for the nucleons in a process similar to what occurs when two bags of marbles collide. Because the nuclei hit off-center, the dumbbell that forms begins to rotate and the neck connecting the two

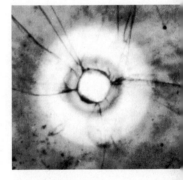

A giant halo surrounds a small grain of the mineral monazite embedded in mica. Bombarded with protons in a 1976 experiment, monazite grains radiated X rays, seeming to indicate the presence of superheavy element 126. Further research attributed the rays to lighter nuclei.

Probing the Heart of the Matter

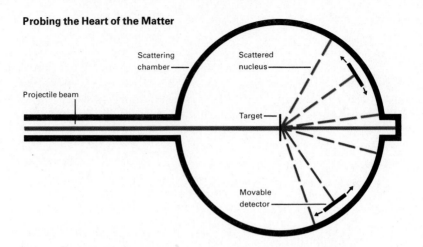

When fast-moving projectile nuclei strike target nuclei, *above,* they produce scattered nuclei whose charges, masses, energies, and angles of emergence reveal the nuclei's nature. Most projectile nuclei miss the target nuclei, *opposite page.* Others interact and veer off course. A few strike off-center, and the nuclei flow together to form an unbalanced dumbbell that rotates slightly before breaking apart in the process of quasifission. A direct hit may yield a single, excited compound nucleus.

ends initially grows thicker because of the nuclear attractive force. However, the electrical repulsion ultimately pushes the two ends apart. The elongation slows down the rate of rotation, just as spinning ice skaters slow down by extending their arms. The dumbbell snaps into two rotating and oscillating pieces that fly apart.

Some nucleons are transferred between the target and projectile; they usually move from the larger nucleus to the smaller one, which tends to make the fragments more nearly equal in size. The number of transferred nucleons is generally greater if the projectile strikes the target close to its center, leading in some cases to fragments that are roughly equal in size. This highly inelastic reaction is often called quasifission because the fragments have properties similar to those of normal fission fragments, but the projectile and target never form a single compound nucleus.

If the projectile strikes the target nearly head-on, the two may coalesce into a single compound nucleus. Such compound nuclei form most easily when the target and projectile contain only a moderate number of nucleons. When they contain many nucleons, the composite nucleus cannot withstand the strong electrical repulsion among protons and breaks apart almost instantaneously.

Such experiments at the new accelerators tell us what happens when two heavy nuclei collide, and theorists are busily calculating what should happen on the basis of models of nuclei and comparing these predictions with the experimental data. They use two approaches that represent extensions of methods that were developed to

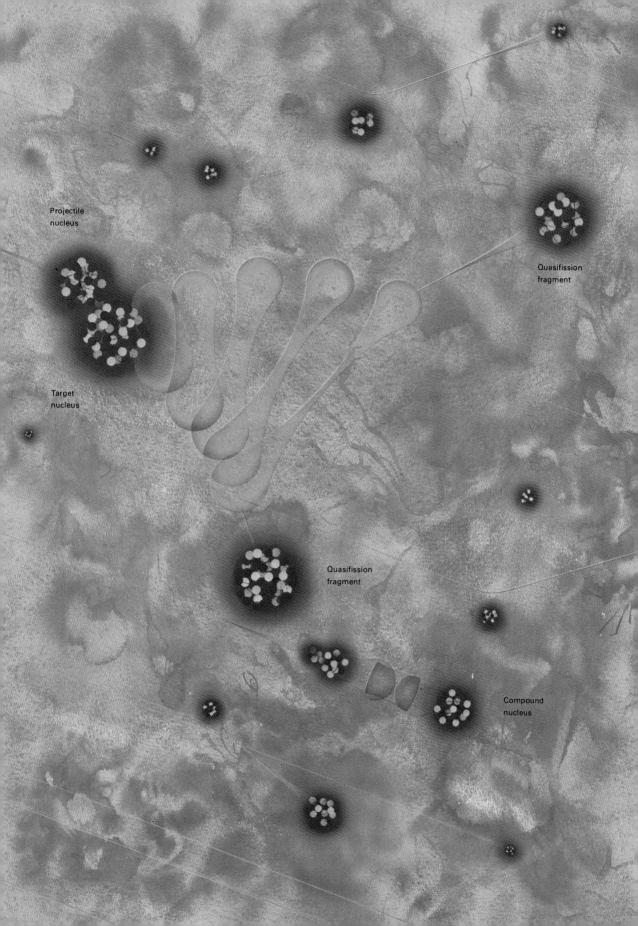

Projectile nucleus

Quasifission fragment

Target nucleus

Quasifission fragment

Compound nucleus

study nuclear fission. In a microscopic approach, the theorist starts with the force between two nucleons and then calculates the position and velocity of every nucleon during the collision. In a macroscopic approach, the theorist ignores individual nucleons and deals with large numbers of particles as though the nuclei were drops of liquid.

Both approaches show the outcome of a collision to be like a tug of war between two groups of competing forces: (1) the nuclear force, which tries to hold the nucleons together; and (2) the electrical repulsion between positively charged protons and the centrifugal force caused by rotation, which try to tear the nuclei apart. Nuclear viscosity, a type of internal friction, also affects the tug of war because some of the overall motion is converted into random motion of the nucleons, a type of nuclear heat. Both experimental results and calculations indicate that, because of quasifission, it is far more difficult to form a compound nucleus than physicists had originally thought.

The difficulty is perhaps best understood by a geographic analogy in which the sum of all the forces–nuclear attraction, electrical repulsion, and centrifugal repulsion–is represented by the steepness of a mountainside. The force at any one place would simply be the downward pull of gravity on a ball placed there. Creating a superheavy nucleus is comparable to driving a golf ball from the base of the imaginary mountain into a hole near the top. An added challenge for the golfer-scientist is that he must adjust his stroke so that about halfway up a broad slope the ball hits and merges with a target ball. Together, the balls must then roll uphill over a pass and into a steep, narrow

A researcher mounts a detector on a new spherical scattering chamber, *left,* at the Lawrence Berkeley Laboratory (LBL) in California. In a different experiment in LBL's streamer chamber, *above,* about 80 fragments leave a spectacular shower of tracks after a collision between an argon nucleus moving at nearly the speed of light and a lead-oxide nucleus.

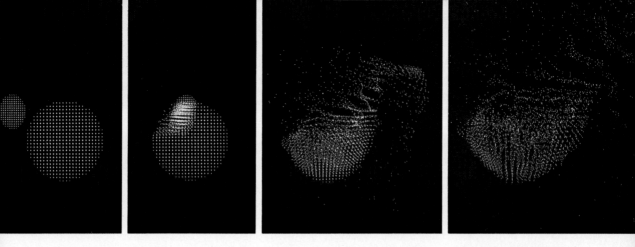

Computer drawings show what may happen when a neon-20 nucleus moving at nearly the speed of light strikes a uranium-238 target nucleus off-center.

canyon that contains the cup. As if the terrain were not tricky enough, nuclear viscosity acts as a kind of thick, wet mat of tall grass on the slope and it slows the balls down. Usually, the two balls separate or break up into several smaller balls that roll back downhill.

Experiments that show quasifission are done with projectiles that have enough energy to overcome the electrical repulsion of target nuclei. However, a new accelerator at the Lawrence Berkeley Laboratory can now accelerate projectiles such as neon to energies hundreds of times as great as this. Moving at nearly the speed of light, these projectiles can literally smash their targets. In grazing collisions, the nucleons in the region where target and projectile overlap are torn away and ejected as individual protons, neutrons, or other light nuclei. The remaining portions of the target and projectile emerge as large fragments. Collisions closer to the center of the target are more violent, and the target and projectile are more thoroughly disintegrated. Such collisions also produce large numbers of pi-mesons, short-lived particles one-seventh as massive as nucleons.

High-energy collisions act somewhat like collisions between individual nucleons and somewhat like collisions between two drops of fluid. When two heavy nuclei collide at high energy, they become compressed. This compression may form shock waves that move through the target, such as those that occur when a supersonic jet plane breaks the sound barrier. With sufficient compression, the nucleons may form a new kind of stable, dense nuclear matter that may also exist in certain stars. Another new kind of nuclear matter containing pi-mesons in addition to nucleons might result. Researchers have yet to create these new forms of nuclear matter, which would be similar to the forms water takes when it freezes into ice or vaporizes into a gas.

We are now setting out on a new journey. By studying collisions between heavy nuclei, we hope to learn more about them, to produce superheavy elements, and to find new forms of nuclear matter. But we should keep in mind that we are entering an unexplored realm of science. Like Christopher Columbus, who set out to find a new route to the Orient and discovered America instead, we are likely to make far more important discoveries than we intended. Wherever we land, we are on one of the most intriguing scientific voyages of the century.

127

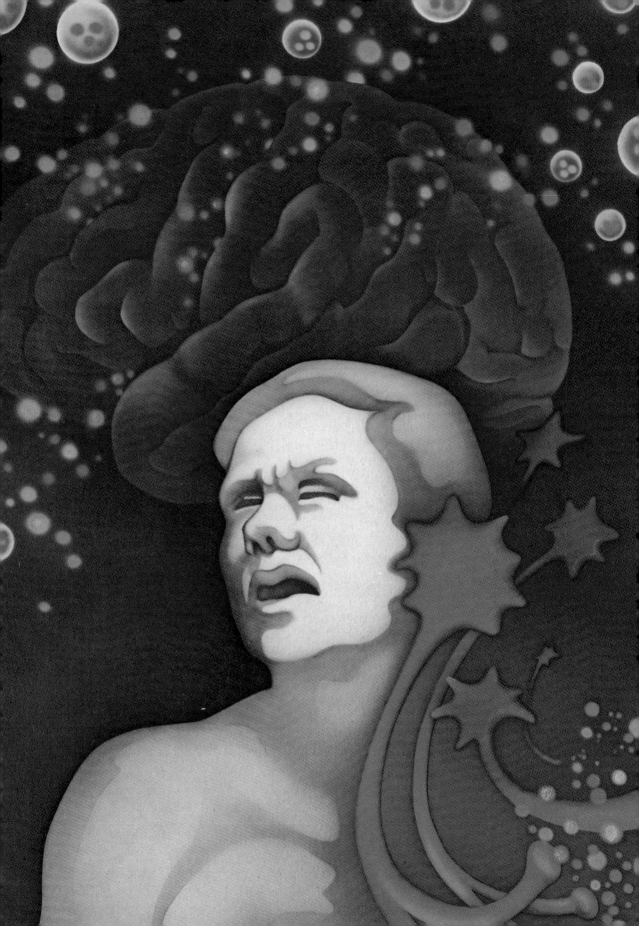

Our Body's Own Narcotics

By Solomon H. Snyder

Our brain and spine contain chemicals that could become revolutionary new drugs for treating both physical and emotional pain

Two substances that seem to have great promise as potent drugs against human pain were discovered in the brain in 1975. By 1977, scientists throughout the world had learned a great deal about these remarkable substances, which are called enkephalins. Evidently they are the body's own narcotics, dulling pain in much the same way that morphine, opium, heroin, and other narcotics do. The enkephalins may also prove to be extremely effective against emotional disorders.

The discovery of enkephalins sprang from a series of events that began with a growing suspicion that the nerve-rich brain or spinal cord, or both, contained narcotic receptors. These are sites on cell surfaces to which the narcotics would have to bind, much as a key fits in a lock, in order to produce their characteristic effects. Scientists assumed that narcotics must act via such highly specific sites, on nerve cells rather than in a general or widespread way, for several reasons.

For example, some narcotics are extraordinarily potent. The powerful drug morphine has a chemical cousin, etorphine, that can relieve the same amount of pain in doses only $\frac{1}{10,000}$ as large. A typical effective dose of such a drug does not provide enough molecules in the body to act in a widespread way. In addition, narcotic antagonists, commonly administered to drug-overdose victims, that stop narcotics from acting are also effective in very low doses. These antagonists, such as the drug naloxone, probably fit the narcotic receptors and bind to them. In this way, an antagonist would block access of narcotics to the receptors.

One way to prove that narcotic receptors exist is to mark a narcotic or an antagonist so that it can be traced in the body. Then we can check to see if it attaches itself to brain or spinal tissue, or both. In 1973, in our laboratory at the Johns Hopkins University School of Medicine in Baltimore, graduate student Candace B. Pert and I marked naloxone by making it highly radioactive. Then we added it to cells taken from rat brains and found that it bound to the cells. Scientists at Sweden's University of Uppsala and at New York University obtained similar results.

In another study using this technique, we found that the receptors concentrated around the synapse, the junction between the taillike nerve ending of one nerve cell and the body of another nerve cell. When nerve cells pass impulses along in communication with one another, one releases a chemical called a neurotransmitter from its nerve ending. The neurotransmitter crosses the synapse and transfers the impulse to the cell body of the other nerve by acting on its outer membrane. The synapse of nerve cells involved in transmitting pain is an ideal place for a narcotic to do its work.

In 1973, we ground up various parts of rat, monkey, and human brain, and added a radioactive narcotic to them. By measuring how much of the narcotic attached to each brain region, we found out how rich in receptors each is.

There were marked differences. The thalamus, a large oblong structure in the back of the forebrain, is a good example. The thalamus is the brain's major center for processing sensory information. It is divided into two parts, the lateral and medial thalamus. The lateral thalamus processes information regarding touch; light pressure to the skin; and sharp, prickly pain; none of which are influenced by narcotics. The medial thalamus processes deep, chronic, burning pain, which is most susceptible to relief by narcotics. We found that the lateral thalamus has few receptors, while the medial thalamus has a great many. Presumably, the receptors in the medial thalamus are at least partially responsible for the ability of narcotics to relieve pain.

The results of these experiments made us wonder just how the narcotic receptors are distributed in detail throughout the brain and spine. To find out, we injected rats and monkeys with radioactive narcotics, sacrificed the animals, and sliced their brains and spines

The author:
Solomon H. Snyder is a professor of psychiatry and pharmacology at the Johns Hopkins University School of Medicine.

into thin sections. Then we placed each of the sections on photographic paper. The radioactive material "developed" the silver grains in the emulsion on the paper, thereby pinpointing the precise location of the receptors in each slice.

This technique revealed tiny areas with extremely dense concentrations of narcotic receptors. The greatest concentrations were in the medial thalamus and other brain parts known to be involved with body functions that are affected by narcotics. For instance, the pretectal nuclei, groups of nerve cells in the brain stem, are loaded with narcotic receptors. These cells regulate the diameter of the pupils of the eyes, so we may have discovered why such drugs as heroin cause the pupils to constrict. The abundance of narcotic receptors in the brain stem's vagus nerve nuclei, where coughing is triggered, may explain why narcotics relieve coughing. In the spinal cord, narcotic receptors are concentrated in a dense band known as layers I and II that runs up the spine like a column. Neurophysiologists have known for some time that this band contains the first synapse points for the nerves that carry pain perception.

We were surprised to find that the amygdala, which has no known role in physical pain, has the greatest concentration of narcotic receptors in the brain. The amygdala, located just under the cerebral cortex, is part of the limbic system, which plays a major role in regulating emotions. Perhaps narcotics trigger the euphoria that they produce through these receptors.

There is no reason why the body should have evolved receptors for narcotics unless it produces some narcoticlike substance of its own. With this in mind, pharmacologists John Hughes and Hans W. Kosterlitz at the University of Aberdeen in Scotland began a series of experiments in 1975 aimed at finding the body's narcotic. They added extracts of pig brain to mouse vas deferens (part of the male reproductive system) and guinea pig intestine. These two tissues contract when electrically stimulated, but the contraction is inhibited by narcotics in direct proportion to the strength and amount of the narcotic. Thus, the scientists were able to test various components of the brain tissue for narcotic activity until they found a narcoticlike material. They then isolated and purified it.

Meanwhile, pharmacologist Gavril W. Pasternak and I, as well as biochemist Lars Terenius at Uppsala University in Sweden, tried another method of finding the natural brain narcotic. We added substances extracted from calves' brains to a mixture of radioactive narcotics and brain tissue containing narcotic receptors. Then we measured the amount of radioactive narcotics that attached to the receptors. Any natural narcotic in the brain extract would compete for receptors with the radioactive material, so the less radioactive material we found, the more natural narcotic was present in the extract. We did find a natural narcotic, and we showed that the relative amount of it varied in proportion to the number of narcotic receptors.

How Nerves Communicate
One nerve cell passes an impulse to another by releasing a chemical substance called neurotransmitter into the synapse, the space between them. This substance relays the impulse by attaching itself to receptors on the surface of the second nerve's cell body.

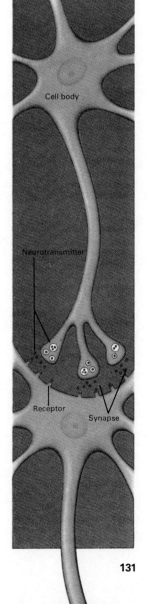

Cell body

Neurotransmitter

Receptor

Synapse

On Dec. 19, 1975, Hughes, Kosterlitz, and their co-workers reported the chemical structure of the newly discovered material from pig brain. It turned out to be two substances, both peptides, which were called enkephalins. A peptide is a chain of amino acids linked to each other. Each of the two brain peptides is only five amino acids long, and they differ just in the identity of one amino acid at one end of the molecule. For one enkephalin, this amino acid is methionine; and for the other it is leucine.

In April 1976, biochemist Rabi Simantov and I isolated and purified the narcoticlike brain substance from calf brains. We also reported its chemical structure to be the same as that of the pig enkephalins. The fact that we had found the enkephalins in the brain and that they acted in the brain and spine seemed strong evidence that they originated in these two organs. But an earlier discovery provided strong evidence to the contrary. Pharmacologist Avram S. Goldstein of Stanford University had shown that substances in the pituitary gland, located just below the skull, could act like narcotics. Furthermore, the sequence of the five amino acids of methionine-enkephalin is also contained within a large 91-amino-acid peptide found in the pituitary gland in 1964. Many scientists reasoned that this large peptide, called beta-lipotropin, might break down to produce the methionine-enkephalin and several other peptides with different functions.

To determine whether beta-lipotropin is the source of brain enkephalins, Goldstein and pharmacologist David Cheung removed the

Locating Narcotic Receptors
Rat brains injected with radioactive narcotics revealed the greatest receptor concentrations (red) in parts that regulate functions affected by narcotics.

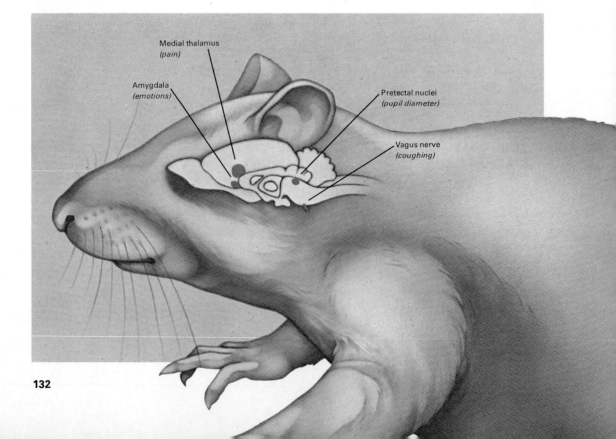

Medial thalamus
(pain)

Amygdala
(emotions)

Pretectal nuclei
(pupil diameter)

Vagus nerve
(coughing)

pituitary glands of rats and checked the enkephalin levels in their brains up to three weeks later. They found no change from normal levels, indicating enkephalins do not originate in the pituitary.

Of course, we need not know where a substance is produced to begin studying it and putting it to practical use. This is particularly true of the enkephalins because their small size makes them easy to synthesize, creating essentially unlimited quantities with which scientists can perform experiments.

Initial experiments quickly pinpointed two major problems that must be solved before the enkephalins can be used effectively as drugs. Although enkephalins injected directly into animal brains are as potent as morphine in killing pain, they are effective for only a very brief time because they break down very rapidly. For this reason, enkephalins are practically useless when they are taken orally or are injected into the bloodstream.

In addition, experiments with animals indicate that the enkephalins are addictive. For example, pharmacologists Eddie Wei and Horace Loh of the University of California, San Francisco, reported in November 1976 on an experiment in which they gave rats constant infusions of enkephalins through a small tube implanted in the brain. Then they measured pain perception by shining a hot light on the animals' tails and observing the rate at which the animals flicked them. The faster the rate, the more pain the rats felt. After a period of time determined partly by the enkephalin-dose level, the animals became tolerant, that is, it took more enkephalins to get pain relief. Tolerance is a typical early stage in addiction. Moreover, when the enkephalin doses were stopped, the rats developed withdrawal symptoms, such as shaking and diarrhea.

Probably enkephalins regulate pain intensity, most likely by modulating the so-called pain threshold, the point at which one begins to perceive a stimulus as painful. If this is true, it may be that narcotic antagonists such as naloxone make a test subject more sensitive to pain. Biologist François Jacob used the tail-flick method at the Pasteur Institute in Paris in 1974 to show that rats become very sensitive to pain when treated with naloxone.

Researchers trying to find out what role enkephalins normally play in human beings are handicapped because they cannot experimentally inject the material into people. But naloxone is known to be a safe drug for human beings: It has been administered to thousands of patients. In October 1976, neurophysiologist Patrick D. Wall and his colleagues at University College, London, gave naloxone to normal, healthy human beings and checked to see if their pain thresholds were altered. The pain measured was created by electric shocks on the forearm. The scientists recorded the number of amperes at which the shock was first perceived and the maximum shock each subject could endure both before and after receiving naloxone. The results seem to indicate that the subjects were no more sensitive to pain after receiving

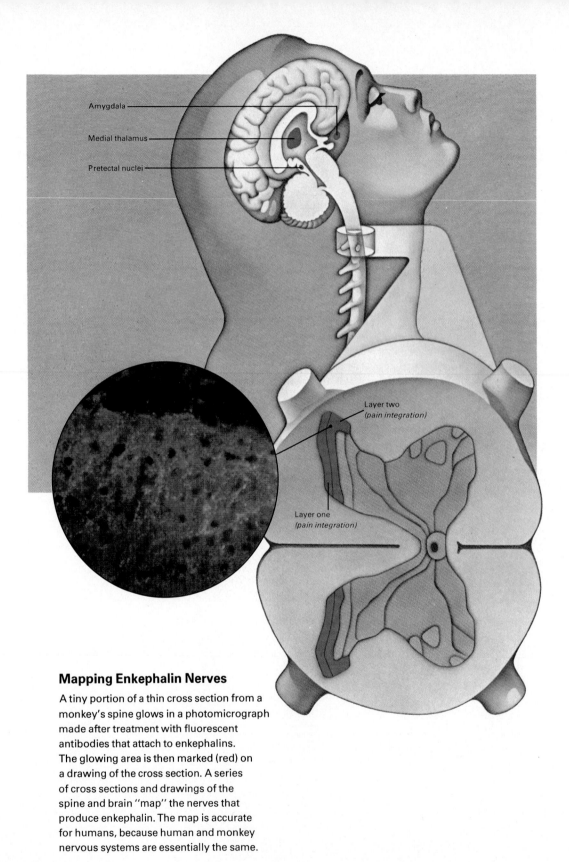

Amygdala

Medial thalamus

Pretectal nuclei

Layer two
(pain integration)

Layer one
(pain integration)

Mapping Enkephalin Nerves

A tiny portion of a thin cross section from a
monkey's spine glows in a photomicrograph
made after treatment with fluorescent
antibodies that attach to enkephalins.
The glowing area is then marked (red) on
a drawing of the cross section. A series
of cross sections and drawings of the
spine and brain "map" the nerves that
produce enkephalin. The map is accurate
for humans, because human and monkey
nervous systems are essentially the same.

the naloxone than before. The researchers suggested that this might indicate that enkephalins dull pain only under unusual psychological conditions. Such conditions might be found in combat, where some soldiers severely wounded in battle seem to be free from pain.

Some researchers are linking the pain relief of acupuncture to enkephalins. Do the precisely placed acupuncture needles trigger the release of enkephalins? Psychologist David J. Mayer explored this question at the Medical College of Virginia in Richmond by determining the influence of naloxone upon acupuncture-induced relief of pain in people's tooth pulp. Tooth pulp is particularly useful in pain studies because pain is virtually the only sensation it transmits to the brain.

In a continuing series of experiments first reported in 1974, Mayer passes an electric current through the tooth pulp via an electrode attached to the tooth surface. He has found that acupuncture alone raises his subjects' pain threshold by an average of 28 per cent, but that acupuncture after an injection of naloxone has no effect. Psychologist Bruce R. Pomeranz of the University of Toronto in Canada described similar experiments with mice in December 1976. Pomeranz reported that he could produce acupuncture analgesia in these small rodents. He measured the effect by the increased amount of time it took the mice to squeak in discomfort when a hot light was focused on them after acupuncture treatment. And, just as Mayer had found in his tests on humans, Pomeranz reported that naloxone blocked acupuncture analgesia in mice.

Even with all this work, scientists had still not found exactly where the enkephalins are produced. Because we had found enkephalins at nerve synapses and because they break down so quickly, we believed they are probably produced at or near the synapse. We also suspected that they are produced by nerve cells not directly involved in transmitting pain. Anatomist Thomas Hokfelt at the Karolinska Institute in Stockholm, Sweden, and also Simantov, pharmacologist Michael J. Kuhar, and I at our lab worked out a way to see enkephalins under the microscope. We inject guinea pigs or rabbits with enkephalins that are chemically coupled to large proteins. The animals' immune systems recognize the dual units as foreign molecules and make antibodies to destroy or neutralize them. In any future reactions, these antibodies will also recognize enkephalins alone as foreign molecules. The blood of the guinea pigs and the rabbits is thus enriched with antibodies against enkephalins.

Isolating the portion of the animals' blood that contains the antibodies, we apply it to thin sections of rat brain and spine tissue mounted on microscope slides. The antibodies bind tightly to the enkephalins in the tissue. To see exactly where this occurs, we treat the tissue with still another antibody. However, this one is fluorescent and binds to the original rabbit or guinea pig antibodies. The entire piggyback arrangement—enkephalins, enkephalin antibodies, and fluorescing antibodies—is now visible under the microscope.

The appearance of fluorescing nerve cells at the synapses of other nerves confirmed our earlier belief that there are enkephalin-producing nerve cells. The concentration of these enkephalin nerves in layers I and II of the spinal cord, in the medial thalamus, and in the amygdala also confirmed our earlier work locating narcotic receptors. Anatomist Carol Lamotte is now using the fluorescent-antibody technique on brain and spine tissue taken from monkeys, which have essentially the same nervous system as do human beings. Lamotte also has found that the enkephalin-producing nerves are concentrated in the areas that are rich in narcotic receptors.

With all these bits and pieces of knowledge about the enkephalins, we can begin to speculate about how they work to alter pain perception. The part of the nervous system where this perception is best understood is the spinal cord. For example, we know that when you burn your hand, free nerve endings of sensory nerves in the skin are stimulated. They send a nerve impulse into a threadlike extension of the nerve ending, up through your arm, and into your spinal cord, where the sensory nerve ends at the synapse with another nerve or several nerves in layers I and II of your spinal cord. The sensory nerve ending then releases a neutrotransmitter that crosses the synapse and triggers an impulse in the other nerve cell or cells. This impulse moves up the spine, ending finally in the brain as a feeling of pain.

We had shown that the enkephalin nerve was also at the synapse. But how did its enkephalin alter the pain impulse passing from the sensory nerve to the others that carried the message to the brain? In May 1976, Lamotte, Pert, and I reported on experiments that help

A Theory of How Enkephalins Work
Enkephalins probably ease the pain from bad burns and other severe injuries. A sensory nerve carries the pain signal into the spinal cord.

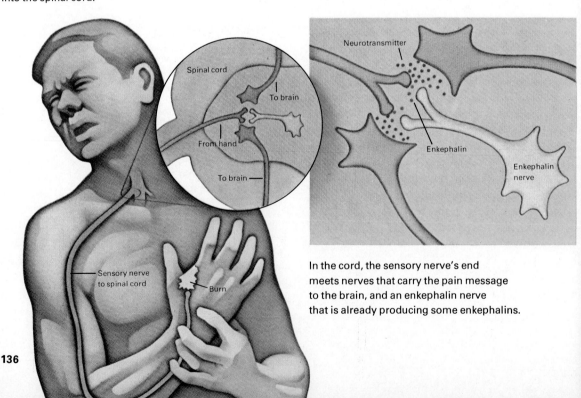

In the cord, the sensory nerve's end meets nerves that carry the pain message to the brain, and an enkephalin nerve that is already producing some enkephalins.

answer this question by precisely locating the narcotic receptor—and, therefore, the enkephalin receptor. We cut the sensory nerves in the necks of monkeys and later examined the area where these nerves are known to end in layers I and II of the spinal cord. We could see that the nerve endings had degenerated. Tests with radioactive naloxone then revealed that narcotic receptors had disappeared along with the sensory-nerve endings, indicating that the receptors are probably normally located on these endings.

We think that the enkephalins, released by the "firing" of the enkephalin nerve, cause a reaction that inhibits the release of the sensory nerve's neutrotransmitter when they bind themselves to the receptors. We do not yet know whether the enkephalin nerve fires regularly and routinely or whether a pain signal triggers it. We suspect, however, that both conditions exist, keeping enkephalins at some normal level at all times, but increasing the level during times of need. Then, with less neurotransmitter moving from the sensory nerves to those nerves that carry pain information to the brain, there will be less pain.

Though we can put together this fairly clear picture of how the enkephalins regulate pain transmission in the spinal cord, precisely how they might act to produce other effects is much less clear. We can assume that, like narcotics, relatively large quantities of enkephalins could cause feelings of well-being. This might explain the role of enkephalins in the amygdala. Perhaps normal quantities of enkephalins in the amygdala act as the body's own tonic against disappointments and losses. This leads us to wonder whether enkephalins play a role in emotional illness. Can a deficiency of enkephalins in those brain re-

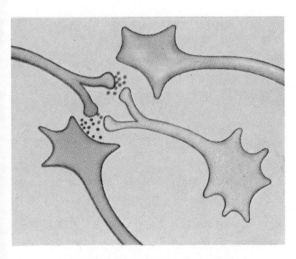

With more enkephalins available to receptors on a portion of the end of the sensory nerve, neurotransmitter production is blocked and less pain is perceived.

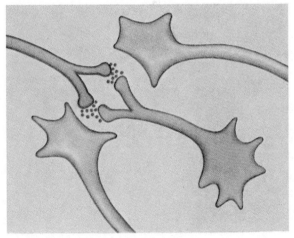

If still more enkephalins are produced, they may block all or nearly all neurotransmitter production, greatly easing or even eliminating the pain.

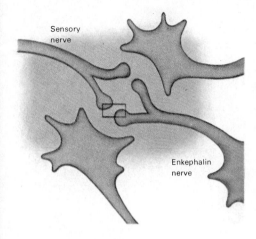

Sensory nerve

Enkephalin nerve

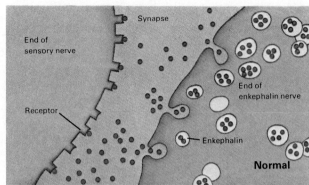

Synapse

End of sensory nerve

Receptor

End of enkephalin nerve

Enkephalin

Normal

Enkephalins and Addiction

Under normal conditions, enkephalins probably fill a specific number of narcotic receptor sites. A feeling of euphoria is experienced when the morphine or some other narcotic fills most of the remaining sites.

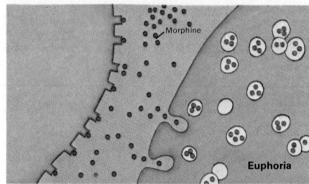

Morphine

Euphoria

Additional morphine overloads the sensory nerve's receptors, and it signals a halt to enkephalin release. Euphoria can then be achieved only by increasing the amount of narcotic that is taken, a phenomenon known as tolerance.

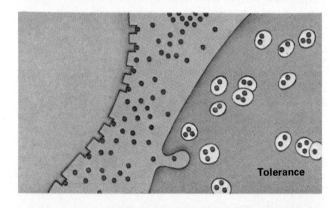

Tolerance

With no enkephalins being released, and no narcotic to fill even the normal number of receptors, the craving that is characteristic of addiction follows. Without the narcotic, receptors stay empty, which results in typical withdrawal symptoms.

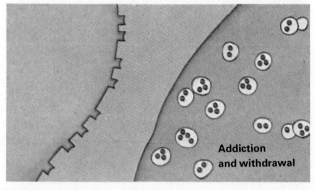

Addiction and withdrawal

gions that are involved in emotions result in increased mental pain and depression? Indeed, physicians and social scientists have long suspected that many narcotic addicts first turned to drugs in an effort to counteract profound emotional pain.

Such possibilities are clearly important. One way to test their validity is to inject narcotic antagonists such as naloxone into test subjects and carefully observe them for psychological changes. Another approach is simply to measure and compare the enkephalin levels in normal people and in those with various emotional disorders. Although, for obvious reasons, we cannot perform experiments in which we sample human brain tissue, we can use spinal taps to measure chemicals in the spinal fluid. Such studies are now underway in our own and other laboratories.

If we assume that enkephalins affect us emotionally in the same way as do narcotics, it helps us understand some of the most puzzling aspects of drug addiction. The major characteristics of addiction are tolerance and physical dependence, and both these characteristics have been difficult to explain. While tolerance refers to the diminishing effect of a drug that is used repeatedly, physical dependence refers to the excruciating withdrawal symptoms that occur when an addict stops taking the drug.

Presumably, the body's narcotic receptors are always exposed to a constant level of enkephalins. When a person receives a substantial dose of morphine, he or she feels euphoric, but the receptors are "overdosed." Then the nerve cells that contain the narcotic receptors probably send a message to the enkephalin-producing nerves saying, in effect, "I am getting too much narcoticlike stuff. Turn off the enkephalin machine." The enkephalin nerves then stop firing, and the narcotic receptors have only morphine with which to contend. When this happens, the person needs even more morphine (to fill the receptor sites left open by the lack of enkephalin) to become euphoric.

If morphine is withdrawn at this point, the receptors not only have no morphine, but they also lack even the normal amount of enkephalins. This condition triggers withdrawal symptoms, which last until the nerve cells that contain the receptors send a new message to the enkephalin nerves, "We don't have any narcoticlike substance. Please begin firing and releasing enkephalin again."

As we learn more about enkephalins, our understanding of narcotic addiction and physical and emotional pain will increase. More important, work to create enkephalin drugs is proceeding at an unprecedented rate. Several drug companies have already synthesized hundreds of enkephalin analogues, slightly altered chemical forms of the material. We hope that some of these will resist breakdown in the body, be as effective as narcotics in relieving pain, and also be nonaddictive. With enkephalin analogues and the knowledge we are gaining, we may soon perfect a host of treatments that will ease much human pain and suffering.

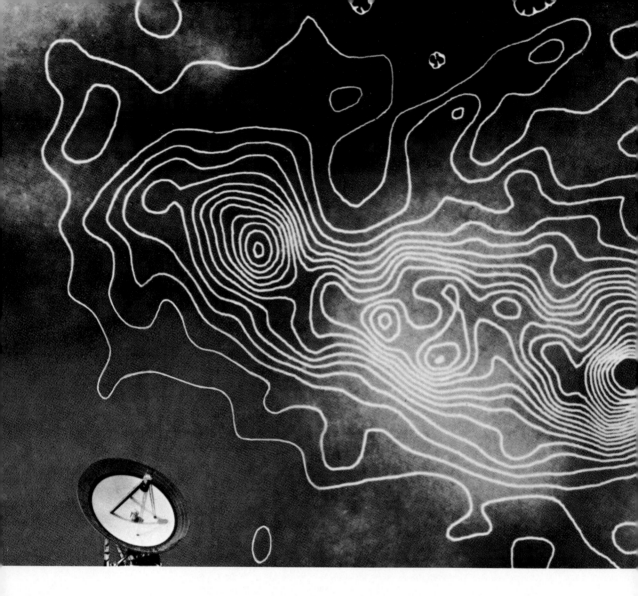

A Sharper Focus On the Universe

By Kenneth I. Kellermann

By linking giant radio telescopes around the world, astronomers can explore the structure of exotic objects in the most distant reaches of space

Nestled deep in the green hills of West Virginia, a dish-shaped telescope keeps an unceasing vigil on the sky. Day and night, through fair weather and foul, it collects weak radio signals coming from distant parts of the universe. Across the continent, in a California desert, a similar telescope points at the same part of the sky. Other radio telescopes, similarly aimed, are located in a remote part of Texas; outside Boston; in the West German countryside south of Bonn; and by the shore of the Black Sea in Russia. When the signals gathered and recorded by each telescope are combined in a computer, they will sketch an image of a distant galaxy with as much detail as if it were produced by a single radio telescope with a diameter nearly as large as the diameter of the earth.

This new technique of radio astronomy is called Very Long Base Line Interferometry (VLBI). Among the exotic objects it is used to study are radio stars; galaxies; pulsars, or pulsating radio stars; quasars, objects that look like stars but emit the power of galaxies; and interstellar masers, clouds that amplify radio waves to many times their original power.

Each day, the earth is bombarded by radiation at many different wavelengths carrying information from throughout the universe. Most wavelengths are absorbed by the earth's atmosphere before they can reach the surface. An earth-based astronomer has only two windows to the rest of the universe – the narrow band of wavelengths that make up visible light, and a restricted range of longer radio waves.

The optical universe is familiar to anyone who looks up into the night sky. But it was not until 1932 that astronomers were able to open the second window to the universe – that of radio astronomy. There was one problem, however. Looking through this window was like looking through frosted glass. Where an optical telescope distinguishes dozens of tiny stars and intricate galactic detail, the image derived from the radio data was a fuzzy blur or blob. One of the most frustrating problems faced by early radio astronomers was the inability to pinpoint accurately the sources of this intriguing radio emission.

Tantalized by the new information coming in about the universe, radio astronomers have struggled to improve the images they received. They have built larger telescopes, worked with combinations of telescopes, and eventually linked telescopes around the world. And with each step, they have discovered new and intriguing celestial objects.

Light from ordinary stars and galaxies comes largely from atoms powered by thermonuclear reactions in the hearts of stars. But the origin of the extremely intense radio, optical, infrared, and X-ray emissions from the more exotic quasars, pulsars, and galactic nuclei has remained a mystery. Detailed investigation of these objects has been restricted by their remarkably small size. Even the largest optical telescopes show them only as fuzzy points of light.

The ability of a telescope to distinguish fine detail, or to see objects that are close to each other as separate images, is called resolution. The resolution of a telescope is determined by the ratio of the wavelength of the radiation received to the diameter of the instrument. Resolution may be increased either by building larger telescopes or by observing radiation at shorter wavelengths. Because radio waves are some 100,000 times longer than visible light wavelengths, astronomers believed for many years that the resolution of radio telescopes had to be many times poorer than that of optical telescopes. For example, radio astronomers often study the 21-centimeter wavelength radiation that is emitted by neutral hydrogen atoms. This is 350,000 times longer than the average wavelength of visible light. At this wavelength, an astronomer would need a radio telescope 53 kilometers (33 miles) in diameter in order to equal the resolution of a 15-centimeter (6-inch) optical telescope.

However, the actual resolution of optical telescopes is usually limited by what astronomers call seeing – that is, the blurring of an image

The author:
Kenneth I. Kellermann is a staff member of the National Radio Astronomy Observatory in Green Bank, W. Va.

caused by the distortion of light waves as they pass through a moving atmosphere. Because of seeing, even a large optical telescope is limited to an angular resolution of about 1 second of arc. A second of arc is $\frac{1}{60}$ of an arc minute, which is $\frac{1}{60}$ of a degree. There are 360 degrees in a circle. Objects of very different actual sizes can have the same angular size, depending on their distance from us. For example, even though the diameter of the sun is 400 times greater than that of the moon, they appear to be roughly the same angular size because the sun is so much farther away from the earth.

The longer wavelengths of radio signals pass through the atmosphere relatively undisturbed. Thus, the resolution of a radio telescope is limited only by its own dimensions. This is limit enough. A conventional radio telescope 100 meters (330 feet) in diameter can produce an angular resolution of only about 1 arc minute, even at its shortest operating wavelength. This is equivalent to the resolution of the human eye and about 100 times poorer than the resolution of an optical telescope under good observing conditions.

To improve resolution, radio astronomers in 1947 began to link two or more separate telescopes to form an array. Each pair of telescopes in the array forms a radio interferometer, which works on the principle of wave interference. If the waves collected by each telescope are exactly in phase—that is, if their crests and troughs match exactly—they combine to make a stronger wave. If they are out of phase, they will cancel each other out. Unless the source is directly overhead and midway between the two telescopes, the distance to one telescope is always slightly greater than to the other. As the rotating earth carries the telescopes along, this difference in distance constantly changes, and the waves received by the telescopes will be alternately in and out of phase. The resulting pattern is called an interference fringe pattern. By studying the changes in this pattern, radio astronomers can glean information about the size and structure of the source. If the telescopes are suitably spaced and the signals from each interferometer are properly combined, a picture of a radio source may be formed with a resolution corresponding to the overall dimensions of the array, rather than those of just one of the telescopes.

Conventional interferometer arrays have diameters up to a few kilometers. The Very Large Array now under construction near Socorro, N. Mex., will have 27 telescopes in a Y-shaped array covering a circle with a diameter of about 35 kilometers (22 miles) when it is completed in 1981. It will be the largest and most sensitive radio telescope array in the world. At its shortest operating wavelength, it will have a resolution considerably better than 1 arc second, better than the largest earthbound optical telescope.

Theoretically, there are no limits to the dimensions of a radio telescope array, but there are practical restrictions. In smaller interferometer arrays, connecting cables carry the signals from each telescope to a central processor where they are combined in a computer. But as the

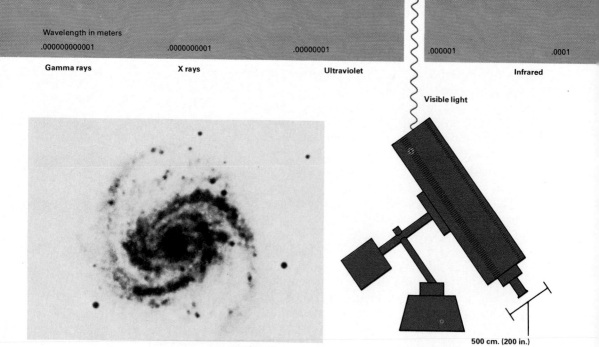

500 cm. (200 in.)

Windows in the Spectrum
Only visible light and some radio waves penetrate the atmosphere and reach the Earth's surface. Compared to an optical telescope, even the largest radio telescope gives poor resolution. Linking radio telescopes in an interferometer array sharpens the image. A Y-shaped array under construction will give resolution similar to the best optical telescope.

arrays expand, the cost of the cable needed to carry the signals to the central processor becomes prohibitive. Natural obstacles such as rivers, mountains, and eventually oceans also limit the dimensions of interferometers connected by cables.

To extend the base lines and obtain even higher angular resolution, astronomers in the early 1960s began using microwave radio links to connect telescopes up to about 100 kilometers apart. This gives a resolution of about 0.1 arc second. In the mid-1960s, however, astronomers began to suspect that much smaller radio sources, with dimensions less than 0.001 arc second, might exist in quasars, in the nuclei of highly energetic galaxies, and in the giant clouds of gas that are interstellar masers.

Resolutions of this order could only be achieved with the transcontinental or intercontinental base lines of VLBI. To cover such distances with microwave radio links would require building many costly amplifier stations, similar to those used by the telephone industry. Instead, scientists developed new techniques, using magnetic tape to record the signals received by widely spaced radio telescopes. To synchronize the tapes precisely, they mark time signals on the tapes with the help of sophisticated atomic clocks that are accurate to within one-millionth of a second per year.

Long radio waves

Radio waves

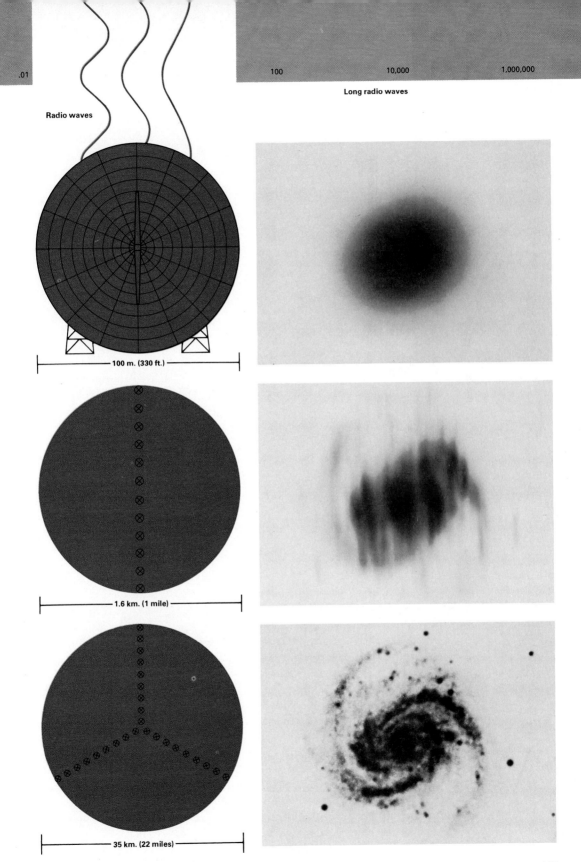

100 m. (330 ft.)

1.6 km. (1 mile)

35 km. (22 miles)

Where the Telescopes Are

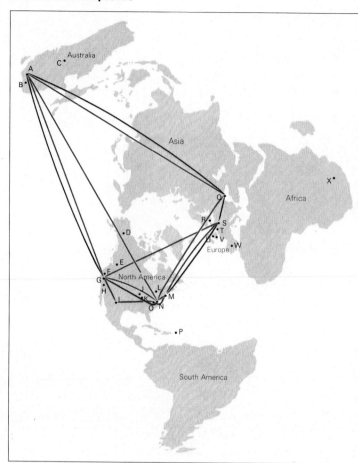

A. Tidbinbilla, Australia
B. Parkes, Australia
C. Woomera, Australia
D. Fairbanks, Alaska
E. Penticton, B.C., Canada
F. Lassen Park, Calif.
G. Owens Valley, Calif.
H. Barstow, Calif.
I. Fort Davis, Texas
J. Iowa City, Iowa
K. Danville, Ill.
L. Algonquin Park, Ontario, Canada
M. Haystack, Mass.
N. Naval Research Laboratory, Maryland Point, Md.
O. Green Bank, W. Va.
P. Arecibo, Puerto Rico
Q. Semeiz, Russia
R. Gothenburg, Sweden
S. Bonn, West Germany
T. Dwingeloo, The Netherlands
U. Jodrell Bank, England
V. Slough, England
W. Madrid, Spain
X. Johannesburg, South Africa

Radio telescopes around the world can be linked in various combinations to study distant sources.

Canadian astronomers began operating the first successful VLBI system in 1967, followed a few months later by a United States link. Within a few years, VLBI systems as large as the earth's diameter had been developed, involving the cooperative efforts of radio astronomers around the world. More than 10 telescopes in the United States, 2 in Canada, and others in Australia, Great Britain, the Netherlands, Russia, South Africa, Spain, Sweden, and West Germany have been joined in various combinations to unravel the radio emission from sources at the limits of the universe.

Many quasars and energetic galaxies are surrounded by vast radio clouds, millions of light-years across, that radiate with the power of 10,000 billion suns. The radio emission is caused by electrons moving through weak cosmic magnetic fields at close to the speed of light. Tremendous energy is required to power these giant radio galaxies and quasars, and the source of such an enormous energy supply and the mechanism by which it is converted into particles and magnetic fields has been one of the major questions of modern astrophysics. For

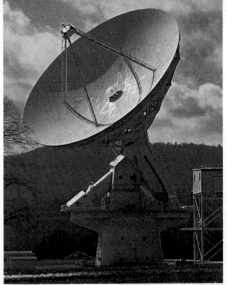

Radio telescopes come in many different designs and are found in remote, widely separated locations. The setting sun silhouettes the telescope near the desert at Goldstone, Calif., *above.* Others nestle in the hills of Green Bank, W. Va., *above right,* and the West German countryside near Bonn, *right.* Dusk settles over the telescope at Onsala, Sweden, *below.*

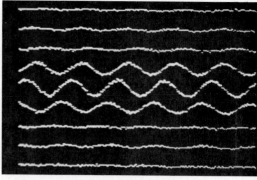

When the signals from two telescopes are precisely synchronized in a computer, wavy interference fringes appear. If the synchronization is off by 500-billionths of a second, the fringes disappear.

many years, astronomers have speculated that the energy supply lies deep within the galactic nucleus or quasar, perhaps in the form of a giant black hole, or perhaps as some yet-unknown physical process. But it has not been possible to study the source of energy in any detail.

Through VLBI observations, radio astronomers have found that some quasars and galactic nuclei cover an angle of less than 0.001 arc second. Even at the immense distances of these objects, this angle is equivalent to a diameter of only a few light-years (a light-year is the distance light travels in a year). By comparison, the Milky Way Galaxy is more than 100,000 light-years across. Yet one of these extraordinarily small quasars or nuclei holds an energy source so fantastically powerful that it can apparently power the giant surrounding radio-emitting clouds for hundreds of millions of years. Understanding this process is a formidable challenge.

The smallest galactic radio nucleus that has been observed so far is the one in the center of our own Galaxy, the Milky Way. The nucleus is only one-fourth the size of our solar system, about 10 astronomical units across. (One astronomical unit equals the distance from the earth to the sun.) With this high-resolution look at the nucleus of our galactic system, astronomers hope to learn about the more powerful energy sources that lie within other galactic nuclei. Other compact radio sources are found scattered throughout our Galaxy. They are associated with double-star systems, X-ray-emitting stars, or pulsars.

VLBI observations show a remarkable amount of detailed structure in some of the quasars and galactic nuclei. Often the radio emission appears to come from two or more regions that are several light-years apart. These two radio blobs are often found to be separating, as if receding from a powerful explosion. In some cases, the measured velocity of the objects seems to be greater than the speed of light—a truly remarkable and unexpected result with profound implications because, according to the laws of physics, nothing can be accelerated to a speed beyond that of light.

Radio astronomers first found evidence of this motion, which they dubbed superlight, in a series of observations made between 1968 and 1970. But they have only recently accumulated enough data to convince themselves that they are not being fooled by inaccurate or inadequate measurements. Since the initial discovery, radio astronomers from the California Institute of Technology, the Massachusetts Institute of Technology, the National Aeronautics and Space Administration (NASA), the National Radio Astronomy Observatory, and West Germany's Max-Planck-Institut für Radioastronomie have collaborated to demonstrate rather convincingly that the double radio components in several quasars and one radio galaxy are indeed separating with apparent speeds between 5 and 10 times the speed of light.

The best-studied object is quasar 3C 345. Transcontinental and intercontinental VLBI measurements have shown that if 3C 345 is as far away as is generally believed, then its components are flying apart at about eight times the speed of light. Similar results have been found in quasars 3C 273 and 3C 279, and radio galaxy 3C 120. Since 1970, the quasar 3C 279 and the radio galaxy 3C 120 have each sent out two pairs of radio-emitting clouds that are separating with apparent superlight velocities. In both cases, the second pair appeared just as the first was fading away, and seemed to follow the same path as the first pair.

According to conventional physics, nothing can be accelerated to travel faster than the speed of light, so how do we explain these strange results? The suggested explanations fall into three categories, some of which challenge our fundamental concepts of physics and cosmology.

The most straightforward and least exotic explanation describes the superlight motion as some sort of illusion. Instead of actual motion, perhaps a series of stationary sources flare up in succession to give the illusion of motion, much as theater marquee lights appear to move. But this type of activity should produce components that appear to be moving toward each other as well as separating, and so far only separations have been observed.

Or the source may be moving at just under the speed of light in a direction toward the observer almost along the line of sight. The signal from such a source would approach the observer only slightly faster than the source itself. The velocity of the sideways motion may then seem to exceed the speed of light. But this, too, would be only an illusion. And because it is improbable that all fast-moving components would be so conveniently aligned, it is difficult to use this interpretation to explain why apparent superlight motion is seen in about half the radio sources that have been regularly studied.

Another possibility is that these objects really are moving faster than the speed of light. Contrary to popular thought, the laws of relativity permit this and physicists have speculated for some years that faster-than-light particles might exist. They call such particles tachyons,

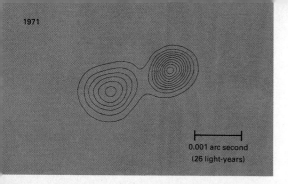

1971

0.001 arc second
(26 light-years)

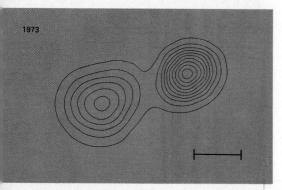

1973

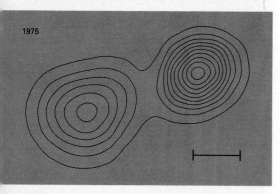

1975

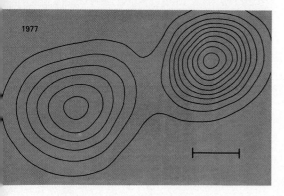

1977

Faster than the Speed of Light
Radio emission from quasar 3C 345 seems to
come from two distinct regions that are several
light-years apart. Observations since 1971
show that the components appear to be flying
apart at about eight times the speed of light.

Tiny regions in the Orion Nebula, *opposite page,*
act as an interstellar maser to amplify the radio
emissions from hydroxyl ions and water vapor.

from the Greek word *tachys,* meaning swift. According to the laws of relativity, ordinary matter cannot
be accelerated past the speed of light because, as its
speed increases, the mass increases as well. Steadily
greater forces are needed to increase the speed further, until, at the speed of light, an infinite force is
required to make the particle go faster. But imagine
the tachyon, a particle that by definition always
travels faster than the speed of light. Then an
infinite force would be needed to slow it down to
less than the speed of light. Thus, the speed of light
is a sort of barrier that cannot be crossed from
either direction. Although tachyons have never
been seen in earth-based laboratories, mathematically they can exist. Many of the now well-known
elementary particles of modern physics were predicted mathematically years before they were
found in the laboratory. Perhaps the high-resolution VLBI systems have detected tachyon motion
deep inside the enigmatic quasars.

Yet another explanation is that the quasars are
not as far from the earth as most astronomers believe. The distance of a galaxy or quasar is calculated from the red shift of its optical spectrum.
When a source of light is receding from the viewer,
the waves of energy it emits appear to be length-
ened so that a specific emission line is shifted toward the red side of the spectrum. The amount of
displacement, or red shift, is proportional to the
object's speed and distance. Quasars have very
large red shifts and the conventional interpretation
of their red shifts places them near the limits of an
expanding universe. However, not all astronomers
agree. Ever since the first quasar was identified in
1963, many astronomers have challenged this interpretation of quasar red shifts, which seem to indicate an inexplicable source of energy to account for
the intense radio, optical, and infrared emission
observed. Indeed, some observers have used the
discovery of apparent superlight motion to support
their arguments that quasars must lie much closer
to the earth. If quasars are, in fact, closer than is
generally assumed, then their components would
be separating at less than the speed of light.

But no other viable explanation for the great quasar red shift has been offered. Any other interpretation of the large red shifts would also imply fundamentally new physical laws.

Quite apart from the broad range of radio emission found in galaxies and quasars is the narrow band of radio emission from giant interstellar clouds of molecules such as formaldehyde (H_2CO), alcohol (CH_3OH), water vapor (H_2O), or ions such as the hydroxyl (OH) radical. Unusual temperature, pressure, and density conditions may be found in interstellar space around areas where new stars are forming or around very old stars radiating at infrared wavelengths. Astronomers theorize that, under these conditions, huge clouds of H_2O or OH may act as giant interstellar masers, greatly amplifying radio waves over a narrow range of frequencies. VLBI studies show that the radio emission from these masers comes not from a single point, but from a number of very tiny regions, some less than a few astronomical units across, and each moving with a somewhat different velocity. These tiny regions may be condensations in the interstellar gas where stars and planetary systems could be forming.

In order to get a closer look at these tiny regions, radio astronomers from Australia, Russia, and the United States teamed up in 1976 to study the 1.3-centimeter emission from interstellar water-vapor clouds. Deep inside a celestial cloud known as W49 where new stars are being born, they found extremely small water-vapor sources, some less than one astronomical unit across. The longest base line in this experiment extended from Maryland Point, Maryland, to Tidbinbilla, Australia, nearly the diameter of the earth. It provided an angular resolution of about 0.0001 arc second, an extremely small angle equivalent to seeing a football on the moon. The best optical telescopes cannot even resolve an entire football field at that distance. Similar resolution was achieved in another joint international experiment by a group of West German, Russian, and U.S. scientists who collaborated to study radio galaxies and quasars.

Although the VLBI systems were developed to study celestial radio sources, they also have important practical applications in such areas as synchronizing global time, measuring distances accurately, predicting earthquakes, and navigating interplanetary spacecraft.

Radio interferometers will work only if the signals from the two telescopes are accurately synchronized. Many experiments require synchronization to within one-millionth of a second. Fortunately, it is not necessary to have each clock set with this degree of accuracy. By manipulating the signals from the two telescopes in a large computer, many different synchronizations may be tested until the correct one is found. By carefully measuring the computer-corrected clock error, scientists expect to synchronize clocks around the world to within one-billionth of a second.

As a by-product of the astronomical observations, the distance between two VLBI telescopes is determined with very great accuracy, perhaps to within a few centimeters. This opens up a wide range of applications. For the first time, it may be possible to measure the motion of continents directly. The accurate measure of intercontinental distances, especially across oceans, is also of strategic military value, because the effectiveness of intercontinental missiles depends on knowing the precise distance between launching site and target.

A more humanistic application could potentially save thousands or even millions of lives by using VLBI to predict catastrophic earthquakes. Geologists believe that, prior to a major earthquake, there are small movements in the earth's crust that might be detected by measuring a small change in the length of an interferometer base line that extends across a fault.

Although many exciting new discoveries have been made with VLBI systems, we have only scratched the surface of what is possible. Existing systems are limited because the individual telescopes are not ideally located to form a clear image. This is because the telescopes are all part of existing observatories, whose locations were chosen long ago for reasons of convenience or freedom from local radio interference. In addition, many of these telescopes are not designed to work well at short wavelengths where the angular resolution is greatest. Astronomers are looking forward to the day when an array of precision telescopes, each equipped with modern sensitive receivers, is constructed at precise locations throughout the world.

Baut transporting the magnetic tape recordings from each telescope to some central location where they can be replayed presents a formidable logistic problem, particularly when many telescopes are used. One solution is to eliminate the tape recordings, and relay the signals to the central processor via a synchronously orbiting satellite—one that always stays above the same point on the earth's surface. Satellites now in common use for transmitting international telephone calls and worldwide television could be used for this purpose. In late 1976, a team of scientists from the United States and Canada developed a satellite-linked VLBI system using the Communications Technology Satellite, a joint project of NASA and the Canadian Department of Communication. Signals received at the 43-meter (140-foot) radio telescope in Green Bank, W. Va., are relayed by this satellite to the Algonquin Radio Observatory in Ontario, Canada, where they are combined in a computer processor with signals received at a 46-meter (150-foot) radio telescope.

By extending the satellite hookup to link many telescopes throughout the world, astronomers will be able to explore pulsars, galaxies, quasars, and interstellar masers with unprecedented resolution, and at the same time learn much about our own earth. Within the next decade, radio astronomers hope to be able to put large antennas into orbit around the sun, extending the base lines to interplanetary dimensions and opening up a new range of resolution with which they can probe the mysteries of the universe.

New Clues to Changing Climate

By W. Lawrence Gates

Scientists use sediments from the ocean floor and high-speed computers to reconstruct climates of the distant past

Americans will long remember the winter of 1976-1977 as one of the most unusual since man began to record the variations of the weather. Snow fell in Miami, Fla., for the first time in memory, while Alaskans basked in unexpected warmth. Much of the Central and Eastern United States shivered through the coldest winter since the founding of the republic, while the Western United States experienced record-breaking drought.

The unusual winter caught most Americans unprepared. Fuel stockpiles that would have been enough for a normal winter were quickly exhausted. The drought dealt a severe blow to agriculture, hydroelectric power generation, and winter sports in the normally rain-drenched valleys and snowpacked mountains of the West.

What caused the severe winter? Could it have been foreseen? Do these changes in weather presage the start of a new ice age? Scientists

Powdered snow on grapefruit in a Florida grove and a dried-up reservoir in Washington indicate unusual weather during the winter of 1976-1977.

cannot yet answer these questions, but they are busily searching for the key to understanding changes in weather or climate.

Daily changes in weather—the usual concern of meteorologists and weather forecasters—contrast sharply with the long-term changes that concern the climatologist. When viewed over seasons, years, decades, and centuries, the average weather, which we call climate, appears to change relatively slowly. The climatologist deals with temperature changes of only a few tenths of a degree over decades and centuries, while the meteorologist is faced with daily changes ten or even hundreds of times greater. In fact, scientists first recognized less than 100 years ago that climate actually does change.

The study of climate and its changes is made difficult by the great complexity of the factors that produce it. In addition to the atmosphere, with its familiar wind, rain, and clouds, climatologists study the earth's oceans, its ice masses, and the differences in its land surfaces. These elements interact in subtle ways that often prevent wide-ranging changes in climate. For example, part of the sun's radiation is absorbed and reradiated by the ground at infrared wavelengths. Air near the surface may absorb the infrared rays and warm up, which may cause surface moisture to evaporate and clouds to form. The clouds tend to intercept or reflect some of the sun's radiation, which in turn causes the air to cool near the ground. This cooling offsets the initial warming and moderates the process.

Many such interactions link the atmosphere, oceans, ice, and land masses. Taken together, these effects tend to regulate the climate and to confine its changes within narrow limits. Unfortunately, these subtle interactions also make it difficult to pinpoint the exact cause of a particular change, such as a severe winter or the great ice ages of the past. It may not be too surprising, then, that the causes of climate changes remain a mystery.

In their search for clues to how the climate may change in the future, however, scientists are studying the earth's past climates more closely than ever before. They have found that our present climate, which most people regard as normal, is, in fact, highly unusual. On the whole, the earth is now warmer than at any time in the last several hundred thousand years. Ice ages, or periods of relative cold when ice sheets covered large land regions, have occurred eight times during the past 700,000 years. These ice ages are also called glaciations and the periods of relative warmth between them are called interglacials.

The study of past climates—paleoclimatology—involves a variety of geologic and biological sources, each of which serves as a natural "tape recorder." For example, sediments found on many parts of the ocean floor show a characteristically layered structure that developed as material that was suspended in the sea gradually sank to the bottom and accumulated there.

Although deep-sea sediment consists mostly of sand and mud deposited from the surrounding continents by the actions of erosion, wind,

The author:
W. Lawrence Gates is a professor and chairman of the Department of Atmospheric Sciences at Oregon State University in Corvallis.

or volcanic eruptions, it also includes the calcium carbonate shell skeletons of marine microorganisms that live in the ocean's surface waters. These shells provide the key to paleoclimate. For example, tiny, one-celled sea animals called foraminifera live only in the upper layers of water. Different types of foraminifera thrive at different temperatures. Consequently, the relative abundance of their shell remnants in the different layers of a deep-sea sediment sample provides a sequential record of the average changes in surface water temperature in that region.

Researchers aboard deep-sea drilling ships have collected thousands of cores, vertical sections of sediment about 8 centimeters (3 inches) in diameter and up to 9 meters (30 feet) long, by drilling a hollow cylindrical pipe into the sea-floor sediment. Paleoclimatologists can determine the age of the various sedimentary layers in a core by radiocarbon dating or a similar radioisotope-dating technique. Thus, deep-sea cores allow scientists to estimate what the sea-surface temperature was at selected times and places in the past.

In recent years, extensive core collections have been built up at Columbia University's Lamont-Doherty Geological Observatory in Palisades, N.Y.; the Scripps Institution of Oceanography in San Diego, Calif.; and Oregon State University (OSU) in Corvallis. Scientists from various countries banded together in the early 1970s to study past climates, especially to mine the rich climatic record in the deep-sea sediment cores. The research group, called Climate: Long-Range Investigation, Mapping and Prediction (CLIMAP), reconstructed the global sea-surface temperature distribution for July and August of about 18,000 years ago. This time is close to the greatest glaciation during the most recent, or Wisconsin, ice age. They used mathematical techniques to correlate the many core samples, and checked the

Complex interactions that link sunlight to the earth's atmosphere, oceans, ice, and land masses tend to keep climate changes very small and gradual.

A geologist collects samples of sediments near McMurdo Sound in Antarctica that were once covered by huge glaciers. The fossil remains of marine organisms found in the samples will give clues to the history of the climate in that area.

calculated temperatures against temperatures known to exist for the most recent sediment at the top of the core. In 1976, CLIMAP scientists constructed maps that showed that ocean surfaces in the last ice age were generally colder than they are today, with the largest differences occurring in the higher latitudes. In the North Atlantic, for example, the warm Gulf Stream flowed almost directly eastward from the vicinity of present-day New York toward Spain, rather than northeastward toward Great Britain as it does now.

The CLIMAP team also estimated how much ice covered the land and seas during the Wisconsin ice age. Huge ice sheets reached their greatest extent a few thousand years apart in different locations. CLIMAP geologists examined previously published reports as well as rocks, soil, and other evidence in the field in order to estimate how far the ice sheets extended and how thick they were. Where evidence was scant, the researchers relied on the best data available.

Much of the northern lands were covered by vast ice sheets that reached a thickness of as much as 3 kilometers (1.8 miles). In North America, the ice extended from Greenland across all of Canada and reached as far south as the Ohio River before it began to recede and melt, leaving freshwater puddles—the Great Lakes. In Europe, a simi-

**Climatic History
Beneath the Sea**
Researchers aboard the
R. V. Vema lower a device to
collect cores from the
ocean floor, *left.* Scientists
use radiocarbon dating to fix
the age of a layer, *below,*
then determine the relative
abundance of shell skeletons
from different types of
foraminifera (sea organisms)
in the layer, *bottom.* Because
each type thrives in water
of a certain temperature,
far left, the data reveal the
temperature of the area
surface at the time the layer
was deposited on the sea floor.

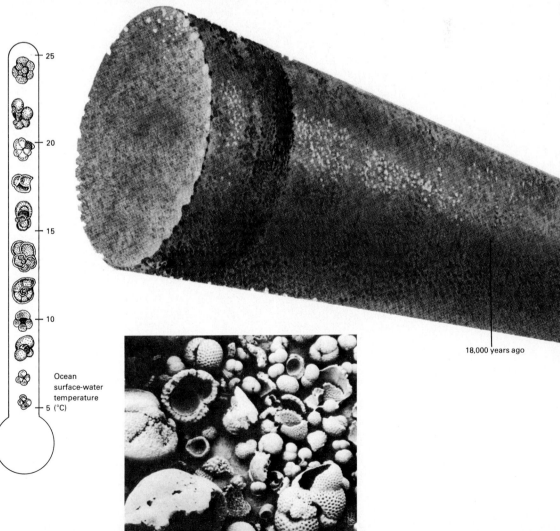

25

20

15

10

Ocean
surface-water
temperature
5 (°C)

18,000 years ago

lar ice sheet covered Scandinavia, the Baltic Sea, and parts of Great Britain, and extended southward and eastward into central Europe. Glaciologists believe that this sheet was connected to the one in North America by a continuous ribbon of sea ice extending across the North Sea from Europe to Iceland and Greenland. A smaller ice sheet covered north-central Siberia, and there were extensive glaciers in the mountains of central Asia, South America, and New Zealand.

The volume of water locked in these ice sheets should have lowered the worldwide sea level about 85 meters (279 feet) below what it is today. Ocean scientists have verified that the sea level was much lower 18,000 years ago by studying submerged coral reefs around Caribbean islands. From 75 to 90 meters (250 to 295 feet) below sea level, they found notches cut in the coral reefs by the pounding of ancient waves. Vast expanses of floating sea ice surrounded the ice sheets then, and the floating mass around Antarctica nearly reached the tip of South America. The ice sheets that cover Greenland and Antarctica now are remnants of this and previous ice ages.

Other members of the CLIMAP team determined what kinds of vegetation or soil covered the land that was not engulfed by the ice sheets. Near the glaciers, melting ice created streams that sculpted sandy plains filled with gravel and other glacial debris. Farther away, where forests once grew, grasslands and vast treeless plains took over because they were better suited to the cooler climate.

The patient detective work of paleoclimatologists can tell us how cold the oceans were, how big the ice sheets were, and what kind of plant life covered the earth 18,000 years ago. But what can such information tell us about the climate then? While the CLIMAP team pieced together its map of the earth's surface, other researchers worked on projects destined to revolutionize the study of ancient climates. They developed mathematical models of the atmosphere that are similar to those that meteorologists, with the aid of computers, use in forecasting the weather. Unfortunately, these general atmospheric circulation models (GCM) produce inaccurate daily weather forecasts. But happily for the paleoclimatologists, GCMs prove accurate in calculating the average weather for a month, a season, or a year—in other words, for calculating the climate. These calculations are based largely on information about the earth's surface such as that assembled by the CLIMAP team.

GCMs are designed to predict atmospheric changes for large regions of the globe. But even the most powerful computer cannot calculate changes at every point in the atmosphere. Instead, climatologists cast an imaginary net on the atmosphere and have the computer determine atmospheric changes—for example, changes in temperature and pressure or air speed and direction—at each point of intersection on this global net. The imaginary three-dimensional network has points spaced 200 to 500 kilometers (125 to 310 miles) apart in the surface directions, and 1 to 5 kilometers (0.6 to 3 miles) vertically.

The Summer of 16,000 B.C.

Land Features

Snow and ice

Sandy deserts and patchy snow

Steppes and semideserts

Grasslands

Forests and other vegetation

A paleoclimatic map produced from research data shows that glaciers (elevations in meters) covered much of North and South America in July about 18,000 years ago. The Gulf Stream ran almost due eastward then and the sea level was about 85 meters (279 feet) lower than today.

A researcher at a computer terminal puts paleoclimatic data into the computer, which then calculates features of past climates that appear on the screen.

We understand these atmospheric changes well enough to describe them in mathematical equations that interrelate the effects of sunlight coming into the atmosphere from above, the infrared radiation that the earth emits, clouds and precipitation, and the complex mixing of heat, moisture, and energy in the atmosphere. In fact, these equations make up the GCMs. In order to start the calculations, the scientists must determine and put into the computer boundary conditions–the values of certain properties at the top and bottom of the atmosphere. For example, they must know the intensity of the solar radiation that hits the top of the model atmosphere, the elevation and character of the land surface, and the temperature of the ocean surface. The computer solves the equations at a given time for each point in the model atmosphere, then solves them again for an hour, a few hours, or even longer periods later. The calculation process can be repeated again and again. When such solutions are averaged over a month, a season, or a year, they represent the climate for that period.

Scientists have fed boundary conditions for today's earth into about a dozen different GCMs to simulate the present average climate, say for a typical January or July. They have determined many features of the atmosphere that agree with observations. For example, their cal-

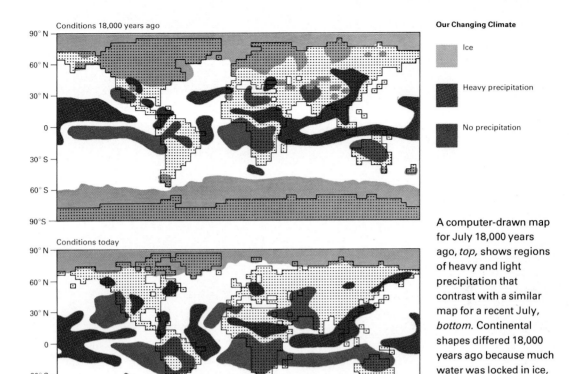

Conditions 18,000 years ago

Conditions today

A computer-drawn map for July 18,000 years ago, *top,* shows regions of heavy and light precipitation that contrast with a similar map for a recent July, *bottom.* Continental shapes differed 18,000 years ago because much water was locked in ice, lowering the sea level.

culations of the average worldwide seasonal temperature, wind, rainfall, and cloud cover agree reasonably well with those actually observed. The models also show such features as the position and strength of the midlatitude jet stream; the zone of maximum precipitation in the equatorial regions; and the zones of maximum storm activity, identified by relatively low atmospheric pressure, in the middle and higher latitudes.

These models can also be used to calculate climates in the distant past if the boundary conditions at the earth's surface can be assembled from paleoclimatic sources. CLIMAP researchers recognized this possibility when they reconstructed the sea-surface temperature and ice-sheet topography for a typical July and August of 18,000 years ago. My colleagues and I first used this CLIMAP data plus estimates of the other boundary conditions in an atmospheric model in 1976 to recreate the global climate of the last ice age.

What kind of climate existed then? The model shows that during the Wisconsin ice age, average temperatures differed most from today's temperatures in regions near the major ice sheets. Near what is now Washington, D.C., for example, the ice age July was about 13°C (23°F.) colder than it is today. The temperature near present-day Los

Angeles and Miami was as much as 11°C (20°F.) colder. Similarly, southern Europe and much of central Asia were markedly cooler. Even in the Southern Hemisphere, in regions far removed from the major ice sheets such as South America, Africa, and Australia, the ice age July was an average of 5°C (9°F.) colder than July is today.

The ice age July was also somewhat drier, especially over the continents of the Northern Hemisphere. Worldwide precipitation was about 14 per cent lower than it is today. The most dramatic changes in ice age precipitation occurred over the Indian Ocean and Southeast Asia, where there was no heavy rainfall as there is now during the summer monsoon. Instead, more rain fell farther south over the open water of the Indian Ocean.

An easterly flow of relatively cool air from the then-colder North Atlantic Ocean was the dominant air-circulation pattern in the Central and Eastern United States 18,000 years ago. This is in sharp contrast to the southerly flow of warm, moist air from the Gulf of Mexico that now characterizes summer climates in these regions. This surface circulation was part of a general counterclockwise flow around a high-pressure cell centered over the North American ice sheet. Thus, if you had lived near what is now Washington, D.C., during the ice age, your summer climate would have resembled that now found in Newfoundland, Canada.

Because most of the CLIMAP and other available paleoclimatic data are used as boundary conditions for the atmospheric model, there is very little data left with which to test the model's picture of the ice age climate. One way to check the model's estimates of ice age temperature and rainfall, however, is to analyze fossil pollen preserved in layers in bog and lake-bottom sediments. The estimated ice age cooling, as indicated by the changes in plant life, agrees with the results of the model. Cross-checks on the amount of rainfall are fewer and less certain. But researchers continue to search for other paleoclimatic evidence to verify the model's simulations.

Meanwhile, the CLIMAP team is assembling a companion set of ice age boundary conditions for a typical December and January 18,000 years ago. These data will permit us to simulate climates of the Northern Hemisphere winter and the Southern Hemisphere summer, so that we may catch a glimpse of the seasonal extremes of ice age climate. My co-workers and I at the OSU Climatic Research Institute are now making such calculations, as are Syukuro Manabe and his co-workers at the Geophysical Fluid Dynamics Laboratory of the National Oceanic and Atmospheric Administration at Princeton, N.J., and scientists at the National Center for Atmospheric Research in Boulder, Colo. By comparing the results of our atmospheric models, we hope to firmly establish the major features of the ice age climate.

Dressed warmly against the chill of summer 18,000 years ago, North American hunters cross rock-strewn terrain left by a receding glacier.

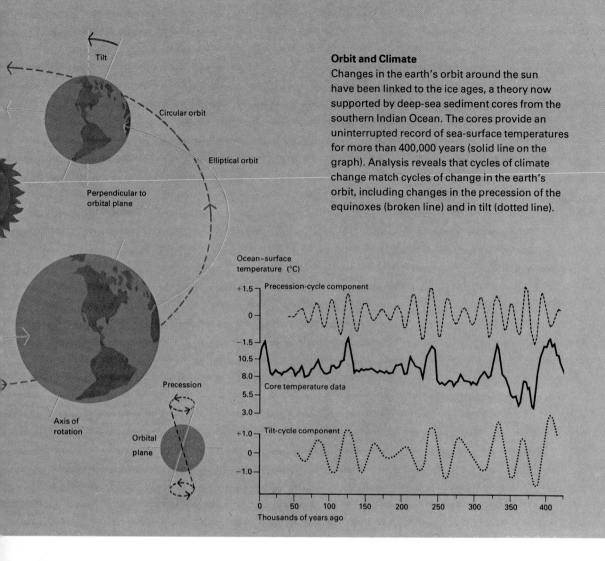

Orbit and Climate
Changes in the earth's orbit around the sun have been linked to the ice ages, a theory now supported by deep-sea sediment cores from the southern Indian Ocean. The cores provide an uninterrupted record of sea-surface temperatures for more than 400,000 years (solid line on the graph). Analysis reveals that cycles of climate change match cycles of change in the earth's orbit, including changes in the precession of the equinoxes (broken line) and in tilt (dotted line).

The use of atmospheric models has revolutionized the study of paleoclimatology. It should enable us to reconstruct not only the average climate of the last ice age, but also the climates of earlier ice ages and interglacial periods. By using atmospheric models together with similar models for the oceans, we may also be able to reconstruct the earth's climate as far back as 100 million, 200 million, or even 300 million years ago. The positions of the continents and the shapes of the oceans then were quite different than they are today, and so, too, were the climates. Because ancient climates influenced what kinds of animals and plants flourished, unraveling the history of these paleoclimates may provide the climatic "missing link" in the story of the evolution of life on earth.

Exciting as this research will be, it also has its problems. Aside from the difficulty of getting the boundary-condition data necessary for the models and to check their results, our present models do not tell us how or why the climates of the past changed. They provide only a "snapshot" of the average seasonal or annual climate under given conditions. In order to trace climate changes through the complete course of the last ice age, for example, we need not only better mathematical models of the atmosphere and ocean, but also models of how the ice sheets evolved. When we make such models and couple them together, we will then have to devise more efficient ways to use computers in order to integrate vast amounts of data and solve the equations in a reasonable amount of time.

Based even with the best models and more powerful computers we may have to look beyond the earth's atmosphere and oceans to learn what causes an ice age. In December 1976, researchers James D. Hays of Lamont-Doherty, John Imbrie of Brown University, and Nicholas J. Shackleton of Cambridge University in England reported strong evidence for the theory that cyclical changes in the earth's orbit around the sun caused the ice ages. They examined in detail two deep-sea sediment cores from the southern Indian Ocean and established an uninterrupted record of sea-surface temperature dating back more than 450,000 years. Analysis of the temperature data revealed three cycles of climate change that match cycles of change in the earth's orbit caused by the gravitational attractions of objects in the solar system other than the sun.

The longest climate cycle, about 100,000 years, matches the period of change in the shape of the earth's orbit around the sun from nearly circular to elliptical and back. The second climatic cycle is about 41,000 years, the same as the period of change in the earth's tilt, or axis of spin with respect to the perpendicular to the earth's orbital plane. The shortest cycle, about 23,000 years, corresponds to the period of precession of the equinoxes, in which the earth's axis slowly traces an imaginary circle in the sky in the same way that the axis of a spinning top slowly drifts in a circle. Consequently, the earth is closest to the sun in January now, but in about 10,000 years from today it will be closest to the sun in July.

Because these orbital changes alter the seasonal and latitudinal distribution of sunlight that strikes the top of the atmosphere, they could trigger the onset of a new ice age or end an existing one. For example, if the summer sun were a little higher in the Northern Hemisphere sky at a time when the earth was also a little closer to the sun, Arctic ice sheets would melt more easily.

A new ice age appears destined to occur sometime in the next few thousand years unless human effort can contribute to a long-term global warming. For example, burning fossil fuels may heat the atmosphere because carbon dioxide, a combustion by-product, acts like a thermal blanket. We hope to quicken the pace of research in order to better understand the earth's climatic machine. If we cannot find ways to prevent the predicted cooling, then, like our ancestors, our descendants may have to learn to live through an ice age.

How Plants Tune In the Sun

By William S. Hillman

In adapting sunlight to their needs, plants seem to be very choosy about which wavelengths of light to use

For thousands of years, humans worshiped the sun as the source of life. In a sense, they also should have been worshiping the world's plants. For it is the complex relationships of sunlight and green plants that provide all of us with the basis of life – the food that we eat and the oxygen in the air that we breathe.

Botanists have been trying to understand the plant-light relationship for many years. Currently, many of them are working on the ways that plants react to specific wavelengths, or colors, within sunlight.

The visible light spectrum – that part of the electromagnetic radiation spectrum that we can see – runs from short-wave, deep-blue light, which has a wavelength of 400 nanometers (nm), through intermediate-length green and yellow light to long-wave red light (700 nm). A nanometer is one-billionth of a meter, or about 0.00000004 inch. It is mainly the wavelengths in the visible spectrum that affect plants, although ultraviolet and far-red light, on either side of the visible spectrum, also affect them.

Of all the interactions between plants and light, the best known—and most studied—is photosynthesis, in which plants use sunlight, carbon dioxide, and water to make carbohydrates, which they use as food, and give off oxygen. In the process, blue and red light is absorbed by molecules of a green pigment called chlorophyll. Plant scientists are working out the fine details of how chlorophyll acts, in part because they hope to design artificial photosynthetic systems to capture and use the sun's energy.

Other botanical scientists are trying to understand two less-well-known plant-light relationships, each involving a different plant pigment. One involves red and far-red light absorbed by the blue pigment phytochrome, and initiates many different plant responses. The other process, phototropism, uses blue light. The active pigment has not yet been identified.

In photosynthesis, light is a source of energy. In the other two processes, light, in effect, provides information. In response to the phytochrome system, a lettuce seed, for example, may germinate or lie dormant; a Japanese morning-glory may flower or produce only leaves. Phototropism helps the plant move and grow so as to obtain the best available light: The sunflower faces the sun and houseplants may bend toward the brightest window.

Almost every event in plant development can be affected in some fashion by the phytochrome system. Its effect on seed germination, which helped lead to its discovery, illustrates the extraordinary way in which it acts. In the early 1950s, plant physiologist Harry A. Borthwick and biochemist Sterling B. Hendricks of the United States Department of Agriculture's (USDA) research station in Beltsville, Md., experimented with the Grand Rapids variety of lettuce to study the effects of light on seeds. Most of the Grand Rapids seeds remained dormant when they were moistened and kept in darkness. However, they germinated within a day or two after receiving a few minutes of daylight, even if they were immediately returned to darkness. In many of their early experiments, Borthwick and Hendricks used an intense carbon-arc light shining through a giant glass prism to provide the light that they required. The prism broke the light up into a rainbow of colored rays so that batches of moistened seeds could be briefly exposed to various wavelengths of light.

A major landmark in modern plant physiology was established by one striking discovery: A few minutes of red light at a wavelength of about 660 nm caused rapid germination in the darkness that followed, but when the seed was exposed to a slightly longer wavelength—the far-red 730 nm—immediately after the red exposure, germination did not occur. Hendricks, Borthwick, and their associates carried this experiment further. They treated batches of seeds with a series of red (R) and far-red (F) light exposures—for example, R,F,R or R,F,R,F—and

The author:
William S. Hillman is senior plant physiologist at the Brookhaven National Laboratory in New York.

A beam of white light is broken into its spectrum and reflected onto seeds at the Argonne National Laboratory near Chicago. Scientists use the apparatus to study how different wavelengths affect seed germination.

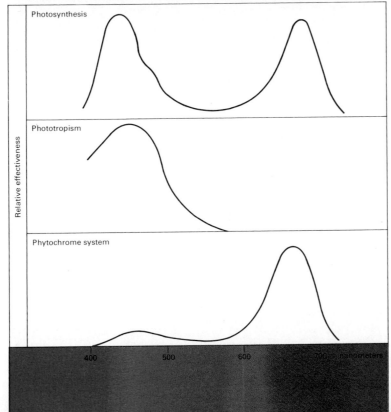

Each plant-light system is most efficient at specific wavelengths. Photosynthesis uses blue and red light. Phototropism uses blue light; the phytochrome system uses red light.

the percentage of germination in subsequent darkness depended entirely on the last exposure. The percentage was high if the last exposure was red (as in R,F,R) and low if it was far-red (as in R,F,R,F).

It is important to realize just how little light is involved in such effects. The intensities are substantially less than that produced by a small refrigerator bulb in an otherwise dark room. And each exposure need last only a few minutes.

Experimental botanists, led by Borthwick, Hendricks, and their co-workers, found during the late 1950s and early 1960s that the red, far-red reversible system also affected germination in other plants. They found that it can be used to control other plant processes, too.

For example, seeds or tubers sprouted and grown in darkness do not develop normally. They have long, weak shoots and small, inconspicuous leaves. In addition, they have no chlorophyll, so they are white or light-yellow in color. This condition, called etiolation, is probably advantageous in nature during early growth, allowing a plant to grow rapidly through soil, or from under rocks or other obstructions, as if reaching for the light. Indeed, etiolation ceases abruptly and normal development takes over as soon as plants reach sunlight.

For many years, scientists assumed that photosynthesis or a process closely related to it caused etiolated plants to begin normal growth. It remained for the Borthwick-Hendricks group at Beltsville to show that this effect is really another function of the red, far-red reversible system, which later came to be called the phytochrome system.

The pale bean sprout so familiar in Oriental food is an example of an etiolated seedling. Its stem is elongated and the tiny true leaves between the two fat cotyledons, or seed leaves, are folded tightly. The top of the bean sprout is bent in a crook, a characteristic that may help it break through the soil under normal circumstances. But if such shoots are exposed to a few minutes of light several times a day instead of being kept in darkness, their growth pattern changes profoundly. The crooks straighten, the leaves open out, and the stem grows thicker instead of longer. The changes can be measured after only a few hours of a given light exposure. Exposure to red light promotes normal development even in darkness while far-red given immediately after the red tends to maintain the etiolated pattern of growth. Here again, when brief red and far-red exposures are followed by darkness, the effective wavelength is the one given last.

Besides affecting the early life of many plants, the phytochrome red, far-red reversible system also can change the processes of mature plants. For example, the phytochrome system participates in photoperiodism, the way in which some plants seem to tell the season of the year by measuring daylength.

W. Wightman Garner and Harry A. Allard coined the term "photoperiodism" in the 1920s to describe their discovery that flowering in certain plants depends on the length of their daily exposure to light. These two agricultural scientists, predecessors of Borthwick and

Hendricks at USDA's Beltsville laboratories, had been investigating why certain varieties of soybean always flowered at the same time each year no matter when they were planted. Since its discovery in plants, photoperiodism has also been found to control processes in many kinds of insects, birds, and mammals.

Of the several types of photoperiodic response in flowering, perhaps the most commonly studied is the short-day type. Short-day plants flower normally if they are exposed to no more than a certain number of hours of light each day. Long-day plants, on the other hand, require no less than a certain número of hours of light. The precise duration varies widely among species and even among varieties of the same species. But when short-day plants receive light for longer than the critical length of time, flowering either fails entirely or is greatly delayed.

Lettuce seeds germinate most effectively when exposed to red light (R) and least effectively when exposed to far-red (F). When R and F are alternated, the seeds' germination depends on the last one they are exposed to.

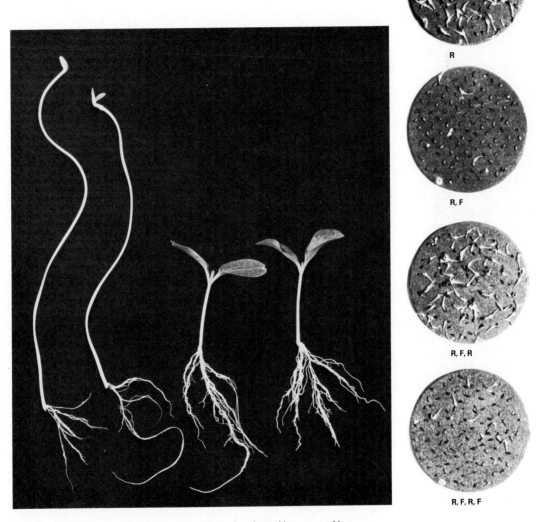

R

R, F

R, F, R

R, F, R, F

Squash seedlings grown in darkness have undeveloped leaves and long, spindly stems compared with plants at right, grown under normal light.

Petunias are long-day plants that flower best when exposed to long hours of light, *below.* Marigolds, *opposite page,* are short-day plants and flower quickly with 8 hours of light daily.

In one variety of morning-glory studied by Shun-Ichiro Imamura, Atsushi Takimoto, and their colleagues at Kyoto University in Japan, the critical daylength is about 15 hours. Plants of this variety that receive more than 15 hours do not flower, no matter how good the other conditions; they simply continue to grow as long, flowerless vines. On the other hand, plants raised from seed with eight hours of light a day flower so rapidly that they never develop into vines. They remain as stubby bushes in which all the potential growing points are converted to flower buds which then develop into flowers. The flowers set seeds and the plant then dies, having completed its life cycle, but not as a normal morning-glory.

For experimental purposes, scientists originally assumed that photoperiodism worked through photosynthesis. For example, excessive

8   12   16   20   24

Hours of Light Per Day

photosynthesis might prevent flowering in short-day plants, while long-day plants might need at least that much photosynthesis. Many kinds of experiments have discredited this concept, but the most important ones were those in which scientists used light-breaks, brief periods of light that interrupted the darkness plants experience in each 24-hour period.

For example, under appropriate laboratory conditions, the Japanese morning-glory produces flowers rapidly on a daily schedule of 10 hours of light alternating with 14 hours of darkness, and completely fails to flower on 16 hours of light and 8 hours of darkness per day. However, the full additional six hours of light is not needed to prevent flowering. In fact, on a schedule of 10 hours of light alternating with 14 hours of darkness, merely interrupting each dark period approxi-

8　　　　**10**　　　　**12**　　　　**14**　　　　**16**　　　　**24**

Hours of Light Per Day

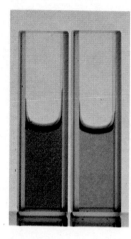

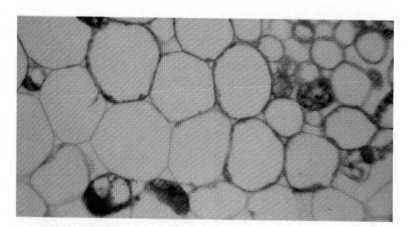

Dark-blue phytochrome, *above,* absorbs red light, which changes it to a lighter-blue form that absorbs far-red light. The two forms of the pigment act differently in plant cells. The red-absorbing pigment, stained brown for microscope study, is distributed throughout plant cells, *above right,* until it is exposed to far-red light. Then the pigment clumps together near the cell wall, *right center.* A sensitive spectrophotometer, *right,* measures the changes in absorption of red and far-red light.

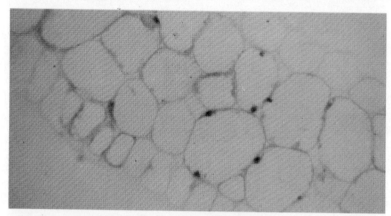

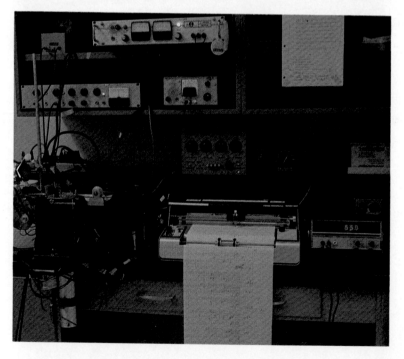

The city-state of Ebla dominated the landscape of what is now the northwestern part of Syria 4,500 years ago.

Italian archaeologist Paolo Matthiae did not at first realize the significance of the few clay tablets that he and his co-workers found at a hot and dusty site in northern Syria in 1974. They had been carefully excavating there for 10 years, uncovering bits of pottery and other artifacts once used by the people of civilizations that had come and gone over the centuries. But these clay tablets were written records, and the first of more than 16,000 they would find. The magnitude of the discovery became apparent only in 1975, when another Italian scholar, Giovanni Pettinato, deciphered and translated some of the writing. It revealed in detail a long-lost and forgotten, but important, civilization. It also documented previously unsubstantiated accounts of some of the earliest places mentioned in the Bible.

Most of the important archaeological finds in this part of the world have been discovered accidentally. For example, the Dead Sea Scrolls were found by a shepherd pursuing a stray goat into a cave. But such was not the case with these clay tablets, the discovery of which is a triumph for patience and carefully planned and executed work.

Matthiae found the tablets exactly where they had been left, in soil layers previously dated from bits of pottery and other artifacts at as far back as 2400 B.C. Many of the tablets are written in a hitherto unknown Semitic dialect that Pettinato has dubbed Eblaite—for Ebla, the name of the long-lost city. Other tablets are written in Sumerian, and some in both Sumerian and Semitic. They describe the activities of a teeming metropolis of some 260,000 persons, whose influence stretched all the way from the Sinai in the south through what is now Israel, Syria, and Lebanon to Cyprus in the west, and into the highlands of what is now Iraq and Iran in the east. Besides providing background for Biblical accounts, the tablets outline diplomatic affairs, sophisticated trade agreements in textiles and finished metals, and laws and administrative orders in this city of Ebla that disappeared some 4,500 years ago.

Matthiae, of the University of Rome, decided in the early 1960s to excavate at a point about 50 kilometers (30 miles) south of Aleppo, near a village called Tell Mardikh. He began digging in 1964 in a 15-meter (50-foot)-high mound of dry, dusty dirt that spread over 56 hectares (140 acres). There was no reason for the mound to be at that place in the Syrian plain. So the archaeologist reasoned that such an obstruction on the landscape hid some man-made construction, probably an abandoned city.

For the first 10 years, this was just another archaeological dig, turning up broken bits of pottery, discarded tools, and stones of various sizes and shapes, some of which were parts of buildings and other man-made structures. By dating the pottery bits, Matthiae and his co-workers determined that the spot had been occupied many times. As they dug down through each level, they found remnants of the relatively recent Arabic period (prior to A.D. 1500); the earlier Byzantine period; the still-earlier late-Roman, Hellenic, and Persian periods; and the Iron Age (beginning about 1200 B.C.). But, when they dug still deeper, their painstaking work gradually uncovered a city of great size that existed during the middle Bronze Age (from about 2000 to 1600 B.C.). One of the first objects found at this level was a small stone statue bearing the name Ibbit-lim, a man described in ancient Sumerian texts as the king of a city-state called Ebla about 2000 B.C. Another was a fragmentary statue of a goddess in Sumerian style and workmanship that the archaeologists dated to the 27th century B.C. These discoveries were but a prelude to the unique tablets they would find as they excavated further.

The importance of the Ebla tablets can hardly be exaggerated. They add greatly to our understanding of the Middle East and the period from which they came. Ebla was in the western part of the fertile crescent, an area whose history was little known. The fertile crescent stretched in a giant arc around the Syrian desert from the

The author:
David Noel Freedman is director of the Albright Institute of Archaeological Research in Jerusalem, Israel, and professor of Near Eastern Studies at the University of Michigan.

Mediterranean Sea to the Persian Gulf. It is called "fertile" not only for agricultural reasons, but also because many great civilizations and cultures began in the area. It was there that the Sumerians developed one of the world's first great civilizations, and the Babylonian, Assyrian, Mitannian, Phoenician, and Hebrew civilizations began.

The team of Italian archaeologists and workers from the village scraped away the debris from a massive gateway, part of a huge wall that had once encircled the city. In 1974, they uncovered public structures such as towers, ceremonial staircases, and a few rooms of a massive temple. They also found the first 42 clay tablets covered with an ancient, wedge-shaped type of writing called cuneiform.

The archaeologists have set up headquarters in a compound on the edge of Tell Mardikh and directly at the foot of the mound. Matthiae is the project's chief archaeologist. His wife, Gabriella Matthiae-Scandone, an Egyptologist, has worked with him at this excavation from the beginning. Pettinato, also from the University of Rome, is an expert in deciphering cuneiform tablets and a specialist in Sumerian and Akkadian languages. The other members of the archaeological team include graduate students from the University of Rome and the Institute of Near Eastern Studies in Rome. The remaining workers live in the village and surrounding areas. From the time they selected the site and obtained Syrian permission to dig there, Matthiae and Pettinato have enjoyed a cordial relationship with the Syrian Department of Antiquities.

Archaeological excavation has a glamorous reputation, but for the most part it is dull, tedious work conducted under trying conditions. Fortunately, the mound near Tell Mardikh is close to the main highway that links Damascus and Aleppo, so the site is easier to reach than many other archaeological digs. The mound rises sharply from the surrounding plain, which it dominates in all directions. Other, somewhat smaller, mounds can be seen nearby.

A typical day on the mound begins about dawn, so the workers can get as much done as possible before the sun rises high and the heat becomes oppressive. They dig in carefully plotted squares, usually four meters (13 feet) on a side, to provide convenient limits within which to classify objects as they are found. Eight or 10 workmen are assigned to each square. They use a pick and shovel for initial groundbreaking, then sift through the dirt with smaller hand tools searching for recoverable objects. They dig slowly; the main purpose is not to see how much and how quickly debris can be removed, but to see how carefully the dirt can be removed while leaving everything else in place.

The artifacts they uncover are placed in baskets marked with the coordinates for the squares in which they were found. They are then taken to the headquarters workroom for identification and examination. Pottery fragments are sorted and classified on the basis of shape, size, texture, decoration, and special features such as rims, handles, and bases. By correlating these with the levels of rubble in which each

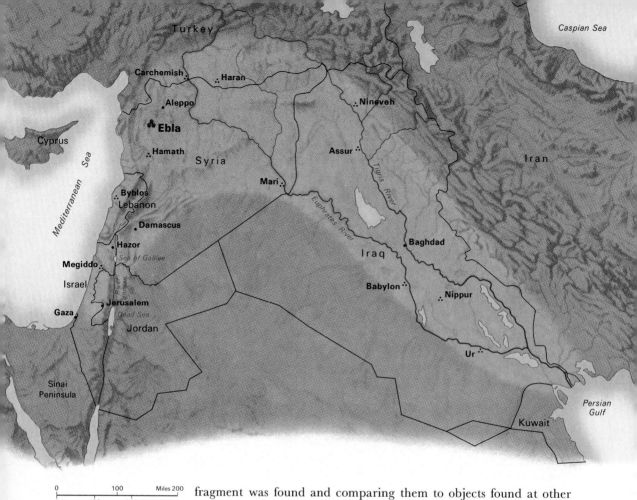

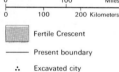

0   100   Miles 200
0   100   200 Kilometers

   Fertile Crescent

—— Present boundary

∴ Excavated city

At the time Ebla was a thriving trade center in the Fertile Crescent, several great Middle Eastern civilizations were beginning to rise to power and dominance.

fragment was found and comparing them to objects found at other sites in Syria and the Middle East, the archaeologists can tell something about the people who lived at Ebla at a particular time.

During the digging season, which extends from August to mid-October, the Italian archaeologists work long hours at the Tell Mardikh headquarters on materials brought from the mound each day. All the objects are registered, and the most important are sketched, photographed, and described. The information is then filed for later selection and publication.

As the work proceeds at the dig, the senior archaeologists go from square to square to observe the progress of the digging and to check unusual features or objects the workers unearth. For example, new pottery may be found in a square different in kind and date from the rest of the material there. This usually indicates either that a pit of a later level has cut through the level being worked on, or that an earlier level has been uncovered. To make sure that they do not mix materials from different levels or periods, the archaeologists must determine the contours of the new pit before they dig further. They can usually determine its outlines by noting changes in the color, the composition, or the density of the soil.

Earlier excavations at other sites in this region had revealed the existence of ancient civilizations with large, well-planned, and heavily fortified cities that had impressive public buildings. Pottery and other artifacts found at these sites indicate that the cities carried on extensive trade among themselves and with peoples in other regions. But without written records, the historical framework and the course of events during this period could not be reconstructed. The newly discovered tablets have already produced an avalanche of historical information that will partially fill this gap.

Matthiae estimates from the size of the palace at Ebla that about 12,000 persons must have lived and worked there. The task of excavating and reconstructing the rooms in the temple began in 1976, and the full extent of its treasures will not be known for many years. Among the first artistic artifacts found in the palace are a staircase inlaid with shells; furniture decorated with carved wooden animals and abstract designs; and wood, stone, and metal sculptures depicting such things as lions attacking bulls and warriors fighting with swords. One prized find is a bull that has a bearded man's head, fashioned out of beaten gold over wood. There are also bowls, urns, jugs, and a number of other household objects.

About 15,000 clay tablets were found in one room, the palace's royal archive, in 1975. They apparently had been stored on wooden shelves that rotted away long ago. In 1976, about 1,600 tablets were found in two other rooms. Since there are many rooms yet to be excavated, more historical riches may still be found in the palace. Perhaps another archive room will be discovered, or even the royal library. Such a library would be distinguished from the royal archive just as the Library of Congress is distinguished from the National Archives, and it might well contain a preponderance of literary texts rather than economic and statistical records.

The sheer volume of tablets that the archaeologists have uncovered is overwhelming. Even the simple tasks of recording, photographing, and numbering them before they are removed from the site became a burden for the overworked team members. In addition to gathering and storing the tablets so that they are safe and readily accessible for examination and study by experts, the archaeologists must register them and the exact spots at which they were found. This makes it possible to reconstruct the original scene in detail for future analysis and interpretation.

The tablets were originally square slabs of clay. The writing was scratched on them while the clay was still moist, and they were then dried in the sun or baked in an oven until they hardened. The writer used a stylus, making wedge-shaped impressions with one end and circle-shaped impressions with the other end. The tablets vary in shape and size. The smallest are about the size of a postage stamp, with barely enough room for one or two cuneiform signs on each side. The largest are rectangular, about 40 by 30 centimeters (16 by 12 inches), with up to 3,000 signs arranged in 30 columns on each side. Many of the large ones have a brief description of their contents inscribed on the edge to mark them for classification. They must have been arranged on the shelves much as file folders or phonograph records are arranged today.

A preliminary reading shows that most of the tablets report the economic accounts of Ebla's rulers over a period of perhaps 100 to 150 years. They cover trade in textiles, wood, and processed metals, as well as tribute collected from other cities. Many tablets deal with diplomacy and foreign relations, internal and domestic affairs, and religious and cultural matters. One tablet states that one of the kings had 38 sons. Another says that the penalty for raping a virgin was death. Still another reports the marriage of the king's daughter to a Mesopotamian king, who received several Ebla-dominated towns as a wedding gift. Other tablets contain hymns and incantations addressed to hundreds of specific gods. There are also vocabulary lists compiled by student scribes, and some school exercises, on which the teacher wrote signs on one side of the tablet and the student copied the teacher's work on the other side.

The records show that war was not as common then as in earlier or later times. One tablet tells of a rare instance of Ebla's involvement in a military campaign. It is a letter from Enna-Dagan, the general of the army, to King Ar-en-num of Ebla, reporting that the general had attacked and captured the city of Mari. Enna-Dagan attacked because of an alleged breach of contract or treaty on the part of Iblul-il, the king of Mari. According to other tablets, the king of Mari was driven out and General Enna-Dagan became the new king. Presumably, relations more in Ebla's interests were then established.

Another tablet contains a treaty between Ebla and the city-state of Asshur, signed by a successor of King Ar-en-num and King Dudiya of Asshur. It is of great interest to scholars that this king has essentially the same name as the first king of Assyria. The Assyrian king lists were compiled almost 2,000 years later, when Assyria was a great empire and the royal family wished to trace the kingdom's history back to its origins. Specific information has always been lacking on the earliest names on the list, and the first 17 are simply described as "kings who dwelt in tents." Because Babylonia's King Hammurabi also traced his ancestry back to these kings, scholars had long thought that this group was simply part of a folk tradition, and that the kings were legendary figures named for tribes and ethnic groups that made up later kingdoms. Now, one of these misty monarchs has emerged as the signer of a formal treaty sometime in the 2600s B.C. It is like finding a document signed by King Arthur, the legendary king of England.

It is fortunate that the tablets were written in two languages that can be compared with each other and that some of the tablets contain bilingual vocabulary lists. This has made it easier to translate the complex and difficult texts. One of the languages is Sumerian, the basic language of the oldest known civilization in Mesopotamia, and the other is an unknown Semitic dialect related to Ugaritic, a Canaanite dialect similar in many respects to Phoenician and Biblical He-

The scribes of Ebla, working in the royal archive, used specially designed tools to press Sumerian and Eblaite texts on the tablets, which were then stored upright on shelves.

A small part of Ebla's vast palace, temples, and other buildings, *above,* has been cleared by archaeologists. The site looks like a low hill, *above right,* when viewed from the road that runs between Damascus and Aleppo.

brew. Knowledge of all these Semitic languages is needed in order to read the text. Though peculiar to Ebla's inhabitants, the dialect was similar to those of many other cities, which made it useful in trade, diplomatic exchanges, and treaties.

The early form of the writing in both languages was pictographs, a system of picture writing in which each sign was a picture of some object, action, or idea. In such a system it is impossible to indicate any grammatical forms, syntactic relationships, or even the distinction between parts of speech. But by the time the Ebla tablets were inscribed, the original pictures had become stylized symbols for certain sounds and meanings and bore little or no resemblance to the objects they were originally supposed to signify.

Besides this—whether as part of an effort to make the symbolic representations effective, or as a natural development in the process of making cuneiform more intelligible and understandable—special signs called determinatives had been devised. These provided specific information when words that had more than one meaning were used. These determinatives were written either before or after the symbols to indicate whether the symbols referred to a city, a deity, a country, or something else.

Determinatives took care of the problems of ambiguity that arose in reading the language. They are not needed in modern alphabet-structured languages because ambiguities of this sort are overcome by the context of the sentence in which the words occur. But standing alone, such words, whether in Eblaite or modern English, are difficult to identify. For example, the English word "dash" can refer to a short line, throwing something against an unyielding surface, a small amount, a showy appearance, a long sound in a telegraph message, or a number of other things.

An even more advanced development in starting writing systems on the road to a full alphabet shows up on the tablets. Symbols, or logograms, were used to represent syllables as well as whole words. Signs were then combined to produce words, the meaning of which had nothing to do with the meanings of the individual signs. In this way syllabic writing became an adjunct of logographic systems, though it never replaced the older method entirely.

Cumbersome as this method of writing was, it was used for almost 3,000 years in this area of the world. It was so complex that few people could read it. It remained the special preserve of the scribal guilds, who jealously guarded their system and its secrets. Many years were devoted to training scribes, who spent the rest of their lives polishing their skills, so there was no need to simplify the system.

Most of the Ebla tablets in Semitic have syllabic writing on them, while logograms were used on both the Sumerian and the Semitic tablets. Syllabic writing would later prove to be more simple because it used from 80 to 90 signs to represent syllabic combinations, compared with the thousands of signs used in the logographic system. The scattered instances of syllabic writing on the tablets identify them as Semitic rather than Sumerian texts.

The translation of the tablets has provided independent background material for Biblical stories in the opening chapters of Genesis. For example, one poetic story about the creation of the world resembles the creation story in Genesis. There is also an account of the great flood that is similar to those found in both the Old Testament and Babylonian literature. In addition, the tablets refer to a place called Urusalima, which scholars agree is Ebla's name for Jerusalem. This is the earliest known reference to the holy city.

Moreover, the tablets contain thousands of names, some of which are practically identical with names of persons in the Old Testament. In no case are the same people involved, but the similarity in names points to a common cultural background. This is actually no surprise. While not mentioning Ebla, the Bible does point to this region as the fatherland of the Israelites. The patriarchs came to Canaan from Haran, where their relatives continued to live long after Abraham and his family had departed. A bride was brought from Haran for Isaac; and Jacob returned to his kinsmen there when he was forced to leave Canaan suddenly. Haran is some distance to the east of Ebla, and is often mentioned in the Ebla texts.

Perhaps the most striking similarity between the tablets and the Bible, at least to Biblical scholars, is the naming of the five cities of the plain—Sodom, Gomorrah, Admah, Zeboiim, and Bela, also called Zoar—on an Eblaite tablet that corresponds closely to the list in Genesis 14:2, and in the same order. The only difference is that the Bible identifies the last city with Zoar ("Bela, which is Zoar"), while another Ebla tablet indicates that the city of Zoar—Za-e-ar in Eblaite—was in the district of Bela. The similarity of the city names—given the rules and restrictions of the Sumerian writing system—and the duplication of the order in which they occur cannot be coincidence. This must point to a common source for both lists. The Ebla tablets do not

The streets of Ebla, *overleaf,* were a focus for a wide variety of daily activities. While merchants sold their wares, metal and pottery workers made tools and cooking pots. Other people gathered to talk and draw water.

mention destruction of these cities, however. On the contrary, the tablets describe them as flourishing commercial centers. Only in the Bible do we have the description of the sudden and violent demise of Sodom and Gomorrah (Genesis 18-19).

Because Abraham and Lot appear in Genesis 14, 18, and 19, it seems quite reasonable to date these patriarchs some time after the Ebla tablets were written. If the height of the Ebla civilization was in the 27th century B.C., then we should place Abraham perhaps in the 25th century B.C. Corroboration of this can be found in correlating Ebla's great King Ebrium with Eber, whom Genesis 10 and 11 identify as the great-great-great-great-grandfather of Abraham. Semantically the names are equivalent, and it is possible that there is some connection between the Eber of Genesis and the great King Ebrium who ruled the city of Ebla.

Assuming that there is such a connection, we can form a list of descendants of Eber that leads to Abraham (based on the genealogical lists in Genesis 10:24-25 and Genesis 11:14-16); it places Abraham exactly 225 years after King Ebrium. Allowing for errors that the scribes may have made in the dating system, we can still conclude that Ebrium was born about 2700 B.C. and reigned later in that century. Abraham was born about 2500 B.C., perhaps as early as 2525 B.C. or as late as 2475 B.C., and the cities of the plain must have been destroyed sometime during the 25th century B.C. It may even be that the violent end of the cities of the plain is associated with the abandonment of Ebla and other middle Bronze Age cities in that part of Syria and Palestine about that time.

Ebla, a city-state, was the capital of a commercial empire, and its kings were successful executives much like the heads of today's multinational corporations. The tablets attest abundantly to the extraordinary range and volume of trade, and the rich rewards amassed by the city's various kings.

During the period covered by the tablets, five kings are known to have reigned at Ebla. The first two, Igris-Halam and Ar-en-num, apparently were not related to each other or to any of the following kings. Then came Ebrium and his son, Ibbi-Sipiš. The oldest son of Ibbi-Sipiš, Dubuhu-Ada, served as crown prince and heir apparent under his father, but never reigned as king. Irkab-Damu, who was not related to any of the previous kings, succeeded Ibbi-Sipiš. During his reign, the city fell to an unknown enemy, and the palace in which the tablets were discovered was sacked and burned. A tablet of a letter from Irkab-Damu to King Zizi of Hamazi describes the impending crisis. Irkab-Damu urgently requests the military assistance of Hamazi's troops, but the help apparently did not arrive or was too little or too late. Ebla fell, and did not regain its prestige and importance until centuries later in the middle Bronze Age.

Ebla's greatest king was Ebrium, who achieved notable diplomatic and commercial success and made Ebla prosperous and secure during

his long and stable reign. He bequeathed to his son a rich and powerful state. But Ibbi-Sipiš was unable to pass the scepter on to his own
son. The story is a partial parallel to the dynasty of David in the Bible.
Like Ebrium, David founded a dynasty and passed a rich and powerful kingdom on to his son, Solomon. However, Solomon could not
deliver his kingdom intact to his son Rehoboam. Unlike Dubuhu-Ada,
Rehoboam managed to hang on to Judah, the southern part of his
father's realm, and the dynasty was able to remain in power for more
than 300 years thereafter.

According to the tablets, Ebla had a highly organized civil service of
nearly 12,000 persons, about 4½ per cent of the total population of
260,000. That would be equivalent in the United States to a bureauc-
racy of more than 9 million persons. Members of the royal family
enthusiastically supported the many gods that were worshiped in this
society, and the tablets list sacrifices that the king, queen, and princes
offered to the various gods of Ebla in their separate temples. In return,
the priests of these temples supported the royal family and contributed
to its survival and success. Again, the similarity to the kingdom of
David and Solomon is noteworthy. The separation of church and state
was unheard of in those days. On the contrary, each was expected to
promote and defend the other, and the king was considered a divinely
appointed symbol of their collaboration and unity.

In addition to the historical information, the Ebla tablets provide new
geographic data. They list thousands of place names, many of which
have remained the same over the millennia, such as Byblos, Sidon,
Hazor, Megiddo, Carmel, Dor, Ashdod, Gaza, Sinai, Aleppo, and
Hama. Of special interest is the reference to Damascus both in the
narrative of Genesis 14 and on the Ebla tablet in which the five cities
of the plain are mentioned. The suggestion in the Genesis passage is
that the five cities are some distance from Damascus. The Ebla tablet,
on the other hand, indicates that the cities may belong to the orbit of
trade and commerce of Damascus. Present-day Damascus is about 185
kilometers (115 miles) from the Dead Sea, the region with which the
five cities have generally been identified. In trying to locate the five
cities of the plain, some archaeologists now consider it wise to look to
the northeast area of the Dead Sea rather than to the southwest area,
where the Genesis text had previously led archaeologists.

The Ebla tablets are a vast resource for increasing our understanding of early urban development in the Near East and the political,
social, and economic history of that period. It is as if we suddenly
learned something about the history and geography of Rome and the
Roman Empire for the first time. Someday, a detailed picture will
emerge of how these people grappled with the problems and possibilities of creating one of the earliest urban trading centers in history in
the Near East. But the main lesson we can learn from this exciting
discovery is that the more deeply we probe into man's past, the more
modern ancient man seems to have been.

Taming
The Winds

By Jack E. Cermak

**Engineers use specially designed wind tunnels
to learn how they can minimize the dangers
of strong winds and maximize their benefits**

Stu Miller, one of baseball's top relief pitchers, strode to the pitcher's mound through swirling clouds of dust in San Francisco's Candlestick Park. The American League was threatening to tie the score in the ninth inning of the 1961 All Star Game and Miller's job was to protect a 3-2 National League lead. As he started to deliver his first pitch, a strong gust of wind blew him off the pitcher's mound. "Balk," shouted umpire Stan Landes, and two American League base runners advanced into scoring position.

Miller's experience was only one of the more unusual embarrassments wrought on talented baseball players over the years by the strong and erratic winds that blow in Candlestick Park, also known as "the cave of winds." In that 1961 game, eventually won by the National League in the 10th inning, baseball's best players of the day committed seven errors, and blamed most of them on the capricious Candlestick winds.

Why was Candlestick Park so windy and gusty? Could the wind be controlled? To answer these questions, the city of San Francisco in 1963 hired Metronics Associates, a group of meteorologists in Palo Alto, Calif., and my Fluid Dynamics and Diffusion Laboratory at Colorado State University in Fort Collins, Colo.

Our investigations, using a small-scale model of the park in a specially designed wind tunnel, showed that local geographic features and the stadium itself create the wind patterns in Candlestick Park. The ball park stands at the southeast end of Bay View Hill, precisely the point at which high-speed winds skirting around the south end of the hill collide with a strong flow of air coming around the other side from the northwest. This collision produces a chaotic swirling flow that moves randomly about the field. Horseshoe-shaped upper stands opening toward the northeast further amplify the gusts.

We discovered three possible ways to reduce the gustiness and speed of the wind in the stadium. In order of cost and effectiveness, they were: building a partial dome roof opening to the northeast; erecting a porous screen on top of the upper stands; and extending the upper stands to form a closed oval around the playing field. Only the third suggestion, the least expensive and least effective, has been followed, somewhat reducing the wind problem.

Perhaps most frustrating of all, our studies showed that the winds are much less gusty only a short distance from Candlestick Park. If the park had been built one playing-field length to the north, much of the wind problem would have been avoided.

Wind continuously interacts with man and his activities. It knocks down buildings, sets tall towers swaying, buffets pedestrians and pummels them with wind-borne debris, and damages crops. Every year, in the United States alone, wind causes an average of 250 deaths, 2,500 injuries, and $500 million in property damage—an unnecessarily high loss rate. On the plus side, wind dilutes and scatters air pollutants, and carries pollen. It can also be harnessed to generate electricity.

Wind engineering integrates traditional fields of science and engineering—meteorology, fluid dynamics, structural mechanics, and statistical analysis—to minimize the unfavorable effects of wind and maximize the favorable ones. Originally, wind engineers focused on how wind affects structures. But our concerns have expanded to include environmental protection, and we are moving toward applications that will convert wind energy to heat and electricity, and improve agricultural productivity.

Most of the effects of wind on buildings, bridges, chimneys, and towers are damaging. As wind moves around and over a building, it slows down, turns, and accelerates again. These changes in air motion result from different pressures caused by the interaction of the building with the wind. The pressure is positive, or greater than atmospheric, on the side facing into the wind, and negative on the top and other sides. The negative pressure helps to suck the wind across the

The author:
Jack E. Cermak is director of Colorado State University's Fluid Dynamics and Diffusion Laboratory. He has been studying wind engineering for more than 20 years.

Wind indicators on a model of Candlestick Park and surrounding area show the variable gusts that blow there.

Revamping Candlestick Park

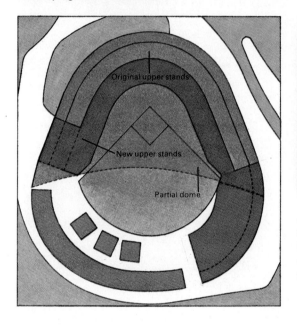

Original upper stands

New upper stands

Partial dome

A partial dome covering half the park, *left,* was suggested as the best solution to the problem of the winds. But a less expensive, and also less effective, suggestion was followed, *above.* The upper stands were extended completely around.

building and down the other side. You can easily sense the forces involved by extending your hand out of a moving automobile. With your hand open and the palm facing straight ahead, you will feel the wind pushing in the direction of the airflow. Now tilt the top of your hand forward at an angle of about 15 degrees. You will feel the wind lifting your hand as well as pushing against it. On a windy day, random gusts will cause these forces to fluctuate even when the automobile is traveling at a constant speed. Wind pressures on a building will vary in a similar way.

When wind pressure causes tall buildings and towers to bend too far, elevators cannot operate and interior walls crack. In extreme cases, a building may become permanently distorted. A few buildings sway so badly that some people in them get motion sickness. Liquids slosh, chandeliers swing, pictures tilt, the building creaks, and televi-

Boarded-up windows in the "Plywood Building," Boston's John Hancock Tower, indicate the extensive damage that strong winds can cause.

sion reception is distorted. Rapid changes in pressure can break windows, dislodge pieces of wall and roofing, and allow air, water, and dirt to seep in through cracks. Different pressures against various sections of tall buildings may disrupt heating, air-conditioning, and ventilation systems, as well as make it difficult to open swinging doors. Outside the building, gusty winds may buffet pedestrians and carry polluted air into building vents. Strong winds can also cause damage to roofs on homes, overturn mobile homes and trailers, and knock down signs and power lines.

Among the benefits of wind is its ability to scatter and dilute air pollution. However, local conditions can vary near pollution sources, and planners must be careful to make the best use of prevailing winds. For example, if the smokestack for a proposed coal or oil power plant is not tall enough, pollutants will be carried to the ground before the wind can scatter them. This problem is complicated when plant buildings or nearby land features channel the smoke into an airstream that is moving downward rather than sideways or upward. To avoid stacks that are ineffective or unnecessarily tall and expensive, builders need to know where stacks should be placed to take advantage of local wind conditions and how tall they should be.

Nuclear power plants, geothermal plants, and storage areas for liquid natural gas present different problems. Nuclear power plants hold radioactive waste gases in huge tanks, allowing the radioactivity to decrease before the gases are released when wind conditions are best suited for rapid dilution. Determining the best time to release such gases from a particular plant and where to locate the emission vents requires detailed information about local winds and how they interact with the plant buildings.

Geothermal power plants, which use steam from deep wells, release substantial amounts of noxious hydrogen sulfide into the atmosphere. Operators must be able to predict where the hydrogen sulfide will go under various weather conditions, and how concentrated it will be in the air. This problem is greatly complicated when the geothermal plants are located in complex terrain, such as that of the geyser fields in northern California.

In storing liquid natural gas, safety considerations depend on knowing the gas concentrations that would result if a tank were to rupture. A dike surrounding the tank would hold the liquid gas, but the liquid would begin to boil immediately upon exposure to the atmosphere, releasing the gas into the air. How far downwind must the gas be carried before its concentration will be dilute enough so that it will not burn? The distance depends on the ability of the wind to mix air with the gas and dilute its concentration.

The atmospheric motion that concerns wind engineers takes place in a thin layer of air covering the earth, called the planetary boundary layer. The thickness of this layer is determined by the height at which surface friction no longer affects the general flow. This height varies from about 300 meters (990 feet) over oceans and large lakes to 600 meters (1,980 feet) above large cities. Usually the motion is turbulent—gusty and swirling, with rapid changes in speed and direction. Average wind speed usually increases with height, while turbulence intensity is usually greatest near the ground and decreases with height. Increasing surface roughness—from sea to grassland to forest to city—also increases turbulence by increasing the surface friction.

Temperature differences within the boundary layer can also profoundly affect average wind speed and turbulence. A layer of cold air below warmer air, called an inversion, tends to discourage up-and-down mixing. If a disturbance carries a parcel of cold air upward into the warmer layer, physical forces push the cold air back to its original position. But when warm air lies below colder air, in what meteorologists call a lapse condition, up-and-down mixing is increased. During a clear day, the surface alternately warms and cools, depending on the amount of sunlight reaching it. This creates a lapse condition from midmorning to sundown and produces an inversion during the night and early morning hours.

Large cities greatly complicate wind action in the planetary boundary layer. Tall buildings penetrate deeply into the layer, and surface roughness and temperature vary widely from place to place. These features cause major changes in turbulence and speed that cannot be predicted by numerical or analytical methods. As Candlestick Park illustrates, local land features can change the wind direction, create jetlike gusts, and generate large-scale swirls. In regions that have many ridges and valleys, such as the geyser fields of northern California, the wind becomes even more complex.

We can get some idea of the complexity of such turbulent motion by watching smoke in the air. Smoke coming out of a stack has two types of motion—general motion in one direction, and turbulent motion indicated by the jagged edge and occasional kinks of the plume. Detailed examination with sensitive wind gauges indicates that the turbulence is a disordered collection of whirls, or vortexes. Because turbulent air moves in a random fashion, efforts to develop a mathematical formula to describe it have had only limited success.

In our attempts to come up with a formula to describe turbulence, my colleagues and I began intensive studies in 1950 of turbulence in the boundary layers that develop inside wind tunnels used to test aircraft. Upon comparing the turbulence measurements from the wind-tunnel boundary layers with measurements obtained in the atmospheric boundary layer, I became convinced that a properly designed wind tunnel could be used to model wind behavior. In 1956, I launched a major effort to design, finance, and construct the first such wind tunnel, at Colorado State University.

Wind tunnels used in testing aircraft are designed with short test sections. The goal is to produce uniform wind speed and temperature throughout the tunnel. But to properly simulate the meteorology of an area, a wind tunnel should have a test section that is about 15 times as long as it is high. Average wind speed, turbulence characteristics, and

Smoke streaming past model of a tall building in a wind tunnel shows how airflow is broken up, creating swirls and turbulence downwind.

How Air Temperature Affects Smoke Flow

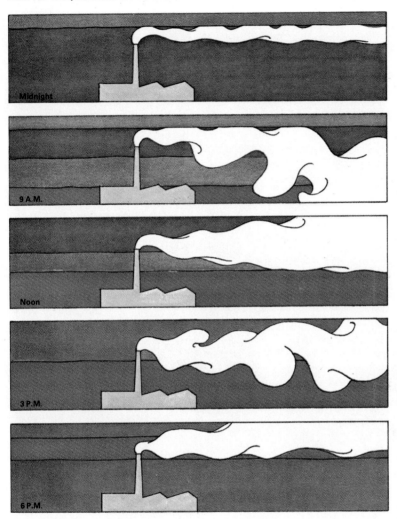

Midnight

9 A.M.

Noon

3 P.M.

6 P.M.

A smoke ribbon drifts at night at the top of a layer of air cooled by the surface. The warming ground heats the air in the morning, stirring up the smoke, which then settles into a cone-shaped plume about noon. Increasing heat from the ground in midafternoon causes strong eddies in the air to break up the smoke. As ground temperatures fall in early evening, the surface layer of air cools again and the smoke disperses above a rising layer of cool air.

Wind-tunnel models show how low stacks, *top,* will send smoke into buildings. Tests show how high to build the stacks so the wind will lift the smoke over the buildings, *above.*

air temperature will vary with height within this meteorological wind tunnel just as they do in the atmosphere for similar surface roughness and temperature conditions. To produce the proper temperature distribution for both land and air, it must be possible to control the temperature of the test-section floor, as well as the air entering the test section. This means that the tunnel must be a closed recirculating system. This wind tunnel is used to study the effects of wind under a variety of atmospheric conditions.

When only the effects of strong winds are significant, a simpler tunnel, which we call an environmental wind tunnel, can be used. With strong winds, intense turbulence churns up and destroys the temperature layers in the atmosphere. Because all levels of the boundary layer are at the same temperature, no heating or cooling is re-

A researcher sets up a model of the Sears Tower in Chicago in an environmental wind tunnel, *above,* used to study winds over large urban areas. The blocks represent outlying areas. A topographical model in the same tunnel helps to study power plant emissions near the Mississippi River, *above right.* A meteorological tunnel is used to study complex temperature effects on winds around a large industrial site in South Africa, *right.*

quired to simulate these conditions in a wind tunnel. The prime requirement of an environmental wind tunnel is that the length of the test section be at least 10 times its height. In our Fluid Dynamics and Diffusion Laboratory, we have built an environmental wind tunnel that is 3.66 meters (12 feet) wide, twice the width of the meteorological wind tunnel. This is wide enough so that we can study models of large areas in the tunnel without reducing the scale excessively. A typical scale would be 1:500 (one meter on the model equal to 500 meters actual size), though it may be as small as 1:4,000, depending on the area under study. A turntable allows us to turn the model so that we can study wind coming from different directions. We use the environmental wind tunnel to study wind characteristics over extensive areas of complex terrain and over proposed urban centers, as well as to study wind pressures on buildings.

We can use these tunnels for both basic and applied studies. The basic studies have two primary objectives. We want to relate boundary-layer wind characteristics to mathematical descriptions of surface roughness and temperature. We are also trying to determine how wind interacts with nonstreamlined bodies. These studies will provide the measurements we need to develop mathematical formulas describing fluid motion in the atmosphere. Eventually, we can then extend the use of computers to investigate wind-engineering problems.

Applied studies deal with specific buildings and surface characteristics, and measure the wind effects needed for design and planning. These investigations cover problems in three broad categories—definition of wind characteristics, wind effects on structures, and distribution of air pollutants. Data obtained from small-scale models in our laboratory have been used in designing or remodeling major structures throughout the United States. These include the World Trade Center Towers in New York City; the Renaissance Center in Detroit; the Peachtree Plaza Hotel in Atlanta, Ga.; the Standard Oil Building and the Sears Tower in Chicago; the Johns-Manville World Headquarters in Denver; the Pennzoil Center and One Houston Center in Houston; the Atlantic-Richfield Towers in Los Angeles; the Bank of

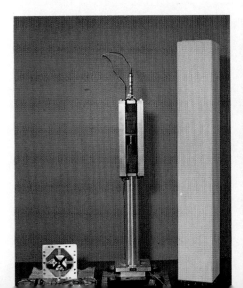
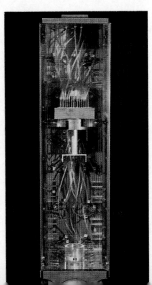

Springs and a damper can be connected to a model of the World Trade Center Tower at right, *far left,* to learn how sway might be modified. Sensors inside a rigid model of the Williams Center in Tulsa, Okla., *left,* measured potential wind pressure on the building prior to its construction.

America World Headquarters in San Francisco; the Garden Grove (Calif.) Community Church; and the Ruck-A-Chucky suspension bridge near Sacramento, Calif.

We have also studied air-pollution problems posed by the Shoreham Nuclear Power Plant on Long Island, New York; offshore nuclear power plants; the Avon Lake Power Plant in Cleveland; and the Harrington Power Station in Amarillo, Tex. And we have studied how to use wind to carry silver iodide to seed snow-producing clouds in the San Juan Mountains of Colorado; to diffuse and dilute natural gas from liquid natural gas spills in Los Angeles; to carry off hydrogen sulfide in the northern California geyser fields; and to carry aerosol pesticides through forests to control tussock moth infestations.

Whenever a model of a proposed building is in the wind tunnel for tests of wind pressure, we also measure wind characteristics in surrounding streets and plazas. This enables architects to consider building changes or to arrange trees, bushes, or other wind screens to protect pedestrians from excessive winds.

Our wind-engineering study of the proposed Yerba Buena Center in San Francisco illustrates the variety of problems that we can analyze with a single small-scale model. This study began in 1971 when the San Francisco architectural firm of McCue, Boone, and Tomsick asked me to investigate the impact of wind on all aspects of their preliminary plans for the center.

The author records data on a clipboard while other engineers observe wind effects created in an environmental wind tunnel. The model is the proposed Yerba Buena Center in San Francisco.

Smoke reveals swirling air under Yerba Buena tower building when the model is set on pillars in a wind-tunnel test of the architect's original design, *far left.* Closed base in modified version deflects the wind, *left.*

Wind also sweeps down the side of the tower building in the original model, *far left,* with possible discomfort and perhaps even danger for pedestrians. Adding a canopy deflects the air over the sidewalk, *left.*

Low exhaust stacks in original Yerba Buena model send fumes into central plaza, *above.* Higher stacks release smoke above pedestrian level, *right.*

We identified several possible problems when we reviewed the architect's drawing to determine the function, shape, and location of each of the 10 proposed buildings. A tower building 115 meters (375 feet) high was to mark the north entrance to a central plaza extending southeast for three blocks through the center. According to the plans, the building was to stand on columns 12 meters (39 feet) above street level. We suspected that this would generate extremely high winds under the building, where pedestrian traffic would be greatest. Furthermore, the tower would reach well above neighboring buildings and thus be exposed to large gusts. The intersections where cross streets entered the central plaza also seemed excellent breeding places for strong gusts. Eight stacks running down the center of the plaza were designed to release exhaust gases from extensive underground-parking facilities. These possibly would send excessive concentrations of carbon monoxide, nitrogen oxides, and hydrocarbons through the plaza and into the air intakes of surrounding buildings.

We built a model to a scale of 1:240. This allowed us to fit a model of the entire 3-block-long center on the turntable of the environmental wind tunnel. Our model, which covered the entire test-section floor, included about four city blocks upwind from the center in each wind direction we studied. We selected the three wind directions most common to the area—south, west, and northwest—for detailed study. We used wind-data measurements obtained at the National Weather Service station a few blocks away from the Yerba Buena site.

With the aid of tracer gases released at strategic locations in the tunnel, we first observed the general character of the air flow through the center. Motion pictures of the smoke movement provided a permanent record, and also aided us in interpreting statistical wind data. Then we measured the average wind speed, turbulence, wind pressure, and concentrations of the tracer gases for a variety of possible changes in the model design. We tested various shapes for the tower, moved several office buildings around, experimented with the location and number of trees, changed the shape of the skylight over the plaza, and varied the height of the underground-garage exhaust stacks.

As a result, we recommended a number of changes, many of which were incorporated into the revised plans. The architects agreed to eliminate the open space beneath the tower after we found that peak wind speeds in this space would be close to four times greater than average speeds measured at the Weather Service building. When the average speed at the Weather Service station was 18 kilometers per hour (kph), or 11 miles per hour (mph), winds would peak at 60 kph (37.5 mph) under the tower. Canopies about 6 meters (20 feet) above the street will deflect the strong winds that would swoop down to street level near the tower. We recommended that trees and bushes be planted to act as windscreens where streets enter the plaza. A transparent skylight covering the plaza will be just below roof level. Two short exhaust stacks that were planned for a height of 4.5 meters (15 feet) will be raised to a height of 19.5 meters (64 feet) to equal the height of the other stacks.

The study provided other benefits. We discovered that the tower needed no special structural changes to withstand wind pressures. This was in great contrast to our findings for the World Trade Center Towers in New York City, where special dampers—visco-elastic connections between floor beams and building columns—were recommended to control the sway. The data gathered on the concentrations of automobile-exhaust fumes helped engineers prepare a comprehensive environmental-impact statement for the Yerba Buena site.

Wind engineers are also applying their expertise to agricultural productivity. Good pollination of hybrid wheat strains results in grain yields up to 20 per cent higher than those of commonly grown strains. But because each hybrid plant is either male or female, pollen must be carried from male plants to female plants. Wind is the best natural carrier. But to get the highest pollination, we must know the best pattern for planting in order to take advantage of the natural wind characteristics of an area. Crop yield also can be increased substantially by making sure carbon dioxide is more evenly distributed to the plants. But how should the rows be spaced and which way should they run to get the best atmospheric distribution of carbon dioxide? These are areas where basic wind research can help.

Among the other emerging applications of wind engineering is the problem of energy conservation. Under certain circumstances, natural ventilation can replace air conditioning. For example, in 1976, we began discussions with the U.S. Navy for a study on how housing units in the Pacific Islands could be placed to use the trade winds—which consistently blow from one direction—for natural cooling.

During the last few years, several federal agencies have supported research on using windmills to convert wind energy to electricity. Wind engineers are working in our laboratory to determine the characteristics of wind over different types of terrain. This information will be combined with wind-tunnel studies of models of different sites to select the best site for setting up windmills. Other questions include how many windmills can be placed on a given field before they start interfering with one another, and how much power can be generated at a particular site based on wind-speed records from Weather Service stations in the area.

Wind engineering is a new, but rapidly developing, discipline that engineers, architects, and meteorologists recognize as necessary to understand and treat local wind problems. So far, physical modeling has been the principal tool of the wind engineer. However, as we obtain fundamental measurements of turbulence and air flow over different kinds of sites, we will eventually be able to use computers to model specific problems. Because the awareness of the impact of wind and its effects on human activities is constantly increasing, the contributions of wind engineering to the welfare of society should continue to grow.

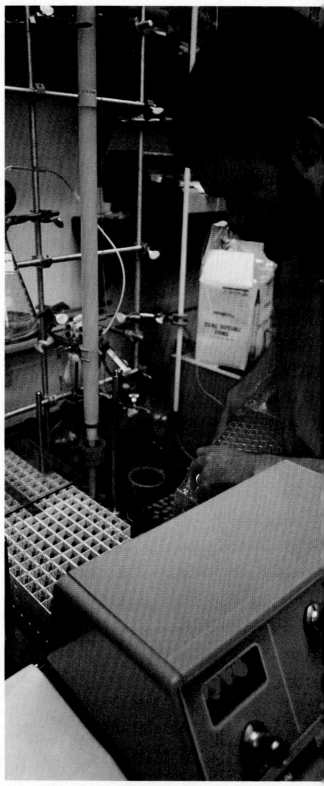

Detectives
Of Disease

By Edward G. Nash

**Scientists from the Center for Disease Control
assemble clues to pinpoint the cause of an
epidemic and find ways that it can be controlled**

Ray Brennan was visibly tired when he arrived home in Athens, a small town in northeastern Pennsylvania, on Saturday, July 24, 1976. He had been attending the four-day American Legion state convention in Philadelphia. The 61-year-old retired Air Force captain soon developed chest pain, fever, and chills. He entered a hospital on Tuesday, July 27, and died that night, his lungs filled with a bloody froth.

All through that week and into the next, the mysterious pneumonia-like illness struck in cities and towns across Pennsylvania. Frank Aveni died in Clearfield, in central Pennsylvania; Louis Byerly in Jeannette, near Pittsburgh. At least four were dead in the Philadelphia area.

By the weekend, physicians noticed an alarming pattern beginning to emerge. On Friday, July 30, Dr. Ernest Campbell of Bloomsburg became disturbed because three of his patients had similar and very serious symptoms. All three had attended the American Legion convention in Philadelphia. Campbell notified state public health authorities. In Williamsport, Dr. Terry A. Belles had five patients with similar pneumonialike symptoms; they, too, were conventioneers. Belles' findings were also relayed to the state Public Health Department in Harrisburg. By Sunday evening, Dr. William Schrack, the Public Health Department epidemiologist responsible for keeping track of the spread of disease in Pennsylvania, was worried. Early on Monday morning, August 2, he compared notes with Edward T. Hoak, state adjutant of the American Legion. Hoak was equally worried, for he had been getting horrifying reports all weekend from legion posts all across the state.

It was a busy day in Harrisburg, as dozens of reports began to trace a ghastly pattern of illness and death that seemed somehow linked to the Philadelphia convention. It was also a busy day in Atlanta, Ga. At 9:15 A.M., Dr. Robert B. Craven of the Bureau of Epidemiology at the Center for Disease Control (CDC) received a call from a Veterans Administration physician in Philadelphia, who described 26 cases of an unusual feverish ailment with many of the symptoms of pneumonia. Already four persons had died. After conferring with Dr. David J. Sencer, who was then CDC's director, and other CDC officials, Dr. Michael B. Gregg, deputy director of the Bureau of Epidemiology, called Harrisburg to offer assistance. The offer was accepted by Dr. Leonard Bachman, the state's secretary of health.

"We had three investigators flying to Pennsylvania before 6 P.M. that day. We had 10 more on the way the next," said Dr. David Fraser, chief of CDC's special pathogens branch, who was in charge of the Center's field operations in Pennsylvania. At its peak, with 27 CDC staff members working in Pennsylvania, the search for the cause of what came to be called "legionnaires' disease" was the largest field operation in CDC's history.

The CDC was founded as the Communicable Disease Center in 1946. It is a direct descendant of a World War II agency set up to control malaria around wartime training camps. Other diseases, such as typhus, were soon added to the agency's responsibilities. After the war, CDC became the part of the U.S. Public Health Service charged with assisting state and local health departments and other agencies in controlling and even eradicating infectious diseases. In recent years, under the Department of Health, Education, and Welfare (HEW), this role has expanded to cover noninfectious environmental and occupational diseases and CDC's army of 3,600 persons—almost half of them working in its 14-building Atlanta headquarters—has become the nation's front-line defense against disease. Its Epidemic Intelligence Service (EIS) has proved to be a valuable training ground for young doctors to learn epidemiology, the study of the causes, distribution, and control of the spread of disease. These investigators are CDC's foot soldiers; they pound pavements, ring doorbells, and ask questions.

The author:
Edward G. Nash is
a senior editor for
Science Year and The
World Book Year Book.

Fraser's Pennsylvania field team was largely staffed by EIS doctors from the Center's Bureau of Epidemiology. Working closely with state and local health authorities, they fanned out across the state and began to fit the pieces of the puzzle together. They collected data on the victims—their personal habits, such as drinking and smoking; the dates they were in Philadelphia; the hotels they stayed in; the meetings they attended; and the food they ate.

As the epidemiologists traced these activities on an almost hour-by-hour basis, a pattern emerged that, along with the clinical symptoms, became the definition of legionnaires' disease. The symptoms would have to have appeared between July 1 and August 18. They had to include a fever of at least 38.8°C (102°F.), coughing, and X-ray evidence of pneumonia. Every victim had to have either attended the American Legion convention or been in the Bellevue Stratford Hotel, the convention headquarters, sometime after July 1. "The symptoms were so nonspecific," Fraser pointed out, "that many illnesses can mimic them. That was one reason we had to be very crisp in our definition of a case. Most serious colds could overlap to some degree with a mild case of legionnaires' disease."

Under this definition, the CDC identified 180 cases of legionnaires' disease. The final death toll was 29 persons, giving the disease a very high death rate of 16 per cent. Curiously, defining legionnaires' disease in this fashion produced 39 cases of "Broad Street pneumonia." These victims had the same symptoms but they claimed they had not been inside the Bellevue Stratford Hotel. However, all had been within one block of the hotel, which is located on Broad Street in Philadelphia, during the appropriate time.

Defining a disease is not the same thing as pinpointing its cause, however. So while the EIS foot soldiers marched across Pennsylvania, the support troops at the Center in Atlanta began their long, thorough, and tedious search for the cause. The epidemiologists collected nasal and throat swabs and blood and tissue samples, and sent them to CDC's Bureau of Laboratories in Atlanta for careful laboratory analysis. The initial fear that this was a swine flu epidemic was quickly allayed. "We had pretty much eliminated influenza by August 3," said Fraser. "It wasn't spreading. That's good evidence that it wasn't influenza. Then by August 5 we had the laboratory reports that definitely ruled it out."

Bacteriologists, virologists, and other scientists at the Atlanta headquarters began eliminating many other possible causes of the disease. They found no bacterial infections and none of the likely viruses. Mycoplasmas and rickettsias, organisms classified between bacteria and viruses, were ruled out. The disease was not histoplasmosis, a fungous infection. Psittacosis, a viral disease carried by birds, was not the answer, either—the pigeons on Philadelphia's Broad Street were found to be relatively healthy. And the toxicologists at CDC could find no evidence of toxins. Paraquat, a chemical used in herbicides, was at one time suspected, as were nickel carbonyl, zinc, and cadmium. But researchers again drew blanks.

There were suggestions that a madman or saboteur had deliberately placed a poison at the convention, perhaps in the hotel's air-conditioning ducts. "If a madman did it," said Fraser, "he certainly was a smart one. He thought of an agent we're having trouble identifying. But our being able to find the cause is really independent of whether or not it was sabotage. The patterns should still be there; the agent should still be there."

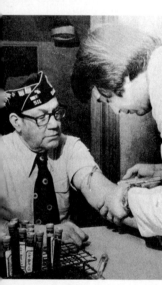
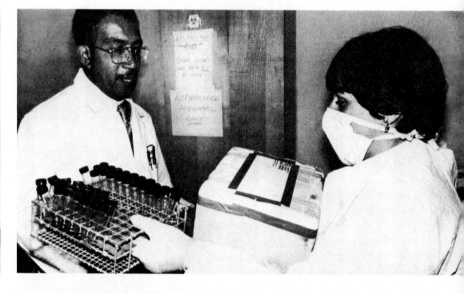

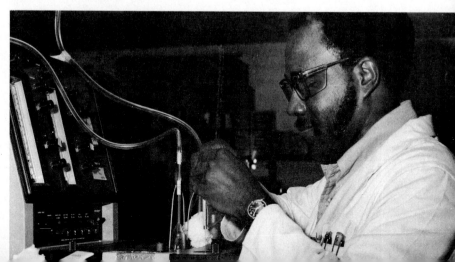

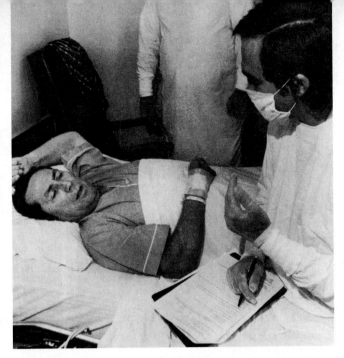

Tracking Down a Disease

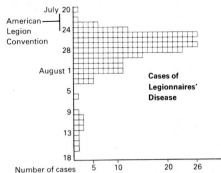

Cases of Legionnaires' Disease

American Legion Convention

Number of cases

Philadelphia's Bellevue Stratford Hotel, *opposite page, left,* hosted Pennsylvania's American Legion meeting. Of one group of conventioneers, *opposite page right,* two died of a strange ailment. Doctors interviewed patients, *left,* and traced the pattern of disease that peaked just after the meeting, *above.*

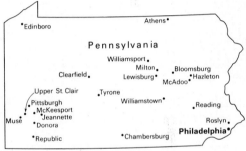

A doctor takes a blood sample from a conventioneer, *opposite page, left.* Samples from the dead and survivors were collected, *opposite page, right,* and tested with sophisticated, computer-linked devices for clues to the cause of the outbreak, *left.* The disease killed 29 persons, including 26 residents of Pennsylvania, *above.*

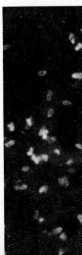

CDC workers handled the dangerous samples in glove boxes, *opposite page, left,* and tested them for toxins, *opposite page, right.* Drs. Joseph E. McDade and Charles Shepard, *far left,* finally found the bacteria that caused the disease, *left.*

As the weeks of fruitless testing stretched into months, criticism grew. Congressional hearings were held in Philadelphia in November. The major criticism was that the epidemiologists were so certain in the beginning that the disease was an infectious one that they failed to consider early enough in the investigation the possibility of toxic substances in the hotel's environment.

The break finally came in late December. Microbiologist Joseph E. McDade of the Center's leprosy and rickettsial branch, still convinced that an infectious organism caused the disease, began looking for a new, unknown organism. Shortly after Christmas, he began re-examining samples taken from guinea pigs that had been inoculated with tissue samples from the disease victims and had developed fever and other suspicious symptoms. The slides had been studied months earlier with no positive results, but now McDade began a more intense scrutiny. This kind of microscopic study, said McDade, "is like looking for a contact lens on a basketball court with your eyes four inches above the floor." McDade found his contact lens—a small group of rod-shaped organisms that had gone unnoticed before.

Then McDade and his chief, Dr. Charles C. Shepard, began the task of proving that this bacterialike organism had caused legionnaires' disease. They injected stored tissue samples from the guinea pigs into chick embryos, all of which died within a week. The researchers found the new organism in many of the embryos. When they tested these with blood-serum samples taken from survivors of the disease, they found that these persons had developed antibodies to the organism. Later they found the organism once again in tissue taken from four patients who died of legionnaires' disease.

So part of the mystery was solved. A bacterium previously unknown to science caused legionnaires' disease. As a bonus, further study implicated the organism in two previous unsolved disease outbreaks, including one in Washington, D.C., that killed 16 persons in 1965. But how did the organism get to Philadelphia, to the Bellevue Stratford Hotel? Why did the disease break out so suddenly and viciously, only to disappear almost as quickly? Most important of all, what can be done to prevent or control future outbreaks? By the summer of 1977, CDC scientists and Pennsylvania health authorities were still trying to answer these questions.

CDC has always worked closely with state and local health authorities. As Sencer said, "We assist rather than do it ourselves. That has really been our way of doing business since the early 1940s." "When the country was established," Dr. J. Donald Millar, director of CDC's Bureau of State Services, explained, "certain powers, such as national defense, were vested in the federal government. But public health became a state responsibility. The legal bases for demanding vaccinations for this or that disease, or for quarantining people under certain circumstances are all state laws." The Bureau of State Services is one of the largest at the Center. Its duties include working with state and

A CDC fieldworker interviews a housewife as part of a field survey that includes talks with relatives and neighbors of the victims of a disease.

local authorities in immunization services, tuberculosis and venereal disease control programs, and environmental problems, such as lead poisoning and rodent control.

Ten regional HEW offices throughout the country help coordinate CDC's assistance to the states. Included in the offices are 750 CDC health workers known as Public Health Advisors who are assigned directly to state and local health departments on a full-time basis.

Many EIS officers also work in the state health departments. Such teamwork has helped to clear up many puzzling disease outbreaks. When 196 of 343 passengers on a chartered jet flying from Tokyo to Copenhagen, Denmark, with a stop at Anchorage, Alaska, were struck by food poisoning in February 1975, epidemiologists were able to trace the outbreak to the infected thumb of a cook in Anchorage who had prepared ham slices served on the plane a few hours before landing in Copenhagen. The investigation involved the work of Danish health authorities, the Greater Anchorage Area Borough Health Department, the Alaska Department of Health and Social Services, the U.S. Food and Drug Administration (FDA), the Department of Agriculture, three EIS doctors in the field, and several CDC divisions.

The CDC receives data on disease occurrence from hospitals, physicians, state and local agencies, and others in the health-care field. Consequently, it can monitor death and disease patterns for the entire

A computer technician searches for a tape in the vast amount of data in CDC's computer room, *above*. A doctor removes a test sample of blood serum from a refrigerated storage area, *above left*. The CDC's extensive collection of biological specimens and its banks of data are important international medical information resources.

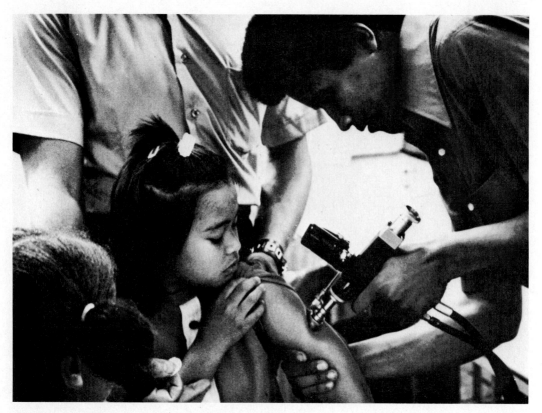

A medical technician inoculates a child after the collapse of South Vietnam in 1975. The CDC coordinated the public-health program to care for Vietnamese refugees in the U.S.

nation. Many of its findings, especially on important and epidemiologically interesting cases, are published in the *Morbidity and Mortality Weekly Report* (*MMWR*). For example, a June 1976 outbreak of psittacosis among workers in a Nebraska poultry-processing plant was traced first to turkeys slaughtered on two particular days, and finally to a specific flock of birds that had come from Texas. The *MMWR* also lists each week's reported cases of such common diseases as chicken pox, measles, and mumps and such rare ones as anthrax, plague, and rabies. Except for venereal disease, infectious diseases have been declining in the United States because of effective state immunization programs developed with the help of CDC and better general health care.

CDC watches one infectious disease very closely–influenza. The virus that causes influenza A, the more severe of the two forms of influenza, mutates periodically, creating new strains that pose a grave health problem. "One can think of flu as the last great plague," said Millar. "It hasn't the same vulnerability that other immunizable diseases have because the virus changes frequently. Therefore, the vaccines are not nearly as effective as smallpox or polio vaccines, for instance. Flu has a tremendous capacity to spread and a short incubation period. It can spread like wildfire." It was just such concern that prompted the controversial swine flu program in 1976.

"The phone rang on February 3," Dr. Gary R. Noble, chief of CDC's respiratory virology branch in the Bureau of Laboratories recalled. "It was Dr. Martin Goldfield of the New Jersey Health Department calling to say that he had a couple of samples that were unusual. He was sending them down for typing. Later, he sent three more. All had come from trainees at the U.S. Army base at Fort Dix." As many as 500 recruits had been infected; one had died. Tests at CDC identified a virus that closely resembled the swine flu virus that killed 20 million persons in 1918 and 1919, including 548,000 Americans. Many feared that another serious epidemic might occur.

After consulting with scientists and health authorities, President Gerald R. Ford on March 24 announced plans to develop a vaccine and mount a massive program to inoculate virtually "every man, woman, and child in the United States." Because 1976 was a presidential election year, the swine flu program became a political issue. Critics charged that the plan was a costly political ploy, and not really necessary. Supporters cited the danger a new strain of influenza posed. They also pointed out that the $135-million cost of the program was far less than the billions of dollars and many thousands of lives lost during the outbreaks of Asian flu in 1957 and Hong Kong flu in 1968. Congress eventually appropriated funds for the immunization program in April. Although other federal agencies such as the National Institute of Allergy and Infectious Diseases and the FDA were involved in developing the vaccine, "the agency that had primary responsibility for seeing that the vaccine was getting into arms was CDC," as EIS officer Dr. Richard J. O'Brien put it.

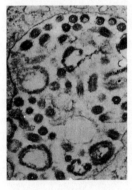

Two deadly African viruses, the Ebola virus, *top,* and the Marburg virus, *above,* have been discovered in recent years. The possibility of their spreading around the world has prompted greater cooperation between the CDC and health agencies in other countries.

The program remained clouded with controversy. Drug companies demanded federal insurance against liability before they would develop and manufacture the vaccine. After immunization began on October 1, there were several brief scares about side effects of the swine flu vaccine. In two months, 45 million persons were vaccinated – the greatest immunization program in U.S. history. The program was halted in December, after some of those immunized developed Guillain-Barré Syndrome (GBS), a neurological affliction usually marked by temporary paralysis. By the end of 1976, 951 cases of GBS had been reported to CDC. Of these, 480 had received influenza vaccine of some kind. As it turned out, the 1976-1977 flu season was relatively mild in the United States and elsewhere. And after the initial Fort Dix cases, the swine flu did not reappear.

If the swine flu program was not exactly a howling success, an earlier massive CDC public health program was. In the spring of 1975, after the collapse of South Vietnam, more than 130,000 Vietnamese refugees entered the United States. There was some fear that such a large influx from a tropical country racked by war might pose a serious public health hazard. Working closely with the military medical services, CDC examined every refugee, and saw that all Vietnamese children from 1 to 5 years of age were inoculated against measles

A mobile quarantine unit is unloaded from a cargo plane for delivery to CDC. The unit will isolate victims of highly contagious diseases being taken to regular medical facilities.

and polio. "It certainly was unprecedented in the number of people to be screened and the short time available," said Dr. William H. Foege, then CDC's assistant director for operations. The initial fears were soon calmed; the Vietnamese proved to be remarkably healthy.

With its investigative expertise and extensive laboratory facilities, the Center for Disease Control has become a major factor in the worldwide fight against disease. It often aids other countries during disease outbreaks and after any natural disasters that might cause widespread disease.

When a devastating earthquake struck Guatemala in February 1976, for example, CDC sent a team there to help in the recovery. "We use the same approach we bring to an epidemic," said Foege. "First, we develop a surveillance system to find out what the real problems are—there's often a lot of confusion and chaos. Second, we analyze the data; third, we respond specifically to the problems. This can be done very quickly. Our people are trained to follow up on rumors. For instance, there were rumors that in one village there were a number of cases of typhoid. Our team found that someone was confusing measles with typhoid. Without that investigative capacity, people and resources sent there to handle typhoid would have been lost for use elsewhere."

Although there has been no smallpox in the United States for many years, CDC's Bureau of Smallpox Eradication has worked with the World Health Organization (WHO) in its drive to wipe out that highly infectious disease. The effort has apparently succeeded. There have been no cases of the virulent Asian smallpox since October 1975. Some cases of the milder African form were reported in Somalia and Ethiopia in May 1977. Smallpox may have already become the first human disease eradicated. See PUBLIC HEALTH, Close-Up.

But other diseases remain to be conquered, some of them deadly and only recently discovered. In 1975, when a young man who had been hiking in Rhodesia died of a raging fever in a hospital in Johannesburg, South Africa, puzzled doctors sent tissue and blood samples to CDC for examination. Using the electron microscope and sophisticated tests in its supersafe "hot lab," CDC scientists identified the Marburg virus in the samples. The virus, which causes green monkey fever, had been seen only once before, in 1967 in Marburg and Frankfurt, West Germany, and Belgrade, Yugoslavia. Thirty-one laboratory and hospital workers in these cities caught the disease; seven died. The researchers apparently caught the disease from African green monkeys being used in laboratory tests, but the monkeys were probably not the virus' natural hosts because they died, too. The identity of the host organism and how the disease spreads to humans are still unknown.

In a similar and even more deadly case, six CDC staff members joined an international team investigating an outbreak of fever in Sudan and Zaire in September 1976. Named Ebola hemorrhagic fever and characterized by very high fevers and internal hemorrhaging, the

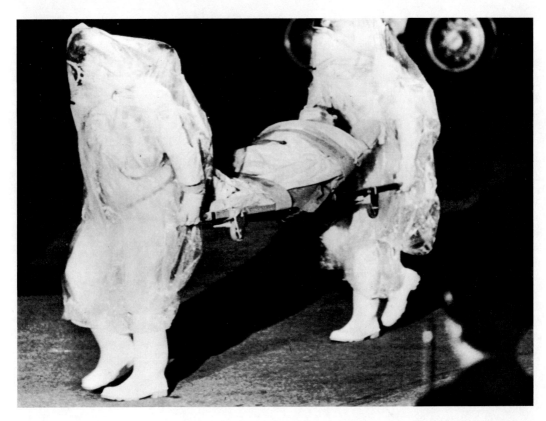

Shrouded in protective clothing, two medical workers carry a victim of Lassa fever, a very dangerous West African disease, from an airplane in Hamburg, West Germany. Modern air travel increases the chances that such a disease, although rare, can break out anywhere in the world.

malady was previously unknown. It is highly infectious–100 of the 600 known deaths were among doctors, nurses, and other health workers treating the victims. Scientists have now identified a virus similar to the Marburg virus as the cause of Ebola hemorrhagic fever. Its natural host is also unknown. "Now comes the time to go and look in nature," said Dr. Patricia Webb, of the viral pathology branch of CDC's Bureau of Laboratories. "It's a really formidable prospect to take on the African flora and fauna because it's so diverse. I think the obvious place to start might be to look at the monkeys." New deadly viruses, unknown 10 years ago, are "an alarming thing," Webb says. "And we are faced with the problem of what to do about travelers who contract these very serious and contagious diseases."

In an age when a jetliner can carry a disease from continent to continent about as fast as a mosquito might carry malaria from house to house, close international cooperation with such organizations as WHO becomes more important. "There's no question in my mind that CDC must become more international," says Foege. "We have more reason perhaps than some other parts of government because diseases are international." In two recent and frightening cases, travelers returning from West Africa–one to London and one to Washington, D.C.–developed Lassa fever, a disease which has a 52 per cent mortality rate in hospitalized cases. As soon as the CDC and British

From its headquarters, a 14-building complex of laboratories and offices in a suburb of Atlanta, Ga., the Center for Disease Control leads the nation's battle against disease.

doctors identified Lassa fever, the patients were isolated and epidemiologists began tracking down all persons who had been in contact with the victims. Luckily, there were no secondary infections and both of the patients survived.

Handling such materials as Lassa fever virus, legionnaires' disease specimens, and other biological and toxicological substances requires strict safety procedures. Hazardous substances are kept in locked freezers in secure storage areas at the Center. CDC has developed a grading system for its scores of laboratories based on the relative danger of the work that can be done in them. These standards were adopted in 1976 by the National Institutes of Health to develop safety guidelines for laboratories conducting recombinant DNA research (see GENES: HANDLE WITH CARE). At one end of the scale are relatively open laboratories designated P-1. At the other end are CDC's two P-4 "hot labs." One of these labs, scheduled to open in 1977, will be among the safest in the world. It will have an air lock and shower room at its entrance in which personnel must wash and change clothing; negative air pressure in its two work areas so that air cannot accidentally escape from the lab; and a double filtration system for vented air. Its "clean room" contains a line of glove boxes, totally enclosed cabinets into which rubber gloves protrude, allowing scientists to work without coming into contact with hazardous material.

Workers in the "dirty room" must wear air suits much like space suits. This room will be used for animal handling and for experiments that cannot be carried out in glove boxes.

Equipment handling and processing is an enormous job at CDC. For instance, each year the laboratories' service section must collect, sterilize, and distribute 15 million pieces of glassware.

The Office of Biosafety is responsible for biological and occupational safety at the Center's various facilities. "My friend and predecessor here, Bob Huffaker, said it was sort of like flying a light plane," John Richardson said of his job as biosafety director. "You have hours of boredom and moments of sheer panic. We have potential problems every day, but we have some very responsible people here."

Richardson's office is also responsible for regulating the interstate and foreign shipment of disease-causing biological agents for diagnosis. The basic shipping container is deceptively simple—a capped tube wrapped in absorbent material, packed in a watertight metal tube, and inserted into a cardboard mailing tube. It is safe enough to be shipped, properly labeled, by ordinary first-class mail.

"The packaging research was done by the U.S. Army at Fort Detrick, Maryland," said Richardson. "They did all sorts of things. They put them in airplanes and crashed them into concrete walls. They dropped them from buildings and dropped heavy metal cylinders on them. It's a remarkably durable container."

A different kind of safety became CDC's responsibility in 1973. The federal government placed the National Institute of Occupational Safety and Health (NIOSH) under CDC's jurisdiction. NIOSH, which has its headquarters in Rockville, Md., operates laboratories in Cincinnati, Ohio; Morgantown, W. Va.; and Salt Lake City, Utah. It is responsible for investigating occupational safety and health problems in business and industry and recommending standards for contaminant or physical hazards to the Department of Labor. The addition of NIOSH and other programs prompts many CDC staff members to ponder the future of the agency.

Sencer thinks that CDC may become more involved in health-care delivery—getting the proper care to those who need it. Millar thinks that primary prevention of infectious disease has been accomplished in the United States except for influenza and venereal disease. "The biggest payoff epidemiologically and economically is in secondary prevention. Diabetes and arthritis are good examples. You can identify the problem and then by certain interventions you can prevent the advancement of the disease or minimize the cost in human suffering and dollars by applying preventive measures."

"If we're really a center for disease control, it's not limited to infectious diseases," said Foege, who was named director of CDC in April 1977. "The big preventive problems in the United States right now concern such things as cigarette smoking, misuse of alcohol, accidents, and improper diet. These are logical areas for CDC's kind of work."

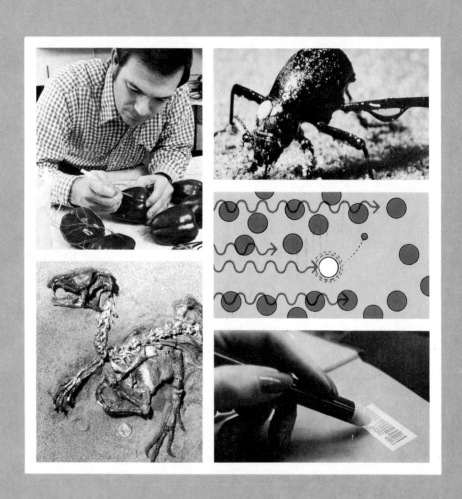

Science File

Science Year contributors report on the year's major developments in their respective fields. The articles in this section are arranged alphabetically by subject matter.

Agriculture
Anthropology
Archaeology
Old World
New World
Astronomy
Planetary
Stellar
High-Energy
Cosmology
Biochemistry
Books of Science
Botany
Chemical Technology
Chemistry
Communications
Drugs
Ecology
Electronics
Energy
Environment
Genetics
Geoscience
Geochemistry
Geology
Geophysics
Paleontology

Immunology
Medicine
Dentistry
Internal
Surgery
Meteorology
Microbiology
Neurology
Nutrition
Oceanography
Physics
Atomic and Molecular
Elementary Particles
Nuclear
Plasma
Solid State
Psychology
Public Health
Science Policy
Space Exploration
Technology
Transportation
Zoology

Agriculture

Triacontanol, a naturally occurring plant-growth regulator that can increase growth by as much as 40 per cent, was extracted from alfalfa by Michigan State University scientists in 1977. Plant scientists Stanley K. Ries, Violet Wert, Charles C. Sweeley, and Richard Leavitt sprayed triacontanol on barley, corn, cucumbers, lettuce, rice, tomatoes, and wheat. Increased growth was distinguishable within a few hours after the growth regulator was applied to the plants. The scientists also discovered that the new compound is effective when sprayed on plants in quantities equivalent to 10 milligrams per hectare (roughly equal to only 1 ounce per 7,000 acres), and that it causes plant growth in darkness as well as in sunlight.

The discovery of triacontanol grew out of the finding two years earlier that chopped alfalfa used as a mulch doubled the yield of a field of tomatoes when compared with an adjacent field growing under identical conditions but without the alfalfa mulch. Ries and his colleagues extracted many substances from the alfalfa plants before they found the triacontanol substance.

Hydroponics. The Nutrient Film Technique (NFT) developed by Allan J. Cooper of the Glasshouse Crops Research Institute in Littlehampton, England, was widely acclaimed as a completely automated system for producing food crops. In this version of a technique known as hydroponics, plants that are tied or propped upright grow without soil. Nutrients are supplied to the roots by a flowing water solution.

The plant roots hang in a channel or gully made of black polyethylene film, and the nutrient solution is continuously recirculated in the channel. This conserves both water and nutrients. In greenhouses, the most beneficial root temperatures can be maintained by controlling the temperature of the solution. Under these conditions, air temperatures can be lowered. This could result in a considerable saving of energy otherwise expended for heating.

New crops. Agronomist Richard Lower of North Carolina State University in Raleigh developed cucumbers

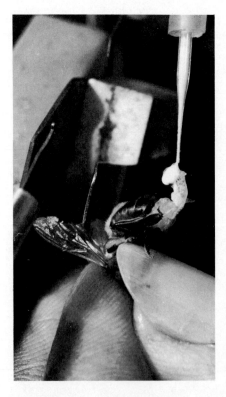

Semen taken from a drone honeybee, *right,* can be frozen and stored in a tank of liquid nitrogen, *far right,* until it is used to artificially inseminate a queen bee.

A scientist inserts thermocouples in bell peppers to check temperature variations in a University of Florida test of how fast peppers can be cooled using various types of refrigeration.

Agriculture

Continued

that grow on bushes rather than vines in 1976. He named his new variety Compact. The Florida Belle, a high-yielding, large-fruited strawberry, was introduced in Florida, and a bronze muscadine-type grape, Dixie, was developed in North Carolina. Scientists at the U.S. Department of Agriculture (USDA) introduced a new pecan, Kiowa, in Texas and Georgia.

The Cheyenne blackberry, an erect-growing plant with large fruit, made its debut in Arkansas, and an Aristocrat pear, highly resistant to all common pear diseases and well adapted to cold, was introduced in Virginia and Kentucky. Other new plants included two squashes developed in Nebraska, the Butternut Ponco and the Butternut Patriot; an Oregon Sugarpod pea; a Floridagold peach; and a Great Northern Star dry bean.

Greenhouses. Theodore Short of Ohio State University in Columbus proved in 1977 that an air-inflated, double-plastic cover reduced greenhouse heating requirements by 57 per cent. Meanwhile, Carl N. Hodges and

fellow scientists at the University of Arizona Environmental Research Laboratories in Tucson developed a new kind of foam insulation that can be pumped into such double-plastic covers or between two separate covers.

The material works as an insulating foam only at night. During daylight, it becomes a liquid through which the sun can shine to warm the greenhouse and provide light needed for photosynthesis. The researchers found that the foam reduces heat loss by 80 per cent.

Merle H. Jensen, also of the University of Arizona Environmental Research Laboratories, designed a greenhouse composed of solar collectors and advanced evaporative cooling devices so that the desired temperature might be maintained with minimum electrical energy. Drip irrigation provides the water and nutrients needed to produce a variety of vegetables – cucumbers, eggplants, lettuce, peppers, and tomatoes. Jensen's greenhouse has a total production area of 23 square meters (250 square feet) and it yields 0.9 kilogram (2 pounds) of vegetables per day.

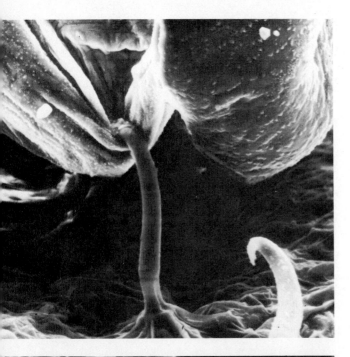

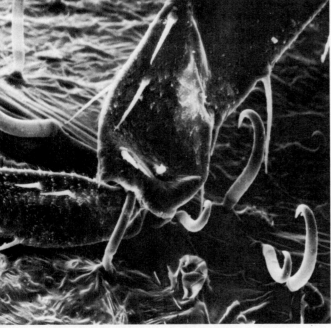

Hooked hairs that provide natural protection to the leaf of a field bean, magnified 700 times, pierce and hold the abdomen of a young insect pest known as a leaf hopper, *top*. Hairs, magnified 350 times, pierce the tissue between the legs of an adult leaf hopper, *bottom*.

Livestock vaccine. A new type of vaccine has been developed to control foot-and-mouth disease, the most dreaded malady of livestock. The vaccine was perfected in 1976 by Howard L. Bachrach, Douglas M. Moore, Peter D. McKercher, and Jerome Polatmik of the Plum Island Animal Disease Center in Greenport, N.Y. It is unique in that it is derived from the coat of protein that surrounds the foot-and-mouth disease virus. All previous vaccines for either livestock or humans have been made from whole killed viruses or weakened live viruses.

The discovery that an animal can be immunized with only the coat of a virus opens new possibilities in vaccine research. It indicates that many virus vaccines may someday be produced without the constant danger that a virus may escape the killing or weakening procedures used today.

Milk and beef production. H. Allen Tucker, a dairy scientist at Michigan State University in East Lansing, found in 1977 that cows produce significantly more milk when the wintertime day is extended to 16 hours with artificial light. Light has a positive effect on the production of prolactin, a hormone relating to milk production. Tucker also reported that calves under extended-light conditions gained more weight than those exposed only to normal daylight. They also had the shorter, finer coats of hair of summer while calves growing in normal daylight had the relatively woolly coats normally produced in winter.

Roger West and Alan Grusby of the University of Florida in Gainesville reported in 1977 that electric shock treatment makes baby beef carcasses more tender because it breaks down adenosine triphosphate (ATP) in cells and thus prevents muscle shortening when the carcasses are put in cold storage. The researchers applied the electrical current to the carcasses about an hour after slaughter. The shock — 5 to 20 seconds of electricity at 320 volts of about 5 amperes — triggered violent muscular reactions that tenderized the beef. The shock treatment had the greatest effect on tougher carcasses, but did not overtenderize already tender beef.

William Chalupa of Applebrook Research Center in West Chester, Pa.,

Tomatoes, more square than round, were developed by scientists at the University of California, Davis. The shape makes them less likely to bruise during harvest and shipping.

experimented with ways to treat feed for cows, sheep, and goats so that it would be digested in the animal's intestine and not in its rumen. The rumen is the first of several compartments in the stomach of cattle, sheep, and goats. High-protein meals provide the greatest benefit in protein when they are digested in the animal's intestine. Chalupa treated soybean meal and other feed high in protein but low in fiber with precise amounts of formaldehyde, causing it to pass quickly through the rumen. Feed high in fiber and low in protein was not treated and was digested mostly in the rumen.

Cold germs. Chemist Russell Schnell of the National Oceanic and Atmospheric Administration laboratory in Boulder, Colo., reported that large concentrations of certain bacteria on plant leaves may reduce a plant's ability to withstand below-freezing temperatures. The reason for this, according to Schnell, is that some species of bacteria are exceptionally efficient ice nuclei, and supercooled water turns to ice around them. On a leaf, at tempera-tures just below the normal freezing point of water, this nucleating efficiency translates into plant-damaging frost, or ice, that would not otherwise have been formed.

Nitrogen fixation. Biochemist Karel Schubert and botanist Harold J. Evans of Oregon State University in Corvallis found in 1976 that much of the energy supplied for nitrogen fixation by beans and other leguminous plants is used to produce hydrogen. This gas apparently serves no function in the biological process of nitrogen fixation (the conversion of atmospheric nitrogen into ammonia). A small amount of hydrogen is used in nitrogen fixation, and Schubert and Evans proposed that ways might be found to use the waste hydrogen in the plants. One suggestion is to inject microorganisms carrying the enzyme hydrogenase into the plants so that the enzymes might act to recycle the molecular hydrogen. Nonleguminous plants, such as the alder, are more efficient in nitrogen fixation because they can reuse the hydrogen they produce. [Sylvan H. Wittwer]

Anthropology

The search for human ancestors continued in 1976 and 1977 with the most significant fossil finds coming from Africa and dating to from half a million to 2 million years ago. The oldest was a skull found in August 1976 at Sterkfontein, near Johannesburg, South Africa, by anthropologists Alun R. Hughes and Phillip V. Tobias of the University of Witwatersrand in Johannesburg.

The skull, found with several stone tools and many animal bones, was dated at from 1.5 to 2 million years ago on the basis of the age of the animal bones. Anthropologists had previously found the bones of many apelike australopithecines at Sterkfontein that could be dated to 2.5 to 3 million years ago. But this skull is the first primitive human found there.

Another skull tentatively identified as that of a *Homo erectus* (Peking man) was discovered at Bodo, Ethiopa, by Paul Whitehead of Yale University. Anthropologist Clifford J. Jolly of New York University reported in April 1977 that the skull consists of most of the face and fragments of the top of the head. Its massive browridges and generally large facial features strongly resemble the Broken Hill skull found in 1921 near Lusaka, Zambia. The skull was found with some stone tools and animal remains that showed signs of butchering. The animal remains suggest that the skull is more than 500,000 years old.

R. J. Clark of Johannesburg in 1976 described part of a skull found in 1973 at Lake Ndutu on the Serengeti Plain of northern Tanzania by Amini Mturi, director of antiquities in Tanzania. A combination of primitive and advanced features suggests that the skull falls somewhere between *Homo erectus* and *Homo sapiens* (modern man) on the tree of man's evolution. It resembles *Homo erectus* in outline and in the thickness of its braincase, but its unusually high forehead is more like that of *Homo sapiens*. The skull resembles one found in 1935 in similar deposits along nearby Lake Eyasi. Many nondescript stone tools, most of them unfinished, were found with the Ndutu skull. A preliminary dating places the specimen at from 500,000 to 600,000 years old.

Anthropology

Continued

Prehuman specimens. An international scientific team, headed by Yale University anthropologist David Pilbeam, reported in April 1977 that they had found 8-million-year-old to 13-million-year-old remains of some 80 prehominid individuals. The specimens were found in the arid badlands of Pakistan's Potwar Plateau. Most of the remains were upper- or lower-jaw fragments of large primates of the genus *Ramapithecus* and *Sivapithecus.*

The Americas. The Old Crow River site in Canada's Yukon Territory is well known for ancient animal-bone deposits and bone tools. In July 1976, William N. Irving of the University of Toronto reported the first human fossil found there. The fossil consists of fragments of an 11-year-old child's mandible (lower jaw). The fossil shows the distinctive coloring of age seen on animal fossils recovered from the site that have been dated to more than 22,000 years of age. The portion of the jaw that attaches to the skull is low and broad, more like that of an Eskimo or an Asiatic than an American Indian.

A human skull found in the Yuha Desert in southern California in 1971 was fully described in April 1977 by Spencer L. Rogers of the San Diego Museum of Man. The skull, tentatively dated to 21,500 years ago, is that of a male about 20 years old with features quite different from those of the Diegueño Indians who now live in the area. The features are similar to those of the coastal La Jollan Indians, but no relationship has been established.

Dental anthropology. Studies of ancient and modern teeth from Japan and China led physical anthropologist Christy G. Turner of Arizona State University in Tempe to conclude that the Japanese people are descendants of Chinese who sailed to Japan about 2,200 years ago. Turner's study of teeth, especially molars, premolars, and incisors, shows that the Jomon people of ancient Japan were closer biologically to the present-day Ainu than to the Japanese. The Ainu are a primitive tribe in northern Japan and Sakhalin, Russia. Many observers have suggested that the light-skinned Ainu are Cauca-

University of Michigan research worker Susan Walker clips hair from an Egyptian mummy while a member of the Cairo Museum staff watches. Analysis of hair and teeth by the school's archaeologists helped to identify the mummy as long-lost Queen Tiy, grandmother of famed King Tutankhamon.

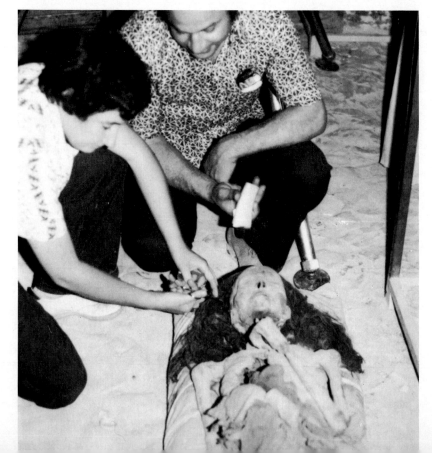

Anthropology

Continued

A copy of an instrument called a Tumi was used by Incan doctors in Peru about 1,200 years ago for head surgery. The hole in the skull, which may be that of an Incan who underwent such an operation, was probably made by the Tumi.

sians, but the dental evidence indicates they descended from Mongoloid stock.

C. Loring Brace of the University of Michigan also used measurements of teeth in April 1977 to determine the biological relationship of the Aborigines of Tasmania, a large island south of Australia. Tasmanians, the last of whom died in 1876, were viewed by many as being quite different from their Australian Aborigine neighbors and more closely related to Negritos, New Caledonians, or even the Dravidians of southern India. Brace's study shows that the Tasmanians are closely allied to the Australian Aborigines, especially those living in south Australia.

Prisons and disease. The discovery of three skeletons on an Arkansas prison farm in 1968 caused political repercussions that significantly aided the cause of Arkansas prison reform in the 1970s. But little attention was paid at the time to the identity of the skeletons, which appeared to be the mutilated remains of murdered inmates.

Eventually, Clyde C. Snow, forensic anthropologist with the Civil Aerome-

dical Institute in Oklahoma City, Okla., discovered that the area had been the burial ground of several hundred convicts who died between 1902, when the prison was established, and 1948. Snow positively identified the skeleton of Thomas "Jaybird" Hamilton, a convict killed by a trusty in 1920, by the teeth and other features of the skeleton. His study, published in April 1977 and dealing with many aspects of anthropology, provides new insight into the nutrition and other physical conditions under which convicts on some Southern prison farms lived.

Michael R. Zimmerman of the University of Pennsylvania reported finding a case of tuberculosis in 1977 in an Egyptian mummy. The victim was a 5-year-old child buried in the Egyptian tomb of Nebwenenef, the first high priest of Amon for Pharaoh Ramses II. Examination showed damage to the lungs and marked twisting of the spine. Zimmerman suggests that human tuberculosis probably evolved from cattle tuberculosis sometime between 3100 and 2700 B.C.　　[Charles F. Merbs]

231

Archaeology

Old World. City life in ancient Rome is well known from years of archaeological excavation but, until recently, little was known about the lives of the farmers who composed most of the population of the Roman Empire. Roman agricultural settlements being excavated in 1976 and 1977 in northern Italy, east-central France, and northern Yugoslavia are yielding information on the everyday lives of the farmers, their political domination by Rome, and their struggles for independence.

Cosa, on the west coast of Italy north of Rome, is being excavated by archaeologists from Wesleyan University of Middletown, Conn., under the direction of Stephen L. Dyson. Cosa was a colony founded by Rome in 273 B.C. to control rebellious Etruscans and to provide Rome with a harbor. The town reached its prime around 100 B.C. and was occupied by the Romans until the Middle Ages. The archaeologists have located 133 house and villa sites of various sizes and periods. Excavation continued in 1977 at the major Le Colonne villa, one of the largest in the Cosa area.

A series of rural valley settlements located in the Serbian part of Yugoslavia are being investigated by Brad Bartel of San Diego State University, Michael Werner of Pennsylvania State University, and Vladimir Kondic of the National Museum in Belgrade, Yugoslavia. They hope to determine the economic relationships of the people in this area, once known as the province of Upper Moesia, before, during, and after Roman colonization. A primary objective is to see how important metallurgy was in their trade relations.

Diggers at one metallurgical site, dating to the A.D. 300s, have found information on the types of smelting furnaces used and the types of metal products made. The evidence indicates that the Romans had little influence among the Illyrian-speaking people of the area, that the population was small, and that there were few Roman military fortifications. This evidence may force us to revise previous archaeological assumptions that the Romans maintained strong forces in Upper Moesia to control barbarian groups.

The Aedui. French excavations in 1976 provided information about the Celts, the major ethnic group in western Europe during the Iron Age. The Celts inhabited parts of Great Britain and all of what is now France. During the second half of the Iron Age, the Romans conquered and subjugated most of the areas inhabited by the Celts, including the territory of the Aedui, a powerful Celtic group who lived between the Saône and Loire rivers. The investigation of Celtic, Gallo-Roman, medieval, and later settlements in the Aeduan territory has been conducted for several years under the direction of Carole L. Crumley of the University of North Carolina at Chapel Hill. The territory, according to Roman records, was an important military and economic center before and after the Roman conquest of Gaul.

Some Aeduan sites, including the impressive capital of Bibracte, were excavated many years ago. The new studies include excavation of Mont Dardon, a mountaintop fortress occupied during the Bronze and Iron ages. Workers there have unearthed data on Celtic agriculture and recovered charred grains and legumes from Celtic cooking fires — including wheat, barley, oats, and Celtic beans. Surveys of other sites between Mont Dardon and the nearby hill fort of Mont Dône have also been conducted in an effort to determine the extent of rural settlement and land use during Celtic times.

Greece. A third season of excavation, survey, and study at Phylakopi, on the island of Milos, Greece, was completed in 1976 under the direction of Colin Renfrew of the University of Southampton in England. The site includes remnants of the early, middle, and late Bronze Age in Greece. Excavations have been concentrated on a late Bronze Age shrine, located in 1975, and on fortification walls of the surrounding village. For many centuries, ancient Milos was an obsidian-mining and Bronze Age trading center.

Archaeological excavations of middle and late Stone Age settlements are being conducted in the Nidzica River Basin of southeastern Poland. Withold Hensel and Jan Machnik of the Institute of the History of Material Culture in Warsaw, Poland, and Sarunas Milisauskas of the State University of New York at Buffalo have uncovered many

A funnel-shaped cup was one of a number of ancient objects found in Stone Age excavations at Bronocice in Kielce province in Poland.

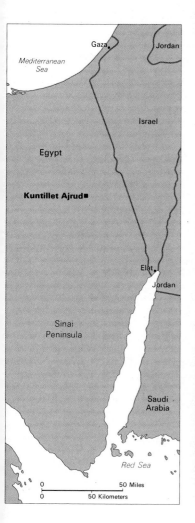

Israeli archaeologists and volunteers excavate
the ruins of Kuntillet Ajrud, a 2,800-year-old
Judean fortress on a hill in the Sinai Desert
between Gaza and Elat. Inscriptions on pottery and
plaster walls indicate that the fortress was built
by King Jehoshaphat of Judea to protect a trade
route linking the Mediterranean and Red seas.

Archaeology
Continued

Baths used during the second half of the Iron Age by the Celts and their conquerors, the Romans, were found in west-central France.

pits, ovens, graves, houses, and ditches at Bronocice in Kielce province.

Sparta houses. A British archaeological team, under the direction of archaeologist H. W. Catling, director of the British School in Athens, Greece, excavated two Bronze Age houses built in the second half of the 15th century B.C. They were found on the side of a steep hill across the Evrotas River from Sparta. The site is believed to be the location of ancient Therapne, mentioned by many early writers as the center of a religious cult established to worship King Menelaus and his wife, Helen of Troy. The archaeologists found offerings inscribed to the couple.

The hill, which has a stunning view of the Evrotas Valley, was first explored by German archaeologists in 1833. British researchers found traces of a Bronze Age building in the early 1900s, but did not realize the importance of their discovery. Catling believes that the houses may have been built around a palace that he hopes to uncover in future seasons of archaeological investigation at the site. [Carole L. Crumley]

New World. George C. Frison and his associates at the University of Wyoming continued in 1976 to excavate several prehistoric sites in the Big Horn area in north-central Wyoming. The Colby site, earliest of these, contains many mammoth bones and some tools used by early men. It was found in what remains of an old arroyo, or dry stream bed, now filled with debris from the river that once flowed there. The site is believed to be between 11,000 and 11,500 years old.

Studies of petrified material indicate that the area was once an extensive grassland where mammoths grazed. The bones are from one adult and five young mammoths. They were apparently placed in two carefully stacked piles after the animals had been butchered by early men. It is not clear how the animals were killed. They may have been trapped in the deep arroyo.

One of the spearpoints used by the hunters was found under one of the bone piles. It is of a type known as a Clovis point. Such points were first found at an archaeological site near

Archaeology

Continued

Clovis, N. Mex. The scientists also found three tools made of bone – a flesher, a chopper, and a possible spear foreshaft; a small stone-flake tool; three other Clovis points; and a hammerstone. Other animal remains found there included those of bison, mountain sheep, and jack rabbit.

The second site excavated by Frison, the Hansen site, was radiocarbon dated at between 9000 and 8000 B.C. It included a number of compacted soil areas believed to have once been the floors of circular lodges. Although little charcoal remained, the diggers found evidence of household fires on top of these floors. They also found considerable evidence that the inhabitants made flint and quartzite implements there, including discarded stone chips and finished tools.

California cave. Louis A. Payen of the University of California, Davis, and Royal E. Taylor of the University of California, Riverside, re-examined and re-evaluated the remains of early man and prehistoric animals at Potter Creek Cave in Shasta County, California. The animal bones and human artifacts in the cave were first scientifically investigated in 1902. Over a period of years, remains of shrub ox, prehistoric horse, mammoth, bison, camelid, and a giant bear were found there. All are now extinct. Many bones were smashed and some were splintered, perforated, and polished. A sharp-edged flake tool made from a river pebble, a freshwater clamshell that was probably used as a tool, and charcoal from a campfire were also recovered.

In 1976, Payen excavated a series of deposits and found other human tools and artifacts. They included spear-throwing devices, dart foreshafts and main shafts, obsidian tools, clamshells, and more charcoal. Radiocarbon dating put the material in the A.D. 100s, but a shrub ox bone from the earlier finds was radiocarbon dated at about 8250 B.C. Payen and Taylor concluded that the extinct animals and the human remains represented two separate occupations of the cave, the first by animals over 8,000 years ago and the second by people about 2,000 years ago.

Mammoth bones 11,000 years old and some stone tools used by early men were found by excavators at the Colby site in Wyoming.

A New Angle in Archaeology

The ancient citadel of Iskanwaya in Bolivia is an impressive pre-Inca complex of masonry ruins on the rugged eastern slopes of the Andes Mountains. Iskanwaya was hidden for centuries by lush forests. Then in 1943, archaeologist Carlos Ponce Sangines, director of Bolivia's National Institute of Archaeology in La Paz, discovered it. He is now in charge of excavating this New World site. The ruins gained international notoriety late in 1976 when an access road to the site was completed.

Archaeologists are particularly interested in Iskanwaya's unusual architecture. The doors and recesses in the walls in all its buildings are trapezoidal in shape, and no one knows why. "This is something completely new in archaeology," says Ponce Sangines. A trapezoid has four sides; only two are parallel.

Archaeologists believe that Iskanwaya, erected and inhabited sometime between A.D. 1200 and 1400, was a major settlement of the Mollo people, some of whose ancestors came from the early highland center of Tiwanaku, 125 kilometers (75 miles) away.

Iskanwaya's ruins cover about 13 hectares (32 acres), and they include the remains of at least 95 separate buildings. Most of the buildings had two rooms, one behind the other, with a trapezoidal doorway connecting them. The first room had no wall directly opposite the trapezoidal door; it was U-shaped and opened onto a large patio. Iskanwaya's buildings are arranged side by side in long rows, all opening onto a common court.

There are several significant, if not fully understood, implications about Iskanwaya. The barrackslike buildings are typical of Inca architecture, and suggest that these ruins were left by ancestors of the Incas. Other Mollo settlements do not show the degree of planning, workmanship, and architectural formality found at Iskanwaya. Therefore, it must have been one of the most important centers of the ancient Mollo civilization.

Iskanwaya is undoubtedly the most impressive ancient Andean site opened since the 1911 discovery of Machu Picchu.   [Michael Edward Moseley]

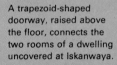

A trapezoid-shaped doorway, raised above the floor, connects the two rooms of a dwelling uncovered at Iskanwaya.

Archaeology

Continued

"Professor, you won't believe this. . . ."

Copper. Kent A. Schneider of the University of Georgia used X-ray fluorescence in 1976 to analyze copper samples from the Great Lakes region and from ore deposits in Tennessee and Georgia. He also examined copper objects found at archaeological sites in Georgia and Mississippi to determine where they came from. The objects were found in Indian mounds and related sites dating from about A.D. 150 to 200. Some specimens found at the Pharr site in northeast Mississippi had Great Lakes copper and others contained Tennessee copper. The Tunnacunhee site in northwest Georgia had Panpipes (tubed musical instruments) and earspools (ornamental plugs worn in the earlobe) made of both Tennessee and Georgia copper.

Sharon I. Goad, also of the University of Georgia, used an optical spectrograph to make similar studies of other copper objects. Her studies indicate that copper objects from the Mandeville site along the Chattahoochee River in southwest Georgia were made primarily from Tennessee copper. Cop-

per earspools and reels from slightly later Copena Indian sites in northern Alabama were made of Great Lakes copper. The work of Schneider and Goad clearly establishes that the Hopewell Indians made objects from copper that came from many different sources, not only Great Lakes copper, as archaeologists previously believed.

Maya dating. Norman Hammond, who began his excavations in the northeastern part of the lowland Maya area under the sponsorship of the British Museum and the University of Cambridge, continued to work in the Rio Hondo and New River areas of Belize (formerly British Honduras) in 1976 and 1977. Hammond believes that the earliest plaster-covered building platforms and pottery of a style known as Swazy may date back to 2500 B.C., which would make this the oldest Early Formative Mayan material found. The oldest of the platforms appears to have been circular with about a 3-meter (9-foot) radius. It was built of earth and once supported a wooden frame building. [James B. Griffin]

Astronomy

Planetary Astronomy. New discoveries about Uranus excited planetary astronomers in 1977. The most important came when several groups of astronomers, watching as the planet occulted, or passed in front of, the faint star SAO 158687 on March 10, 1977, discovered that it is surrounded by a system of rings similar to Saturn's famous ring system.

Astronomers from Cornell University in Ithaca, N.Y., led by James Elliot, observed the occultation from the National Aeronautics and Space Administration's (NASA) G. P. Kuiper Airborne Observatory, a C-141 aircraft flying over the Indian Ocean at 12,500 meters (41,000 feet). They saw the light from the star dim sharply five times in the 40 minutes before Uranus was due to pass in front of it. Only one of the dimmings lasted for more than 1 second. A similar sequence of sharp decreases in the star's light occurred as it emerged on the other side of Uranus. The astronomers concluded that Uranus is surrounded by an extensive system of orbiting particles grouped in at least five distinct rings. Other astronomers, observing from the Earth's surface, confirmed the observations.

The Uranus rings differ from Saturn's in that each is extremely narrow. The four inner rings are only 20 to 30 kilometers (12 to 18 miles) wide, while the outer ring is between 100 and 200 kilometers (60 to 125 miles) wide. As with Saturn, most of the particles in Uranus' ring system orbit about one planetary radius above the planet's surface. Each particle in the rings moves around the planet at about 42,000 kilometers (26,000 miles) per hour, completing the circle in about seven hours. The Cornell observers determined that the particles must be less than 6 kilometers (4 miles) across.

Uranus' rotation remeasured. Two new sets of measurements show that Uranus rotates much more slowly than the 10-hour-and-49-minute period accepted since 1911. Lawrence Trafton, observing with the 2.7-meter (106-inch) telescope at McDonald Observatory at Mount Locke, Texas, found a rotation period of 23 hours. Sethanne

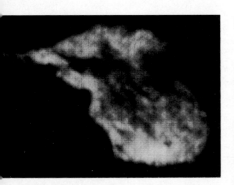

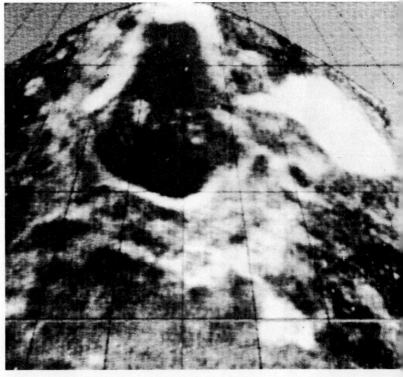

Maxwell, a bright area in a radar picture of part of Venus, *above*, may be the site of a large lava flow. A huge basin in the center of a picture of the northern hemisphere, *right*, may be the result of a meteor impact.

Hayes and Michael J. S. Belton, using the 4-meter (158-inch) Mayall telescope at Kitt Peak National Observatory near Tucson, Ariz., measured it at 24 hours. Both groups estimated the uncertainties due to experimental error at plus or minus three hours.

The Kitt Peak observers also found a new rotation period of 22 hours for Neptune, much longer than the 15 hours and 48 minutes established in 1928. Uranus and Neptune appear to rotate at rates much closer to those of the Earth and Mars than to the rapid spin of their sister giant planets Jupiter and Saturn. This may be related to differences in the way the planets formed.

A magnetic field for Uranus? Following a careful analysis of radio noise recorded by NASA's IMP-6 spacecraft since 1971, Larry W. Brown of NASA's Goddard Space Flight Center in Greenbelt, Md., reported in August 1976 that Uranus probably is emitting bursts of radio waves. If these bursts do come from the planet, Uranus must be surrounded by an active magnetosphere and must have a reasonably strong magnetic field.

Night glow on Venus. The Russian Venera 9 and 10 missions that arrived at Venus in October 1975 included two orbiter spacecraft as well as the two that landed. Instruments on both orbiters measured relatively bright atmospheric emissions on the night side of the planet. The Russian scientists analyzing the data found that the light arose primarily from a single type of molecule in the upper atmosphere, but they could not identify the molecule.

George M. Lawrence and his colleagues at the University of Colorado in Boulder reported in February 1977 that they had solved the mystery. They showed in laboratory experiments that the light comes from newly formed oxygen molecules. During the Venus day, ultraviolet light from the Sun breaks up carbon dioxide molecules and forms large quantities of oxygen atoms as a by-product. These atoms recombine in pairs during the night to form molecules of oxygen gas. Lawrence and his associates showed that when this occurs in the presence of large amounts of carbon dioxide, as on Venus, light is emitted with the same characteristics found by the Russians.

Groovy Phobos. Television cameras on Viking Orbiter 2 obtained remarkably clear photographs of the Mars satellite Phobos as the orbiter swung by the tiny moon on Sept. 18, 1976. Thomas Duxbury of the Jet Propulsion Laboratory (JPL) in Pasadena, Calif., and Joseph Veverka of Cornell University processed a picture of unsurpassed clarity that shows surface features only 40 meters (130 feet) across.

The photograph shows that the surface is heavily cratered, presumably as a result of past collisions with large bodies. Long strings of irregular, elongated craters are aligned roughly with the orbital motion of the satellite around Mars. The surface is also covered by sets of parallel grooves inclined at a large angle to the orbital plane.

The strings of craters, some of which create a herringbone pattern, may have been caused by groups of objects that were knocked off Phobos in a large impact and then collided with the satellite on later orbits. The scientists cannot explain the intriguing grooves. Clearly, we will need to know more about their properties if we are to fully understand the structure and origin of Phobos.

Saturn's rings. Saturn has three primary rings in its system — "A," the most distant ring; "B," the central, bright ring; and "C." A weak "D" ring also appears to exist very close to the surface of the planet.

H. J. Reitsema and Reta F. Beebe of New Mexico State University in Las Cruces and Bradford Smith of the University of Arizona in Tucson reported on a new property of the A ring in late 1976. Their precise analysis of ring photographs shows that the A ring's brightness varies periodically around its surface. More sunlight is reflected from the part of the ring that is about to move behind the planet than from the part that has just emerged from behind the planet. A similar variation occurs where particles pass in front of Saturn.

Three astronomers — Giuseppe Colombo of Harvard University in Cambridge, Mass.; Peter Goldreich of the California Institute of Technology; and Alan W. Harris of JPL — suggest that sheared gravitational waves in the rings cause the phenomenon. A balance between the force of gravity and centrifugal force holds the ring particles in

239

Five narrow rings were discovered surrounding the planet Uranus in 1977. They circle the planet's equator and are perpendicular to the plane of its orbit.

Astronomy

Continued

orbit around Saturn. But this balance is never exact, and one force usually overpowers the other briefly. As a result, the rings oscillate much like a disturbed spring moving back and forth, causing the particles to bunch up in some places and spread out in others. Because the particles in the outer part of the ring orbit Saturn more slowly than do the inner particles, Colombo and his associates note that the oscillations of the ring will become sheared or develop a spiral pattern. All of the rings are expected to share this kind of oscillation, but the effect is expected to be seen clearly only in the fainter, less dense ring.

Saturn's ring particles are predominantly water ice, which breaks down to form hydrogen and the hydroxyl radical (OH). Thus, astronomers expected the planet to be surrounded by an extensive doughnut-shaped ring of orbiting hydrogen atoms, formed when the ice decomposes. Heinz Weiser, Robert C. Vitz, and H. Warren Moos of Johns Hopkins University in Baltimore announced in January 1977 that they had detected such a ring using a

device that measures hydrogen emissions in ultraviolet light from a high-altitude Aerobee rocket. The hydrogen atmosphere above Saturn itself was measured at the expected intensity. The emission from the ring system was about 20 times brighter than expected, however. The experimenters believe this is because the ring particles are slightly hotter than previously thought and therefore evaporate more rapidly.

Also in January, four astronomers—Gordon H. Pettingill and Steven J. Ostro of the Massachusetts Institute of Technology; Donald B. Campbell of the National Astronomy and Ionosphere Center, Arecibo, Puerto Rico; and Richard M. Goldstein of JPL—reported that radar echoes from Saturn's rings indicate there is a substantial amount of material in the D ring. This ring, which has been convincingly photographed only two or three times, is exceedingly faint visually but gives off strong radar echoes. This suggests that large amounts of material exist in objects at least 100 centimeters (40 inches) across.   [Michael J. S. Belton]

Astronomy

Continued

Stellar Astronomy. Astronomers made major progress in the study of pulsars (objects that emit pulsed radio signals), supernovae (exploding stars), and related phenomena in 1976 and 1977. They also investigated our Galaxy, the Milky Way, and made new discoveries about other galaxies.

Pulses of light. Although astronomers have located more than 150 radio pulsars, only the one in the Crab Nebula was known to produce visible-light pulses. But in January 1977, a team of astronomers headed by P. T. Wallace of the Anglo-Australian Observatory used the 3.9-meter (154-inch) telescope in Siding Spring, Australia, to detect visible pulses from PSR 0833-45, a pulsar in the constellation Vela. The quality of their observations, which involved elaborate photoelectric and electronic signal-analyzing equipment, makes the results virtually certain.

The Crab pulsar produces two pulses, a main pulse and an interpulse, at both visible and radio wavelengths, and each pulse is emitted at the same time at both wavelengths. The Vela pulsar,

however, has two optical pulses and two gamma-ray pulses, but just one radio pulse. The pulses do not occur at the same time at the different wavelengths, and the timing between the optical pulse pair differs from the timing of the gamma-ray pair.

Supernova traces. The Crab Nebula is the remnant of a bright supernova observed in A.D. 1054. In 1006, Arab and Swiss astronomers observed an even brighter supernova in the southern constellation Lupus. Sidney van den Bergh of the University of Toronto in Canada reported in August 1976 that he had found the wispy remains of the Lupus supernova on photographs taken with the 4-meter (158-inch) telescope at Cerro Tololo Inter-American Observatory in Chile. The Lupus remnant is much fainter than the Crab Nebula, and Van den Bergh described it as "delicate wisps of filamentary nebulosity." The filaments are expanding into space at a speed of roughly 5,000 kilometers (3,000 miles) per second.

The Milky Way. Astronomers studying the infrared and radio waves that

Polarization map of the light from the Sombrero Galaxy obtained by an electronographic camera is superimposed on a photo of the galaxy. The pattern, probably caused by light that is scattered by dust grains that are magnetically aligned, represents the largest uniform magnetic field observed in space.

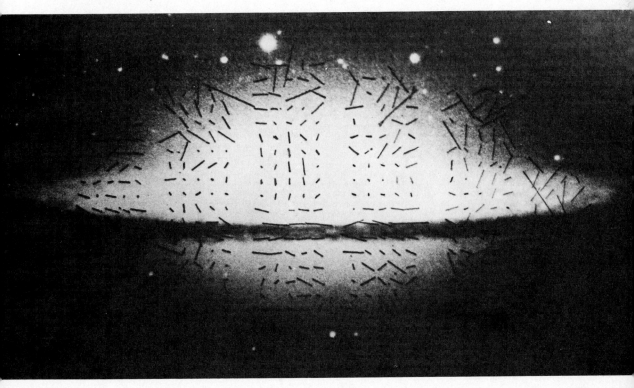

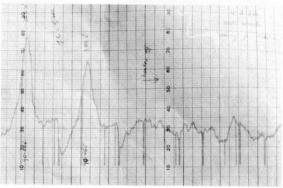

Chemists at Monash University in Melbourne, Australia, *top,* check the microwave spectrum of a compound they produced that helped to identify the spectrum of "X-ogen"—a strange interstellar molecule that could not be found in spectral catalogs. The peaks on the left, *above,* that matched the interstellar spectrum were produced by the highly reactive HNC radical, *right,* which survives uncombined on earth for only a fraction of a second.

penetrate the interstellar dust clouds screening the central region of our Galaxy reported new information about the galactic center in November 1976. Led by Richard R. Treffers, University of Arizona scientists analyzed the spectrum of an infrared source at the galactic center. By studying carbon monoxide absorption bands in the spectrum and comparing the spectrum with those of known stars, they confirmed that the source is a red supergiant star. The star seems to be moving rapidly, perhaps in an orbit around the densely concentrated stars of the galactic nucleus.

Working at Jodrell Bank Observatory near Manchester, England, Rodney D. Davies and his colleagues from the University of Manchester found that a radio source in the nucleus has an unexpectedly small diameter. It may well be a recently produced supernova remnant, or even a pulsar whose pulses are smeared into a continuous signal by the effects of interstellar electrons on the radio waves. The astronomers believe that this object may be the very center of the galactic nucleus.

New nebulae. Allan R. Sandage of Hale Observatories in California reported in November 1976 the first results of a new photographic survey with the 1.2-meter (48-inch) Schmidt telescope on Palomar Mountain. The photographs showed extensive faint nebulae high above the Milky Way. Two prominent nebulae, consisting of dust, are located about 300 light-years above the plane formed by the Galaxy. (One light-year is the distance light travels in a year.) Ordinary reflection nebulae shine by reflecting the light of nearby stars, but these nebulae are far enough above the Galaxy that they reflect the light of the whole galactic plane.

Astronomers usually assume that interstellar dust is closely confined to the galactic plane. Sandage speculates that the dust in the new-found nebulae may have been carried to its current location by the expansion of supernovae that exploded closer to the galactic plane.

Other galaxies. Physicists from the University of Durham in England and the Massachusetts Institute of Technology in Cambridge, under S. Michael Scarrott, reported a landmark study of another galaxy in January 1977. They used the 100-centimeter (40-inch) tele-

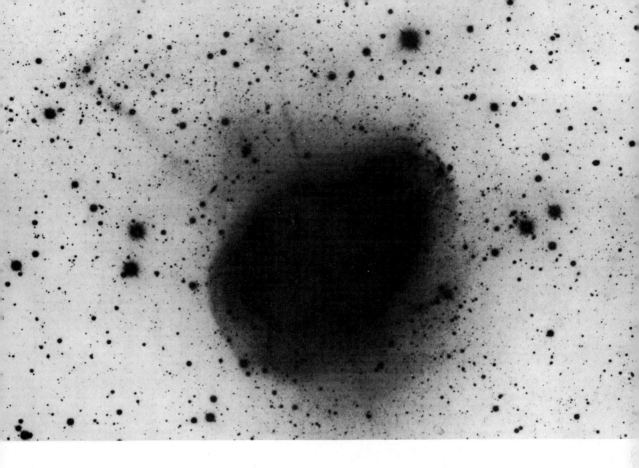

Astronomy

Continued

Jets of material appear to have been ejected from the nucleus of the spiral galaxy, NGC 1097, the first normal galaxy seen with such ejections. All lie in the galactic plane, and two point in exactly opposite directions.

scope at the Wise Observatory in Israel to study the polarization of light across the visible surface of the Sombrero Galaxy, a spiral galaxy that is seen nearly edge-on. Polarization causes light waves to vibrate more in one plane than in any other. The Sombrero's light polarization is presumably produced by dust grains aligned in the galaxy's magnetic field. It shows an overall systematic pattern in the galactic plane and the northern hemisphere of the galaxy that constitutes the largest ordered magnetic field yet observed in outer space.

Vera C. Rubin of the Carnegie Institution of Washington, D.C., and three colleagues presented evidence in February that confirmed a new astronomical interpretation of NGC 1275. This peculiar object is surrounded by extensive filaments. Astronomers once thought it resulted from a collision of two galaxies. However, Rubin's spectroscopic and photographic evidence confirms that it is a single galaxy that exploded to produce the filaments. What seemed to be the second galaxy is apparently an unrelated object moving at high speed across the foreground.

Hydrocarbon discovery. Stephen T. Ridgway of Kitt Peak National Observatory in Tucson, Ariz., and his associates announced in November 1976 that they had discovered acetylene (C_2H_2) in the shell surrounding the infrared source IRC 10216. The acetylene molecule appears to be a principal component of the star's shell.

New telescopes. The University of Massachusetts dedicated the Five College Radio Observatory near Amherst on October 15. It includes a 14-meter (45-foot) radio telescope that is the most sensitive instrument available for observations of radiation at the 1-millimeter wavelength. It is especially suited for the study of molecular clouds in the Milky Way.

Astronomers at the European Southern Observatory at La Silla, Chile, had their first look through a new 3.6-meter (142-inch) telescope on November 7. After eliminating distortions, they obtained excellent photographs of distant galaxies.   [Stephen P. Maran]

Astronomy

Continued

High-Energy Astronomy. Interest in high-energy astronomy deepened in 1976 and 1977. Major areas of interest were the continuing mystery surrounding the nature of globular clusters and the sources of X-ray bursts, the discovery of many new extragalactic X-ray sources associated with Seyfert galaxies, and new detailed information on X-ray sources associated with clusters of galaxies. At the same time, X-ray astronomy stood on the threshold of great observational advances with the launch of the first High Energy Astronomy Observatory (HEAO-A) satellite in the summer of 1977.

X-ray observatory. Equipped with specially designed counters to count X-ray pulses, HEAO-A will search for X-ray sources only one-tenth as intense as those that formerly could be detected. A counter array designed by a group at the Naval Research Laboratory in Washington, D.C., under the leadership of Herbert Friedman is the most sensitive instrument to be used in the survey. The instrument will also be used to detect rapid pulsations from X-ray sources in the center of galaxies. It is especially suited to study the rapid flickering of Cygnus X-1, the leading candidate for containing a black hole – an object so dense that its gravity prevents even light from escaping. The instrument can resolve flickers only 0.00001 second apart, a resolution that should be sufficient to answer specific questions about the predicted behavior of different types of black holes.

Another instrument, the scanning modulation collimator, will locate all known X-ray sources to within a few arc seconds of their exact positions. (One arc second equals 1/60 arc minute, which in turn equals 1/60 degree of a circle.)

A proportional counter with a large field of view, which is sensitive to X-ray emission in the range of 200 to 60,000 electron volts (60 kev), will study fluctuations in background X-ray emission. The space observatory also carries a hard X-ray, low-energy gamma-ray instrument that will study the high-energy portion of the X-ray emission.

Another X-ray satellite observatory, HEAO-B, scheduled to be launched in 1978, will carry the first imaging X-ray telescope designed to study galactic and extragalactic objects. The telescope will be able to detect sources a thousand times weaker than those currently known. It will also determine the position of point sources to within 1 arc second, and will examine the structure of extended sources.

X-ray burst sources. The number of known X-ray burst sources has grown to about 25, mainly due to the efforts of observers analyzing data from the eighth Orbiting Solar Observatory (OSO-8) and the third Small Astronomy Satellite (SAS-C). Burst sources show rapid increases in X-ray emission, reaching a peak in less than one second and then fading back to their normal levels in several tens of seconds.

In most sources, bursts occur every few hours. In the extreme case of MXB 1730-335, bursts occur irregularly at intervals of from 6 to 450 seconds. SAS-C recorded an intriguing relationship between burst intervals and the intensity of the following burst in MXB 1730-335 – the stronger the burst, the longer the interval until the next burst. OSO-8 observers discovered that the behavior of some burst sources can be interpreted as the rapid heating of a ball of gas at very high temperatures, with subsequent cooling.

The controversy among supporters of different theoretical models to explain the bursts continues because there is no definite evidence in favor of any one of them. At a symposium in Boston in December 1976, Jonathan Grindlay of the Center for Astrophysics in Cambridge, Mass., and Jeremiah Ostriker of Princeton University argued in favor of the accreting massive black hole hypothesis. This theory tries to explain burst sources, globular cluster sources, and intense galactic center sources by associating the X rays with phenomena that would be produced by a massive black hole – about 100 times the mass of the sun – in the core of an existing or evaporated globular cluster.

On the other hand, Edison Park-Tak Liang of Michigan State University in East Lansing and Fred K. Lamb of the University of Illinois in Urbana put forth competing theories based on instabilities in the flow of material from one star to another in a binary, or double-star, system. However, no evidence has yet been found to show that

Technicians check over the first High Energy Astronomy Observatory before its launching in mid-1977. Several times more sensitive than its predecessors, it will help to locate extraterrestrial X-ray sources more precisely.

Astronomy

Continued

any of these X-ray sources exist in a binary system.

Seyfert galaxies have extremely bright nuclei and peculiar spectra. The optical and infrared emission of a Seyfert nucleus is believed to be caused not by stars as in a normal galactic nucleus, but by energetic processes occurring in a tiny region surrounding a massive compact object, possibly a black hole.

Early observations from the Uhuru satellite in 1971 revealed extremely intense X-ray emission from the Seyfert galaxy NGC 4151. No other Seyfert had been observed with this kind of emission, leading astronomers to conclude that perhaps NGC 4151 was unusually bright.

Then in 1976, Herbert Schnopper of the Center for Astrophysics and his colleagues, working with the SAS-C satellite, found X-ray emission from 3C 120, one of the most distant known Seyferts, at an even higher energy level. Kenneth A. Pounds of the University of Leicester in England and his coworkers, working with the Ariel-5 satellite, have since detected a dozen more

Seyferts, with X-ray luminosities ranging between those measured for 3C 120 and NGC 4151.

Schnopper and his colleagues theorize that both the X-ray and optical radiation in 3C 120 originate in the same region. The X-ray emission would be caused by bremsstrahlung radiation, given off when hot plasma particles cause speeding electrons to change direction. The optical radiation would be produced by the interaction of the electrons with the radio or infrared field.

Galaxy clusters. Much information on the production of X rays in clusters of galaxies has come from the Ariel-5 and OSO-8 satellite observations. In particular, the detection of ionized iron lines in the spectrum at around 6 kev seems to prove conclusively that the observed X rays come from a highly ionized, extremely hot gas filling the space between galaxies. The detection of the iron lines makes it possible to measure the distance to these objects for the first time by X-ray observations alone.   [Riccardo Giacconi]

Cosmology. Quasars continued to bedevil astronomers in 1976 and 1977. *Quasar* is an abbreviation for quasi-stellar radio source, a name given because radio astronomers first noticed them as intense emitters of radio waves.

Optical spectra of many of these faint, inconspicuous, starlike objects show emission and absorption lines that are considerably shifted toward the red side of the spectrum. The emission lines, produced by the quasar, usually have a larger red shift than the absorption lines, which are produced by some intervening material. Most astronomers believe that quasar red shifts are related to the cosmological expansion of the universe. This would mean that quasars are located at immense distances from the earth and that they are actually extremely luminous despite their apparent faintness. But a few astronomers maintain that quasars are relatively close, perhaps associated with visible galaxies, and that local conditions, such as intense gravity, could slow down the light waves enough to produce the same amount of red shift.

Peter Strittmatter and his co-workers at the University of Arizona in Tucson and at Lick Observatory on Mount Hamilton in California reported in July 1976 that they had found evidence of line locking, or ratios between the absorption and emission red shifts in the spectra of quasars that showed certain coincidences. The results suggested that the material producing the absorption lines had been violently ejected from the quasar at extremely high speeds. If the absorbing gas remained in the vicinity of the quasar, it would be expected to produce a red shift similar to the quasar's own emission red shift. But the difference in speed between the quasar and the ejected gas could cause the difference in absorption and emission red shifts.

Conflicting evidence came in August from Robert Williams and Ray Weymann, also at the University of Arizona. Studying a different quasar, they found a set of absorption lines whose red shift exceeded that of the quasar's emission lines. They interpreted this to mean that the absorbing gas was ap-

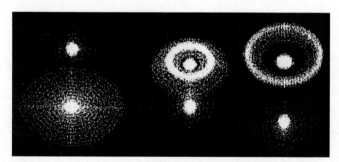

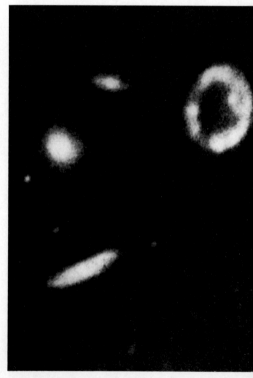

A computer simulation, *above*, shows how a rare ring galaxy may be created when an intruder galaxy pierces a flat disk galaxy, ripping matter from its center. The collision stirs dust and gases in the disk into a ring of dust and stars that expands. The remnant of the intruder combines with material from the center of the ring to form a companion, such as the cigar-shaped galaxy at lower left, *right*.

proaching the quasar and was most likely an intergalactic gas cloud, rather than material ejected by the quasar.

Halton C. Arp of the Hale Observatories in Pasadena, Calif., is the most persistent supporter of the idea that quasars are relatively close, within the same distance as most visible galaxies. In December 1976, he reported finding a quasar near a companion galaxy, with a luminous extension pointing from the galaxy toward the quasar. This could indicate that the quasar was ejected from the galaxy.

In September 1976, Richard Green and Douglas Richstone, also of Hale Observatories, showed conclusively that the distribution of quasar emission red shifts was completely random. Some earlier investigations had found a surprising number of quasar red shifts that seemed to cluster at specific values. If valid, the earlier studies would have cast doubt on the cosmological interpretation of quasar red shifts.

Closing the universe. Debate continued on whether the universe is open and will expand forever, or closed and destined to eventually collapse. Its fate depends on whether there is enough material in the universe to close it. After reanalyzing the mean density of luminous material in the universe, J. Richard Gott III and Edwin Turner of Hale Observatories concluded in October 1976 that the data showed that a closed universe was not possible.

However, their study covered only the matter we can see within galaxies. In March 1977, Marc Davis of Harvard University in Cambridge, Mass., in collaboration with P. James E. Peebles and Edward Groth of Princeton University in New Jersey, reported the results of another study based on an analysis of the distribution of galaxies on photographic plates covering large regions of the sky. Davis and his colleagues concluded that the distribution could be explained only if the universe contains at least one-third of the material needed to close it.

Age of the universe. One characteristic of an open universe is that it must be older than a closed universe. John C. Browne and Barry L. Berman at the Lawrence Livermore Laboratory of the University of California, Berkeley, offered evidence in July 1976 that in-creased the calculated age of the universe to about 20 billion years. Their method involved the radioactive decay rate of rhenium into osmium. A closed universe would be about 14 billion years old, so the new calculations appear to support an open universe.

Galaxy formation. Some cosmologists believe that gravity enhanced small differences in density in the early universe, causing galaxies to form. Others argue that galaxies formed from the fragmentation, or breakup, of massive primordial gas clouds.

The fragmentation theory received an important boost in early 1977. Independent studies by Joseph Silk of the University of California, Berkeley, in January and by Martin Rees of Cambridge University in England and Jeremiah Ostriker of Princeton in May showed that galaxies produced in the fragmentation process would fall within certain fundamental scales of size and mass. The studies indicated that the galaxies we see fall within these scales, and thus could have resulted from the fragmentation of gas clouds as large as entire clusters of galaxies.

The fundamental constants. From time to time, cosmological theories have been proposed that require certain fundamental constants of nature to vary with time. One of these is Planck's constant — the ratio of energy to frequency for a photon, or light particle.

Two groups performed similar experiments in October 1976 to test whether Planck's constant has varied with time. William Baum of Lowell Observatory in Flagstaff, Ariz., with Ralph Florentin-Nielsen of Copenhagen Observatory in Denmark and Jan-Erik Solheim and his collaborators at the University of Texas in Austin compared light from distant galaxies with that from nearby galaxies, using a technique that distinguished photons of different energies. Because distant galaxies are seen as they were at some point in the past, any difference in light would indicate Planck's constant had changed over time. The investigators concluded that the constant has not changed by more than 1 part in 1-trillion per year. This supports the interpretation that large red shifts are related to distance and speed in an expanding universe.   [Joseph Silk]

Biochemistry

In August 1976, Har Gobind Khorana and his co-workers at the Massachusetts Institute of Technology in Cambridge announced that they had synthesized a gene that functions in a living organism. This was a major technical achievement in the field of biochemistry. See Close-Up.

Viral genes decoded. In a second extraordinary achievement, Frederick Sanger, winner of the Nobel prize for chemistry in 1958, and his co-workers at the Medical Research Council Laboratories in Cambridge, England, published the complete nucleotide sequence for the genetic material of Phi X174, a virus, in February 1977. The genetic material of the virus is a single circular strand of deoxyribonucleic acid (DNA), the same substance that makes up the genes of most other living organisms. All DNA is composed of a string of chemical units called nucleotides, whose order of appearance determines the genes' functions. With the new work, the functions of an organism's complete set of genes can be interpreted in terms of their structure.

The key to the accomplishment was a technique worked out by Sanger and his colleagues to determine the nucleotide sequence of DNA by sections. The viral DNA was used as a template, or pattern, upon which fragments of DNA were synthesized using laboratory chemicals, including some radioactive nucleotides. The scientists then analyzed these fragments and the positions of their radioactive nucleotides to determine the complete sequence of the approximately 5,400 nucleotides in the virus' DNA strand.

Previous studies had shown that the DNA strand contained nine genes. The order of the genes had also been determined, but a chromosome map, showing their exact locations, had not been worked out. With the complete sequence available, however, researchers recognized certain nucleotide sequences that mark the beginnings and ends of the virus' genes, and thereby located them exactly.

In addition, confirmation of the presence of several special nucleotide sequences associated with translation of

Drawing by H. Martin © 1977 The New Yorker Magazine, Inc.

"Well, Carpenter, this does it! You and O'Callaghan are through as a team!"

The Man-Made Gene

In 1976, after nine years' work, organic chemist Har Gobind Khorana and his research team at the Massachusetts Institute of Technology completed construction of an artificial gene that works. Khorana announced in August that his group had synthesized the gene that produces tyrosine transfer ribonucleic acid (RNA), which functions in protein synthesis in the intestinal bacterium *Escherichia coli.*

Using gene-transplantation techniques, the scientists inserted the artificial gene into the virus lambda, which infects cells of *E. coli* (see GENES: HANDLE WITH CARE). The virus was then able to grow in the bacteria, indicating that the gene was functioning.

Like almost all genes, the artificial gene is a double-stranded molecule of deoxyribonucleic acid (DNA). A DNA molecule resembles two long ticker tapes, aligned face to face. It has two strands of deoxyribose sugar units, each sugar chemically linked to the next and carrying any one of four bases— adenine (A), guanine (G), thymine (T), or cytosine (C). The bases protrude from the parallel strands and meet in the middle. Because of chemical attractions between the bases, an A in one strand must match up with a complementary T in the other, and a G in one strand must match up with a complementary C in the other. The sequence of bases carries the gene's hereditary information, which is expressed as the gene functions.

In transcription, the first step in gene expression, the two strands of a DNA molecule separate and a molecule of RNA is produced that is complementary in base sequence to one of the DNA strands. RNA is chemically similar to DNA, but contains ribose instead of deoxyribose sugar and the base uracil instead of thymine.

Most RNA molecules produced in transcription then instruct cell structures called ribosomes to hook together a particular sequence of amino acids to make a protein. But transfer RNA functions directly in protein synthesis. After transcription and an elaborate processing during which many of the bases are chemically modified, transfer RNAs deliver certain amino acids to ribosomes for insertion into the growing proteins. Tyrosine transfer RNA delivers the amino acid tyrosine. The gene for this RNA is 207 bases long.

In 1967, molecular biologists John Smith and Sidney Altman of the Medical Research Council in Cambridge, England, discovered the correct base sequence for the main part of the natural gene, a 126-base portion that codes for the transfer RNA itself. Khorana and his associates completed the laborious synthesis of this portion, piece by painstaking piece, in 1973. But to get the gene to work—to be transcribed— certain controlling base sequences had to be added at each end. One end had to have a "promoter" sequence so transcription could start; the other end had to have a "terminator" sequence to stop transcription. Neither the promoter nor terminator sequences were known.

Khorana's group announced in 1974 that they had determined the sequence of both the naturally occurring promoter, which was 56 bases long, and the terminator, 25 bases long. Then they laboriously synthesized these sequences from commercially available chemicals and attached them to the main part of the gene. The researchers made the artificial gene in 19 separate pieces, each from 10 to 15 bases long with short, single-stranded ends complementary in base sequence to single-stranded regions in the adjacent pieces. The complementary ends served as connecting splints when the 19 pieces were finally joined together.

Khorana's achievement opens up the prospect of using artificial genes for the mass production of such scarce natural products as insulin. It also provides a new tool for research on how gene activity is controlled. Molecular biologists are particularly pleased because the work thoroughly proves that their ideas on gene structure and function, developed over 20 years, are correct.

The success of Khorana's group does not imply that genes can be synthesized on demand. Tyrosine transfer RNA comes from a rather small gene—only 207 nucleotides altogether, including the promoter and terminator sequences. In higher organisms, genes are typically 1,000 to 3,000 nucleotides long, not including the regulatory sequences. Synthesis of such large genes would require decades with present laboratory techniques.[Daniel L. Hartl]

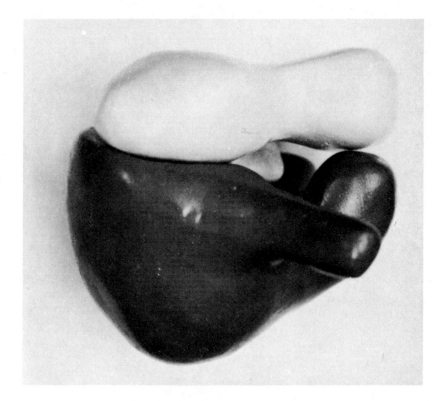

Most accurate ribosome model yet was proposed by James A. Lake of the University of California, Los Angeles, who used a technique in which individual molecules on the surface of the ribosome are marked. This greatly improved the interpretation of the ribosome when viewed through the electron microscope.

Biochemistry

Continued

genes into the protein molecules for which they "code" were found in the Phi X174. For example, preceding each gene is a short nucleotide sequence that is identical to a sequence in a molecule within the ribosomes, the structures that "read" the gene's code and help build the protein.

This is significant because a ribosome does not read a gene's nucleotide sequence directly. Rather, it reads the nucleotide sequence of a substance called messenger RNA (mRNA) that is a complementary copy of the gene. The mRNA, then, also contains a short sequence that is complementary to a sequence in the molecule within the ribosome. Thus, the discovery of the DNA nucleotide sequence that is identical to the sequence of the ribosome molecule confirms a theory that the initial physical contact between a ribosome and mRNA is made through mutual recognition of a complementary sequence.

Analysis of the Phi X174 DNA sequence has already led to one surprise — genes within genes. Most genes act by producing a protein. Each

gene's nucleotides are arranged in a particular sequence. When that sequence is copied as mRNA and read, three at a time, by a ribosome, a particular protein is built.

There was a discrepancy between the calculated number of proteins that the Phi X174 DNA should be able to produce and the number actually produced. More virus proteins had been recognized than could be accounted for by the number of nucleotides (about 5,500) in the DNA strand. This mystery has now been solved by the discovery of two cases in which one gene contains another in the Phi X174 DNA. Scientists had previously thought that genes could not overlap.

The new finding raises a number of interesting points. For example, any two genes sharing DNA nucleotides are, in a sense, competing with each other because, if mRNA is being made for one gene, it might interfere with the production of mRNA for the other. Also, the evolution of such "piggyback" genes must have occurred under unusual constraints, because a chance

Biochemistry

mutation in the region common to both genes would probably have to produce a positive effect on both genes to survive. See GENETICS.

Ribosome structure. James A. Lake of the Molecular Biology Unit at the University of California, Los Angeles, reported in July 1976 on his analysis of ribosome structure. He concluded that each of the two subunits of a ribosome – known simply as the large and the small subunits – are asymmetric.

In order to understand the molecular events through which a gene's nucleotide message becomes a specific protein, we must determine how the various components of protein synthesis interact with each other. We know a great deal about the chemistry of some of the interactions. However, the detailed mechanics of exactly where and how they occur is a mystery, largely because we know little about the three-dimensional structure of ribosomes, the "machine shops" where the mRNA copy of a gene's message is read and the protein coded for in the message is forged.

Several groups of researchers have worked out detailed three-dimensional ribosome models that are based on electron microscope studies. Because of the microscope's limitations, ribosomes appear only as fuzzy blobs through them. Scientists studying ribosomes view many individual ones lying in different positions and then carefully interpret and design their models according to what they have seen.

Lake's model conflicts with one proposed by Gilbert Tischendorf and his colleagues at the Max Planck Institute in Berlin in December 1975. In the German model, each subunit was symmetrical. But, since the 50 different proteins and 3 RNA molecules of which the ribosomes are composed are of different sizes and shapes, it is reasonable to expect that the ribosome would be asymmetric, as in Lake's model.

Ribosome models provide a basis for speculation on function but, more important, they are a "map" upon which scientists are beginning to chart the exact location of each protein molecule that makes up the ribosome.

To locate a ribosomal protein, the scientists make antibodies that will attach to it. Then they add some of the antibodies to ribosomes and examine the mixture with an electron microscope. They can locate the position of the ribosomal protein with reasonable precision by noting where the antibody has attached to the ribosome. For example, Lake, in collaboration with Lawrence Kahan of the University of Wisconsin at Madison, found that protein S13 of the small subunit is on the top of that subunit, near where it connects with the large subunit.

Through studies of this type, the Lake and Tischendorf groups, in spite of disagreements about the ribosome's shape, have come to agree upon the location of many of the molecules on its surface. This work, along with a final ribosome model, promises to provide a complete topographical description of ribosomes that will give a clearer picture of their role in making proteins.

DNA replication. In November 1976, Martin Gellert and his co-workers in the Laboratory of Molecular Biology at the National Institutes of Health in Bethesda, Md., reported discovering an enzyme that changes the physical form of DNA and plays a role in DNA replication, or reproduction.

DNA replicates when its cell divides. This assures each new cell of a complete complement of genes. But biochemists have been puzzled as to the exact form of the DNA that is necessary for replication. Normal, double-helix DNA is spiral, like a loosely coiled spring. But DNA has also been isolated from cells in a superhelical form in which the helix is twisted around itself. Gellert and his co-workers found that the enzyme they discovered changed normal DNA into superhelical DNA. They named the enzyme gyrase.

Evidence that superhelical DNA and, therefore, gyrase are required for DNA replication came from experiments in which Gellert and his co-workers found that antibiotics, such as novobiocin and coumermycin, which inhibit DNA replication, also inhibit gyrase's ability to convert double-helix DNA into superhelical DNA in a test tube. In addition, some gyrase was extracted from bacteria resistant to these antibiotics – that is, bacteria that can replicate their DNA and reproduce in the presence of the antibiotics. This gyrase was not inhibited in similar test-tube experiments. [Julian Davies]

Books of Science

Here are 38 outstanding new science books suitable for general readers. The director of libraries of the Smithsonian Institution selected them from books published in 1976 and 1977.

Anthropology. *The Last Primitive Peoples* by Robert Brain is a survey of life in small self-contained communities that retain their unique cultures independent of the dominant culture groups that surround them. Such communities exist on every continent and include the Lapps, Basques, Amish, American Indians, Appalachian mountain dwellers, and Eskimos. (Crown, 1976. 288 pp. illus. $14.95)

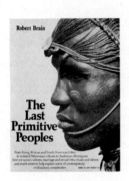

Archaeology. *Archaeology in the Making* by Philip Cleator is an archaeologist's view of the processes and impact of his profession. The author tells of the search for places and societies depicted in mythology and in the Bible, and explains the analytical and interpretative methods of modern archaeological research. (St. Martin's, 1976. 238 pp. illus. $8.95)

The Atlas of Early Man by Jacquetta Hawkes divides the 40,000 years of human history prior to A.D. 500 into eight time periods and describes what happened around the globe during each period in relation to the arts, crafts, and other elements of culture. The author focuses on centers of innovation and achievement in each epoch. (St. Martin's, 1976. 255 pp. illus. $15)

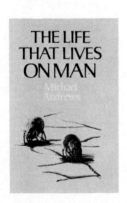

Earth Magic by Francis Hitching is an investigation of the lost world of stone-monument builders who left evidence more than 5,000 years ago of a sophisticated culture seemingly too advanced for its time. Hitching speculates on why these people disappeared. (Morrow, 1977. 320 pp. illus. $10)

Astronomy. *The Collapsing Universe* by Isaac Asimov. An astounding concept, the black hole, is theorized to be a concentration of matter so dense that even light cannot escape its gravitation. Asimov explores this idea and its implications for knowledge of small particles and large galaxies. (Walker, 1977. 204 pp. illus. $8.95)

Comets by Patrick Moore describes the origin and physics of comets in the solar system. Several well-known comets are treated in detail, including some that have now vanished. (Scribner's, 1976. 149 pp. illus. $7.95)

Meteorites: Stones from the Sky by R. V. Fodor presents many basic facts about three types of meteorites including their origin and content and the results of their impact on earth. (Dodd-Mead, 1976. 47 pp. illus. $4.25)

The Universe by Robin Kerrod is a highly abbreviated mixture of information about astronomy and its findings. The author includes information about artificial satellites, space exploration, and mysteries of the stars. (Franklin Watts, 1976. 160 pp. illus. $5.95)

Biology. *The Life That Lives on Man* by Michael Andrews looks at our body surfaces as the habitat for animals and plants such as mites, ticks, fleas, lice, bacteria, viruses, and yeasts. Andrews examines the ecology of this environment and describes the life cycles, effects, and processes of breeding and survival of these forms of life. (Taplinger, 1977. 183 pp. illus. $9.95)

Prolongevity by Alfred Rosenfeld. The science editor of *Saturday Review* presents a cohesive story of the wide range of current theories and experiments on aging and death and the potential for increasing life expectancy. He also discusses the possible social and psychological effects of prolongevity. (Knopf, 1976. 250 pp. $8.95)

Energy. *Energy: The Continuing Crisis* by Norman Metzger examines the gap between energy supplies and needs as mankind heads toward the new millennium. Metzger offers an appraisal of the various options for closing the gap — nuclear fission, including the breeder reactor; making coal into gas and oil; fusion; solar energy in its various forms; geothermal power; and conservation. Included in these appraisals are comments on the background and history of each of these forms of power, on the environmental impact of different energy supplies, and on the costs of using future energy resources. (T. Y. Crowell, 1977. 242 pp. $12.95)

Going, Going, Gone? by James Marshall is a very general examination of the use of energy resources in the United States and why we are experiencing an energy crisis. The book includes suggestions on how to save energy and descriptions of alternative sources of energy that might be developed. (Coward, McCann and Geohegan, 1976. 94 pp. illus. $5.49)

Underground Furnaces: The Story of Geothermal Energy by Irene Kiefer is a brief but thorough examination of heat resources in the earth and how they have been and might still be used as sources of energy. The author assesses the geothermal-energy reserve and the problems and limitations in tapping it. (Morrow, 1976. 63 pp. illus. $4.95)

Wind-Catchers: American Windmills of Yesterday and Tomorrow by Volta Torrey is a collection of facts about windmills and their uses. Torrey describes the problems inherent in wind-power devices and tells of the work being done to use windmills for the production of energy. (Stephen Greene, 1976. 226 pp. illus. $12.95)

Exploration. *From Vinland to Mars: A Thousand Years of Exploration* by Richard Lewis is a journalistic summary of major reconnaissance expeditions that have given us knowledge of the earth. It includes a detailed account of the U.S. space-exploration program and the significant technical developments that led to its success. (Quadrangle, 1976. 436 pp. illus. $15)

Forestry. *Trees and Man* by Herbert Edlin is a comprehensive, descriptive study of trees, ranging from the botanist's knowledge of the tree as an organism to the industrialist's view of trees as useful products. The ecological processes that affect trees are featured, including the conditions that create and destroy forests. (Columbia University Press, 1976. 269 pp. illus. $25)

Insects. *Butterflies* by Jo Brewer and Kjell Sandved blends biological information about butterflies with close-up photographs. It features the morphology of the butterfly's wing, and describes the butterfly's protective devices. (Abrams, 1976. 176 pp. illus. $18.50)

The World You Never See: Insect Life by Theodore Rowland-Entwistle is an integrated pictorial and textual presentation of the insect world that displays these creatures and their ability to reproduce, fly, find food, and camouflage themselves. Special attention is given to the social insects, such as ants and bees, and the insect environment. (Rand McNally, 1976. 127 pp. illus. $9.95)

Medicine and Health. *Epidemics* by Geoffrey Marks and William Beatty describes great epidemics in ancient and modern civilization, ranging from the first recorded epidemic in Egypt in 3180 B.C. to the 1968 Hong Kong influenza outbreak. (Scribner's, 1976. 323 pp. $9.95)

The Greatest Battle by Ronald Glasser portrays cancer as a cellular disease. Glasser traces the development of our understanding of cancer and focuses on carcinogens, including radiation, and how they seem to create cancer. (Random House, 1976. 180 pp. $6.95)

Plagues and People by William McNeill demonstrates how pestilence has affected the course of history, from the first recorded plague in the Athenian Army to the worldwide influenza epidemic after World War I. (Doubleday, 1976. 369 pp. $10)

Microbiology. *Microbe Power: Tomorrow's Revolution* by Brian Ford focuses on the microbe world as a valuable resource in food, drug, and energy production, and in pollution control. The author describes humans as the "sea" designed to provide for the sustenance and transmission to the future of the microbe called a sperm. (Stein and Day, 1976. 181 pp. illus. $10)

Natural Disasters. *The Forces of Nature* by Vivian Fuchs includes articles by scientists about the causes and results of avalanches, blizzards, dam bursts, droughts, earthquakes, floods, lightning, and volcanoes. It cites examples of ancient and recent natural disasters and comments on studies of prediction, disaster prevention, and recovery. (Holt, Rinehart and Winston, 1977. 303 pp. illus. $25)

Nature at War by Hal Butler reports on 13 natural disasters that have occurred in the Western Hemisphere since 1811, including earthquakes, hurricanes, volcanic eruptions, floods, blizzards, and tornadoes. (Regnery, 1976. 240 pp. $6.95)

Oceanography. *The Coral Seas* by Andrew Campbell tells how corals grow; describes their environment and the life that exists among them; and comments on how man affects, studies, and protects them. (Putnam's, 1976. 128 pp. illus. $12.95)

The Forests of the Sea: Life and Death on the Continental Shelf follows the life cycles of creatures that live on the North American continental shelf. The text describes the complex relationship among the plants and animals and the harm done them by modern man. (Sierra Club, 1976. 290 pp. illus. $9.95)

The Living Sea: An Illustrated Encyclopedia of Marine Life by Robert Burton, Carole Devaney, and Tony Long depicts in text and color photographs the basic elements of life in and around the sea among invertebrates, fish, reptiles, sea birds, and mammals. The book provides basic science and detailed descriptions of the classes of species and their relationships to each other and to humans. (Putnam's, 1976. 240 pp. illus. $20)

Physics. *Fabric of the Universe* by Denis Postle uses analogies and parallels from everyday life to explain particle physics and the fundamental laws governing the behavior of matter. Postle also explains what particle physics reveals about the world. (Crown, 1976. 208 pp. illus. $5.95)

Supercold—Superhot: Cryogenics and Controlled Thermonuclear Fusion by Gail Hines explains how scientists create conditions of extreme heat and cold and the abnormal things that happen to elements and materials at both ends of the heat scale. The author also speculates on how these conditions can be used in the fields of medicine, technology, and energy. (Franklin Watts, 1976. 81 pp. illus. $4.33)

Space Travel. *The High Frontier: Human Colonies in Space* by Gerard O'Neill is a speculative description, based on contemporary scientific and technological trends, of a space community. The author provides details on how the colony might be built and populated, and he explains some of the science that might make this possible. (Morrow, 1977. 288 pp. illus. $8.95)

Technology. *Chopper! The Illustrated Story of Helicopters in Action* by Bern Keating is a brief history of the invention and development of helicopters. It includes vivid accounts of such aircraft with photographs of their use in agriculture, industry, science, transportation, and war. (Rand McNally, 1976. 232 pp. illus. $14.95)

The Fire of Genius by Ernest Heyn. The former editor in chief of *Popular Science* tells the stories of prominent inventors and their innovations from 1872 to the present. He includes "tremendous trifles" such as zippers, paper bags, and aerosol spray cans, as well as

inventions that revolutionized our life styles, such as airplanes, photography, the radio, and the telephone. (Doubleday, 1976. 340 pp. illus. $12.95)

How Scientists Find Out About Matter, Time, Space, Energy by Herman Schneider. In words and diagrams, this book explains the workings and applications of various instruments that scientists use to measure phenomena and to extend the human senses. Schneider relates the instruments to ordinary household objects and describes a series of simple experiments for the reader to perform. (McGraw-Hill, 1976. 133 pp. illus. $6.95)

New Trailblazers of Technology by Harland Manchester. Describing devices from electrical soot precipitators to FM radios and cable television, the author recounts the stories of 10 modern inventors whose creative imagination led to the products that now affect our lives. Included are discussions of the laser and instant photography. (Scribner's, 1976. 214 pp. illus. $7.95)

Volcanoes. *Volcanoes* by Alfred and Loredana Rittmann presents brief but succinct information about the mechanics of volcanoes with examples of the most spectacular and well-known volcanoes and volcanic phenomena. (Putnam's, 1976. 128 pp. illus. $12.95)

Zoology. *The Natural History of Marine Mammals* by Victor Scheffer describes the life cycles of the 111 species of marine mammals. Scheffer tells how these unusual beasts descended from six zoological groups and how they then adapted to a hostile environment. (Scribner's, 1976. 157 pp. illus. $7.95)

A Natural History of Zebras by Dorcas MacClintock is a brief but thorough study of the three species of zebras from prehistory to modern times, based on paleontological evidence and field and zoo studies. The author includes information on the zebra's social organization and family life. (Scribner's, 1976. 134 pp. illus. $7.95)

The World of the Bat by Charles Mohr describes the life of the bat through all the seasons of the year. It includes brief zoological and historical information on the relationship of bats and humans, with special comments on how the bat's sonar techniques are used to aid the blind. (Lippincott, 1976. 162 pp. illus. $8.95)   [Russell Shank]

Botany

An artificial system that mimics the natural photosynthesis that provides energy and food for green plants was described by biochemist Thomas R. Janson of the Argonne (Ill.) National Laboratory in a March 1977 report on a 15-year project by Argonne scientists.

With physical chemists Joseph J. Katz, James R. Norris, Michael R. Wasilewski, and others, Janson has been closely studying the physical and chemical changes that occur during photosynthesis. Janson himself produced an artificial leaf device that can convert light to chemical energy much as a real leaf does.

In this complex process, certain chlorophyll molecules called antenna molecules absorb light energy, which is then transferred to a pair of special chlorophyll molecules called reaction centers. After receiving the light energy, the reaction centers give off electrons, which are used to create chemical energy to provide food for the plant.

Janson's artificial leaf is a metal disk coated on one side with chlorophyll and a chemical that receives electrons. The other side is treated with a chemical that gives off electrons. Light shining on the disk creates an electron flow, mediated by the chlorophyll, from the donor chemical to the receiver.

Lasered bacteria. Work on later stages of the photosynthetic process was described by biochemist-physicist Kenneth J. Kaufmann of the University of Illinois in Champaign-Urbana. He reported in February 1977 on his work with collaborators P. Leslie Dutton of the University of Pittsburgh and Peter M. Rentzepis of Bell Telephone Laboratories in Murray Hill, N.J.

The scientists worked with photosynthetic bacteria rather than leafy plants, because the tiny one-celled organisms are simpler in structure. They isolated the photosynthetic centers in the bacteria and transplanted them to nonphotosynthetic bacteria, where the centers functioned when exposed to light.

Using a neodymium-glass laser to time events that occur in picoseconds or trillionths of a second, they discovered two intermediate steps between the ejection of the electron by a chlorophyll

The male flower of the sea nymph, *right*, a water plant, contains coiled pollen grains, *far right*. After they are released, water currents straighten them into long strands and carry them to the female flowers for fertilization. The unusual pollen shape helps the plant to reproduce underwater.

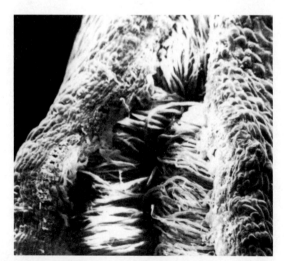

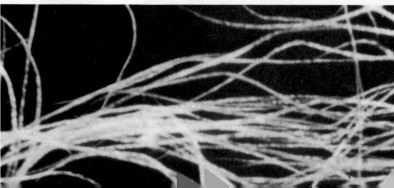

Botany

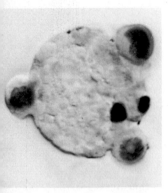

A tobacco-human cell was made by cell fusion. The small dark circles are tobacco cell nuclei. The larger dark circles are the nuclei of human tumor cells. The human nucleus at top right is part of a human cell that has not fused with the tobacco cell.

molecule and its pickup by a molecule of quinone, a previously known step. Researchers say that the ejection of the electron creates an electron-hole pair in which the electron is still bound to the hole created when it was ejected. To prevent it from falling back into the hole, pheophytin, a molecule similar to chlorophyll, quickly picks up the electron and passes it along to quinone.

Such studies of the photosynthetic process may lead to ways to convert sunlight directly into useful energy.

Winter efficiency. The photosynthetic efficiency of *Clamissonia claviformis,* an annual winter primrose found in Death Valley, California, was reported by plant scientists Harold A. Mooney and James Ehleringer of Stanford University in California and Joseph A. Berry of the Carnegie Institution of Washington at Stanford in October 1976. Working in a mobile laboratory, they measured the plant's ability to absorb sunlight and convert it into chemical energy, by the exchange of water vapor and carbon dioxide in the plant's leaves.

The researchers found that the plant leaf converts 8.5 per cent of the light falling on it into chemical energy. This is an unusually large amount, about 80 per cent of the theoretical limit for plants. Most plants can use only about 1 to 3 per cent of the available light.

Mooney and his associates suggest that one reason that the plant has become so efficient is that rain falls infrequently in the desert and a plant has a great advantage if it can grow rapidly and complete its life cycle before the summer heat and drought arrive.

Planting petroleum. Chemist Melvin Calvin of the University of California, Berkeley, suggested in October 1976 that petroleum could be produced from plants. Calvin, who won the Nobel prize for chemistry in 1961 for his work on photosynthesis, recommended two shrubs, *Euphorbia lathyrus* and its relative *E. tirucalli,* that produce an emulsion of hydrocarbons in water. Available technology could separate the hydrocarbons, which are similar to those in petroleum.

The plants could be harvested by simply cutting them off close to the ground. The stumps regrow, so replanting might not be necessary for many years. In addition, the two shrubs grow in areas unsuited to other crop plants. Production estimates range from 25 to 125 barrels of petroleum per hectare (10 to 50 barrels per acre). At 100 barrels per hectare (40 barrels per acre) an area as big as Arizona could supply all the gasoline the United States requires.

Nitrogen fixation. Biologist Rex L. Smith and seven co-workers at the University of Florida in Gainesville in September 1976 demonstrated nitrogen fixation in grasses after inoculating them with nitrogen-fixing bacteria. Plantings of pearl millet and guinea grass, inoculated with the bacterium *Spirillum lipoferum,* produced significantly greater amounts of plant matter than did uninoculated plantings.

Understanding how some plants can fix nitrogen is a prime goal for botanists because large quantities of energy are required to convert atmospheric nitrogen to commercial nitrogen fertilizers. Nitrogen fixation by bacteria associated with legumes is well known, and a few grasses have been found that associate naturally with nitrogen-fixing bacteria. But converting such grasses as wheat to nitrogen fixers would be a major feat.

Other gains were made in 1976 in understanding and controlling nitrogen fixation. For example, cell biologist W. J. Broughton and co-workers at the University of Malaya in Kuala Lumpur observed nitrogen fixation in naked protoplasts (isolated cells with membranes but no walls) from legume root nodules. And geneticists Andrew W. B. Johnston and John E. Beringer of the John Innes Institute in Norwich, England, developed pea-root nodules containing more than one species of nitrogen-fixing bacteria.

An unusual hybrid was reported in July between a plant and an animal. Using special enzymes, biologist Harold H. Smith and co-workers at the Brookhaven National Laboratory on Long Island, N.Y., stripped away the cell wall from isolated tobacco cells and fused the resulting protoplasts with HeLa cells, which were derived more than 20 years ago from a human tumor. The resulting "plantimal" cells, showing characteristics of both the tobacco and the HeLa cells, lived in culture for up to six days.   [Frank B. Salisbury]

Chemical Technology

Seafood processors, faced with stringent pollution-control restrictions on the dumping of untreated wastes from shrimp and crab processing, sought commercial uses for the animals' shells. The research was funded in 1976 by the U.S. Sea Grant Program.

Researchers found that chemically treating chitin, the major substance in the shells, produces a material called chitosan, which can serve as a food thickener or extender and can also be used in making fibers and moisture-proof coatings and films. Chitin also can replace cellulose in some industrial uses. Both chitin and chitosan are nontoxic and biodegradable.

Marine Commodities International of Brownsville, Tex., can produce about 218 metric tons (240 short tons) of the two materials per year, and larger quantities are turned out by Japanese shellfish processors.

Producing fertilizer from the sludge by-product of municipal waste-treatment plants is another goal of chemical technologists. The city of Houston developed a scheme and tested it so suc-cessfully that in 1976 the facility was expanded from a capacity of 82 metric tons (90 short tons) to 127 metric tons (140 short tons) per day. The city plans to increase capacity again to 272 metric tons (300 short tons) by 1981. In the process, the sludge is flash-dried in a furnace to form a material called Hou-Actinite, which is sold to agricultural users such as Texas rice farmers and Florida citrus growers for about $30 a short ton.

Heating and cooling. Researchers at Michigan Technological University in Houghton reported in March 1977 that they had discovered a family of materials that freeze and thaw at room temperatures. The materials, called clathrates, are composed of latticelike molecules that enclose another molecule, such as water. They release or take up almost as much heat as does water when it either freezes or thaws. The clathrate-water compounds have a freezing point between 10°C and 29.1°C (50°F. and 85°F.). Sealed in building walls, automobile panels, and other such places, the materials could

A replacement for part of a monkey's hipbone cools after being coated with bioglass, *right,* a glasslike substance that binds well to living bone. Scientists implant a bioglass-coated object in a baboon's jaw, *far right,* to test the substance's durability.

Chemical Technology

Continued

Flames make no impression on material made of new ceramic fibers developed by the 3M Company. The fibers may replace asbestos for some uses.

help to warm or cool interiors by solidifying to release heat when it turns cold outside and melting to absorb heat when it becomes warm.

Experiments indicate that an ultrathin membrane developed by the General Electric Company for improving burner combustion promises to save energy. The solid polymer film, 0.0005 centimeter (0.0002 inch) thick, allows more oxygen than nitrogen to pass through it, thereby increasing the oxygen content of air around a flame from 21 to 50 per cent. Such enriched air, when used in fuel combustion, produces a hotter flame and, in certain cases, can reduce fuel consumption as much as 50 per cent.

University of Texas researchers in April 1977 reported that they had created a refrigerator compressor with no moving parts other than valves. The device will be used in National Aeronautics and Space Administration (NASA) satellite mappers, which use highly sensitive crystal detectors that must be cooled to $-184°C$ ($-300°F.$). The detectors are too sensitive to withstand the vibration of conventional compressors.

The new compressor works on the basis of a clay mineral known as zeolite (sodium and aluminum silicates) that absorbs up to 25 per cent of its weight in gas, much as a sponge takes up water. In a sealed container, the gas-containing zeolite is heated by an electrical resistance coil. This drives the gas from the zeolite and raises the pressure in the container, producing compressed gas. The compressor also has potential for use in saving energy in home air conditioning. This type of compressor system could use solar rather than electrical energy as the heat source.

Fibers and fabrics. Courtalds Limited of London in April 1977 reported developing a tubular viscose fiber that can absorb as much perspiration and other forms of moisture as wool. The new material is expected to replace cotton in polyester blends. Hollow fibers soak up 120 to 130 per cent of their weight in water, compared with 100 per cent for conventional viscose fibers. Clothes made of the new material can soak up 13 per cent of their weight in moisture before the wearer perceives any dampness.

West Germany's Bayer A. G. in December 1976 reported the development of a lightweight, moisture-absorbing acrylic fiber that is structurally unique. The material, code-named ATF-1017, contains numerous capillaries that branch off into fine canals through which moisture can pass from the outside into a porous fiber core. Unlike wool and cotton, ATF-1017 has good ventilation because it swells very little when it absorbs water. When fibers swell, they fill the air spaces and thus reduce ventilation. Persons wearing ATF-1017 do not sense dampness until the fabric has absorbed almost 19 per cent of its weight in water. ATF-1017 is 25 to 30 per cent lighter than conventional acrylic fibers.

E. I. du Pont de Nemours & Company of Wilmington, Del., in 1976 began marketing a new rubber-toughened acrylic resin for use in manufacturing molded-plastic items. The company claims the new resin is several times tougher than standard acrylics and is as transparent as Du Pont's optical-grade Lucite. The resin will be used for industrial windows, light lenses, skylights, and outdoor signs.

The U.S. Department of Agriculture (USDA) increased the absorbability of its super slurper almost threefold. The newest modified version of this substance soaks up 5,000 times its weight of water. The material could be useful in such items as diapers or as a seed coating to promote germination. The super slurpers consist of copolymers made from cornstarch. But USDA researchers are working on ways to improve this material by experimenting with new polymers made by substituting flour for cornstarch.

Scientists at the USDA's Southern Regional Research Center in New Orleans in August 1976 revealed a process for making fabrics resistant to disease- and odor-causing germs. They treated various fabrics with solutions containing hydrogen peroxide, which is germicidal, and zirconyl or zinc acetate, which bind the peroxide to the fibers and release it very slowly. The researchers reported that cotton treated in this manner was germ resistant after 50 launderings.

The most obvious use for antibacterial fabrics is in hospital sheets and towels. However, they could also be used for fungi-inhibiting hosiery, in clothing to prevent body odor, and in a wide variety of soil-resistant items.

Papermaking. A new pulp-making process, called thermomechanical pulping, was put into operation in October 1976 by F. F. Soucy, Incorporated, of Rivière-du-Loup, Canada, to make newsprint. Ordinarily, wood being converted to wood pulp for papermaking requires chemical treatment. In the new method, the papermaker preconditions a wide assortment of wood chips – including bark – with steam. This softens the relatively brittle lignin, which along with cellulose forms the major part of all wood, and eventually results in longer fibers. The wood chips are then ground, screened, and cleaned to produce pulp that is suitable for papermaking.

The new system lowers construction costs, reduces waste because it uses the whole tree, and causes virtually no air pollution. However, it uses slightly more energy than other pulp-grinding operations. Despite this drawback, researchers are experimenting with adding bleaching chemicals just before the refining step to produce higher paper grades, such as magazine stock.

Microwave technology. The same electromagnetic radiant energy that quickly cooks food in household microwave ovens may find a far-reaching application in toxic-waste disposal. The Lockheed Missiles and Space Corporation of Sunnyvale, Calif., under a contract with the U.S. Environmental Protection Agency, in June 1976 announced development of a microwave system that can break down any organic material into carbon monoxide, carbon dioxide, water, and other elements. Thus, extremely dangerous compounds that cannot be burned or used in landfill safely can be destroyed by literally dismantling them.

Other new products. Zoecon Corporation of Palo Alto, Calif., tested mosquito-killing briquettes during the summer of 1976. The briquettes, which contain methoprene, dissolve slowly in water and can be tossed into septic ditches, drains, and other breeding sites that cannot be sprayed. Each 6-gram (0.2-ounce) briquette will kill mosquito larvae in a 9.3-square-meter (100-square-foot) area of 15-centimeter (6-inch)-deep water for 30 days.

Stuart Pharmaceuticals of Wilmington, Del., in 1976 began marketing in the United States an antimicrobial scrub cleanser for combating infections in hospitals. It will replace hexachlorophene products, severely restricted since 1972. The new product, called Hibiclens, is effective against various bacteria, fungi, and mold. Its active agent is chlorhexidine gluconate.

Norsk Hydro of Oslo, Norway, discovered a way of turning straw into edible fodder. High-fiber, low-protein straw is baled and wrapped in plastic sheeting, into which anhydrous liquid ammonia is injected. After a few weeks, the treated straw develops into an animal feed containing up to 12.5 per cent protein, only slightly less than hay.

Pertwee Industrial, Limited, of Colchester, England, in 1976 began marketing a bag weighing 5.5 kilograms (5 pounds) that can hold 0.9 metric ton (1 short ton) of dry, free-flowing, bulk material. The bags, made of woven polypropylene, can be reused up to 10 times. They can be moved around by fork lifts and emptied in 20 to 30 seconds by releasing a tie-off cord at the bag's base.

University of Alabama scientists in March 1977 reported developing exterior house paints that protect against mildew. An organic fungicide chemically bonded to the polymer paint base prevents the fungicide from evaporating or washing out of the thin film of applied paint.

Allied Chemical Corporation of Morris Township, N.J., reported in January 1977 that its researchers are developing electrochemical devices that can produce chemical compounds with the aid of light. The devices, called photochemical diodes, are sandwichlike wafers of two electrodes. When researchers placed the diodes in acidic water and exposed them to light, hydrogen and oxygen bubbles collected on the diodes. The diodes can be made in microscopically small sizes, and Allied researchers believe it may be possible to suspend them in solution. They could be used to recover hydrogen and sulfur from hydrogen sulfide, as catalysts for chemical reactions, or in solar-energy systems.   [Frederick C. Price]

Chemistry

Roger Guillemin and his co-workers Roger Burgus, Larry Lazarus, and Nicholas Ling at the Salk Institute for Biological Studies in San Diego reported in August 1976 that chemicals extracted from brain tissue may help scientists to understand the chemical basis of schizophrenia and other mental diseases. These chemicals also form a new class of analgesics, or painkillers. Termed endorphins, the compounds were extracted in minute amounts from the hypothalami of 250,000 sheep.

Endorphins are short polypeptides, or chains of amino acids, of varying length. They include two very short peptides, called enkephalins. See OUR BODY'S OWN NARCOTICS.

The three endorphins given the prefixes α-, β-, and γ-, appear to be parts of the larger beta-lipotropin molecule, a hormone isolated from the pituitary gland in 1964 by Choh Li of the Hormone Research Laboratory at the University of California, San Francisco. Amino acids 61 through 76 in the beta-lipotropin molecule correspond to α-endorphin's 16 amino acids. And β-endorphin, which has 31 amino acids, has the same structure as the 61 through 91 amino-acid sequence. With 17 amino acids, γ-endorphin is identical to the 61 to 77 sequence of beta-lipotropin.

When Guillemin's group tested the endorphins on laboratory rats, each endorphin produced a characteristic but different physiological response. Injections of α-endorphin acted as a tranquilizer and a mild analgesic. In contrast, γ-endorphin caused violently aggressive behavior in the animals. The most highly active of the three brain chemicals, β-endorphin caused a catatonic state that left the rats looking as if they had been frozen stiff. However, the rats awoke and became active soon after they received the morphine antagonist naloxone, which blocks the action of morphine. This suggests that the endorphins occupy the same molecular receptor sites in rat brains as do morphine and similar analgesics.

Guillemin believes that the rats' catatonia is similar to that observed in many humans suffering from schizo-

"All I know is he claims he can spin straw into oil!"

phrenia. He suggests that endorphins may hold the key to understanding some of the biochemical mechanisms involved in certain types of mental disease. Preliminary medical studies by Lars Terenius and his associates at the University of Uppsala in Sweden indicate that schizophrenic patients have abnormally high amounts of endorphins in their spinal fluid.

Li and his associates announced in the fall of 1976 that they had isolated β-endorphin from human pituitary glands. Earlier, they had found an almost identical polypeptide in camel brains. Both the camel and human β-endorphin molecules are very effective painkillers, in agreement with Guillemin's work on sheep brain β-endorphin.

Selective chemistry. Laser light, intense, coherent radiation of a single wavelength, is being used to develop a brand new field—laser-induced selective chemistry. The laser technique allows a chemist to select which bond to break in a molecule. It may lead to new methods of separating isotopes, chemical elements that differ only in the number of neutrons they contain.

Chemists Ernest M. Grunwald, Kenneth J. Olszyna, David F. Dever, Philip Keehn, and Dana Garcia of Brandeis University in Waltham, Mass., reported in late 1976 that the excitation energy provided by an infrared laser remained confined to a particular bond in certain chlorofluorocarbon molecules. Moreover, the molecules broke up into the same molecules as they do when they are simply heated.

Normally, gas molecules not only move around in a container, but they also rotate and vibrate in one of many possible energy states. Energy added in the form of heat or light intensifies the motions. But Grunwald's group found that the chlorofluorocarbon molecules behave as if all the laser energy excites only one vibrating bond. If enough energy is added, the molecule vibrates so much that the bond breaks. Previously, chemists would have predicted that the energy would be shared among translational, rotational, and vibrational motion before breaking a bond.

The researchers found similar results when the laser beam strikes chlorofluorocarbons in the liquid state. During the brief laser flash, the molecules collide hundreds of times in the liquid. Thus, the compounds would have been expected to dissipate the extra energy as heat by molecular collisions. Instead, the laser energy remained localized in the vibrational form.

Because a laser beam can be tuned to match exactly the vibrational frequency of a specified bond in a molecule, chemists should be able to select which bond to break in a chemical reaction. In addition, laser energy is absorbed more efficiently than the thermal energy from normal heating.

Chemists Robert R. Karl and K. Keith Innes of the State University of New York at Binghamton separated the isotopes of nitrogen and carbon by selective laser irradiation of symmetric tetrazine, a ringlike molecule that contains two carbon and four nitrogen atoms. When irradiated, tetrazine decomposes into two molecules of molecular nitrogen and two molecules of hydrogen cyanide.

In another experiment, chemists Robin M. Hochstrasser and David S. King of the University of Pennsylvania reported in September 1976 that laser-selective irradiation of symmetric tetrazine crystals at temperatures near absolute zero resulted in a 10,000-fold increase in the relative number of symmetric tetrazine molecules that contained nitrogen 15 and carbon 13. They used a laser to decompose selectively the symmetric tetrazine molecules containing the more abundant isotopes nitrogen 14 and carbon 12.

Ilana Glatt and Amnon Yogev of the Weizmann Institute in Rehovot, Israel, reported a novel application of laser-induced selective chemistry in October 1976. Using an infrared laser, they rearranged molecules of 1,5-hexadiene, which look like the letter "C," into their mirror images, which are shaped like a backward "C." Both the starting molecules and their end-product images are chemically identical. The chemists were able to tell them apart by replacing a hydrogen atom with the hydrogen isotope deuterium in one form of the molecules. By selective irradiation of a bond in one of the two differently labeled 1,5-hexadiene molecules, the chemists were able to rearrange only that molecule.

Chemistry

Continued

New prostaglandin. Prostaglandins, potent, hormonelike substances found in the body, again made news in late 1976 with the announcement that prostacyclin, a new prostaglandin, may prove useful in treating heart attack and stroke. Prostacyclin appears to play a crucial role in the aggregation, or gathering, of blood platelets. The clumping together of platelets in human blood can cause thrombi, or clots, that may result in heart attack and stroke.

Prostacyclin was synthesized and the mystery of its biological activity unraveled by an international effort on the part of John R. Vane and his colleagues at the Wellcome Research Laboratories in England and a research group led by Roy A. Johnson at the Upjohn Company in Kalamazoo, Mich.

The discovery of prostacyclin has a fascinating history that partly involves work by Bengt Samuelsson and his associates at the Karolinska Institute in Stockholm, Sweden. In 1973, the Karolinska researchers and other groups found two endoperoxides, short-lived molecules that convert naturally to thromboxane A_2 and a related molecule, thromboxane B_2. Samuelsson's group found that the thromboxanes cause blood-platelet aggregation.

The Wellcome-Upjohn research indicates that both prostacyclin and the thromboxanes are formed in the body from the same precursor molecules, the prostaglandin endoperoxides. Amazingly, the physiological actions of prostacyclin and the two thromboxanes are almost exactly opposite. Prostacyclin inhibits blood-platelet aggregation while the thromboxanes induce it. Vane announced that prostacyclin is the most potent inhibitor of platelet clumping of all the known prostaglandins and is also a powerful vasodilator, causing blood vessels to dilate. Prostacyclin broke up platelet clumps in test-tube studies.

These discoveries provide important insights into how normal blood vessels avoid clot formation and how platelet aggregation often leads to clots in diseased or damaged blood vessels. Vane believes that prostaglandin endoperoxides are released when platelets brush up against the inner lining of normal blood vessels. This triggers the release of an enzyme that converts the endo-peroxides to prostacyclin, and prevents the formation of clots.

Vane also suggests that no enzyme synthesis occurs and the prostaglandin endoperoxides are converted to the clot-promoting thromboxane A_2 if the platelets brush up against damaged sections of blood vessels. The discovery of prostacyclin and how it acts may result in the synthesis of new drugs that can prevent the clots that too often cause heart attack and stroke.

Singlet oxygen. An international team of chemists led by Nicholas J. Turro of Columbia University and Adolph Krebs of the University of Hamburg's Organic Chemistry Institute in West Germany discovered in October 1976 that heating oxygen from the air in the presence of a special cyclic acetylene catalyst formed a highly reactive type of oxygen. Called singlet oxygen, it may provide insights into such diverse areas as biological processes and how plastics decompose. The new process allows chemists to make singlet oxygen easily.

Normal molecular oxygen (O_2) is in the so-called triplet electronic state in which there is one unpaired electron on each atom of the O_2 molecule. Furthermore, the electrons spin in the same direction. In singlet oxygen, however, both electrons are not only on the same atom, but they also spin in opposite directions. This electronic arrangement makes singlet oxygen highly reactive.

Turro and Krebs propose that the triple bond of the acetylene catalyst and triplet oxygen form a weak complex in which two electrons with the same spin can exist on the same oxygen atom. One of the electrons may then change its spin and pair up with the other to form singlet oxygen, which seeks electron pairs in reactions.

Organic molecules in space. In April 1977, astronomers Lorne W. Avery, Norman A. Broten, John M. McLeod, and Takeshi Oka of the Canadian National Research Council's Herzberg Institute in Ottawa reported finding the largest and heaviest organic molecule in space yet, cyanotriacetylene (C_7HN). It has a molecular weight of 99. The scientists detected the cyanotriacetylene in a cloud in the constellation Taurus. This heavyweight of space molecules tipped its hand by continu-

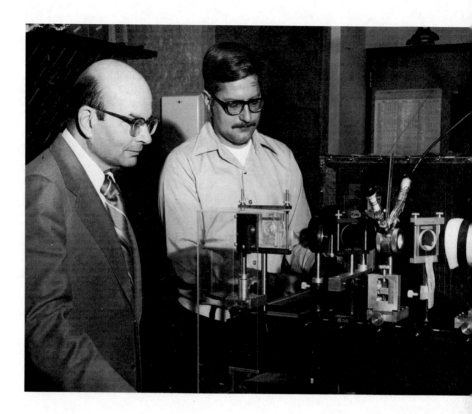

Brandeis University chemists adjust an infrared laser that they used to break specified molecular bonds to initiate chemical reactions.

Chemistry

Continued

ously emitting radiation over a range of microwave frequencies. Astronomers measured and compared these frequencies to those emitted by a sample of C_7HN synthesized in the laboratory by chemist Harry Kroto and his group at the University of Sussex in England.

The C_7HN molecule is so highly reactive that the Sussex workers could trap it for only short periods of time under special conditions. It has a much longer lifetime in the vastness of interstellar space where the probability of its encountering another molecule to react with is very small. Cyanodiacetylene (C_5HN) had been detected previously in space, and scientists now wonder if the interstellar search will yield the next higher compound, C_9HN.

The simplest molecule containing a carbon-carbon triple bond, acetylene (C_2H_2) was found in mid-1976 by a team from the Kitt Peak National Observatory near Tucson, Ariz., and the Massachusetts Institute of Technology in Cambridge, led by Don Hall and Stephen T. Ridgway. See ASTRONOMY (Stellar Astronomy).

Another new molecule found in space late in 1976 was ethyl cyanide (C_3H_5N). A team from Columbia University in New York City, Harvard University, and the National Bureau of Standards detected it in the constellation Orion. Ethyl cyanide is a hydrogenated version of cyanoacetylene, and it may have been formed in a hydrogen-rich atmosphere.

Another important space-chemistry discovery was that of an ionized molecule containing deuterium instead of hydrogen. The deuterated formyl ion, DCO^+, was found in several gas and dust clouds in our own Galaxy, the Milky Way, by a team headed by radio astronomer Lewis E. Snyder of the University of Illinois in Champaign-Urbana. Previously, the only molecules containing deuterium detected in interstellar space were the neutral molecules HD, HDO, and DCN. The discovery of deuterated formyl ions gives scientists additional clues with which to formulate theories about the origin of the universe.   [Lawrence Verbit]

See also CORES IN COLLISION.

Communications

New techniques in 1976 made it possible to increase the amount of detail visible in a television picture transmitted over a communications channel.

Nippon Electric Company Limited of Japan in the summer of 1976 began demonstrating a signal encoder that allows commercial-quality television to be broadcast on one-fourth the bandwidth, or range of frequencies normally required. This is achieved by reducing the amount of information that must be transmitted.

The system does this by comparing each frame of the television transmission with the one that preceded it and transmitting only signals that represent changes between the two frames. Most change is caused by movement; much of the picture, such as background, remains unchanged. When the system detects significant change in a frame, it sends a signal representing the change, plus a code that tells where change has occurred in the picture. This updates the information about the frame, and the process starts over for the next one.

The television signals are digitally encoded — changed into a series of bits, or 0's and 1's. This allows the signal to be transmitted through many signal-relaying devices without noticeable changes in its quality. The system converts a 5-megahertz (MHz) television signal into a bit stream carrying from 16 million to 32 million bits per second. At the receiving end of the system, the bits are converted back into video signals to produce a television picture. Previously developed digital systems required about 125 million bits per second for commercial-quality television transmission and reception.

The amount of detail in a picture depends on the bandwidth used to transmit it. The broader the bandwidth, the more information that can be transmitted. Bandwidths in commercial systems vary from the low-frequency narrow AM radiobroadcast band to the broad, ultrahigh-frequency television and microwave bands.

Television systems, which transmit 30 picture frames per second, must operate on a broad electromagnetic spectrum. But operating on higher bandwidths is expensive because broadcasting on high frequencies requires very sophisticated technology. So there has always been a trade-off between speed, quality, and cost.

With another Nippon system, a standard U.S. color television signal — suitable for conferences or videophones — can be transmitted on an even narrower bandwidth at a rate of only 6.3 million bits per second.

Better copies. Similar improvements took place in transmitting facsimile, copies of documents sent over communications channels. Nearly all facsimiles are sent over long-distance telephone circuits, which have a bandwidth of less than 4 kilohertz (kHz). Although the narrow bandwidth is economical, it takes four to six minutes to transmit a clear copy of an ordinary business letter.

But in April 1977, the 3M Company of St. Paul, Minn., introduced a digital facsimile machine that can send a 300-word letter over ordinary telephone lines in less than 30 seconds. An electronic scanner goes over the document, picking out the black-and-white areas. The scan generates a stream of digital bits which is then encoded to identify and locate only the black letters, numbers, and lines, so that no transmission time is devoted to sending blank space, such as margins and spaces between lines and paragraphs.

Optical-fiber communications. The first practical optical-fiber communications systems began operating in early 1977. Optical fibers, thin strands of glass that use light waves to transmit signals, were used to link a U.S. Air Force wind-tunnel test facility with a computer complex 1.6 kilometers (1 mile) away in Dayton, Ohio. Tests with telephone service using optical fibers began in Long Beach, Calif.; Chicago; and near London.

The General Telephone & Electronics installation at Long Beach is a low-capacity system, handling only 24 conversations over 9 kilometers (5.6 miles) of fiber. It uses light-emitting diodes (LEDs) in its transmitters and two amplifying devices that strengthen signals traveling long distances. The Bell Laboratories installation in Chicago handles Picturephone data transmissions along with more than 600 voice channels. It uses lasers to beam the light signals over its 9 kilometers (5.6 miles) of fiber. The Standard Tele-

Technicians compare the intensity of light traveling along two types of optical fibers. A fiber developed in West Germany, right, transmits far more light over its 200-meter (656-foot) length than does a fiber made of lower-quality glass that is only about 30 meters (98 feet) long.

Communications

Continued

phone Laboratories system in Great Britain is about the same length as the California system and has two relay devices. However, it also uses lasers instead of LEDs. The British system has the highest capacity of the three, with the ability to handle 1,920 conversations over a single fiber.

Some of the most difficult problems technicians had to overcome to make optical systems practical involved handling and protecting the hair-thin glass fibers. They developed cables to protect and isolate the fibers from the normal stresses that metal cables take in stride.

Another major problem was how to make strong splices that would minimally affect the signals traveling over such fibers. For a good splice, the ends of the two fibers must be cut straight and flat so that they will fit together perfectly. The greater the deviation from this ideal, the greater the loss in transmission caused by the splice.

For the new telephone service, the companies all developed mechanical splice techniques that are reasonably practicable. Transmission losses range

from 0.2 to about 1 decibel per splice. General Cable Corporation of Stamford, Conn., announced in May that it was working on a splicing technique in which the fiber ends are fused or welded. Losses with this technique were less than 0.1 decibel in laboratory tests.

Magnetic bubble memories, a decade after their invention at Bell Laboratories, found practical applications in 1977. The "bubbles" are actually magnetic islands in a thin film of garnet crystal. They are polarized in an opposite direction from the rest of the crystal. In digital language, the presence of a bubble means 1, the absence of a bubble means 0. Unlike memories such as tapes that move on reels, bubble memories have no moving parts that can wear out.

Bell Systems began using bubble memories in early 1977 for recorded telephone announcements. The sound of a voice is converted into digital bits and coded into the bubble. The bubble memories for these recordings can store 325,000 bits, enough for messages 13 seconds long. [Albin R. Meier]

Drugs

Scientists tested a new drug to relieve gallstones during 1976 and 1977 — chenodeoxycholic acid, or chenic acid. The substance is a natural constituent of bile acids, which are secreted by the liver and stored in the gall bladder, a pear-sized pouch under the liver.

Each year, a million Americans develop gallstones, pebblelike crystals of cholesterol that form in the gall bladder. Scientists do not know why this happens in certain people. Perhaps poor functioning of the gall bladder allows the cholesterol to collect and crystallize.

Normally, the gall bladder contracts and squeezes bile acids into the small intestine to help digest fatty foods after a meal. When stones are present, however, the contraction wedges them in the small ducts leading to the intestine, causing pain and sometimes nausea and vomiting. When the symptoms become chronic and severe, the only treatment is surgical removal of the gall bladder with the stones. This operation is now the sixth leading reason for admission to a hospital. See MEDICINE (Surgery).

Researchers at 10 medical centers are recruiting 1,000 patients to study the long-term effects of chenic acid extracted from the bile of cattle and hogs. Researcher Leslie J. Schoenfield of Cedars-Sinai Medical Center in Los Angeles believes that, while the acid's ability to dissolve gallstones is unquestionable, this study will evaluate its long-term effectiveness and safety. There has been some concern about the drug's toxicity because experiments with rhesus monkeys and baboons produced instances of liver damage. But liver damage has not been seen in humans participating in chenic acid tests, perhaps because they metabolize the acid differently.

If the study proves the drug's safety and effectiveness, some of the almost 500,000 persons who would otherwise have their gall bladders removed each year may be able to take a capsule every day and avoid surgery. An effective and safe drug would also provide relief for the 15 to 20 million who suffer occasional gallstone attacks.

Saccharin out. The Food and Drug Administration (FDA) announced on March 9, 1977, that it would revoke a regulation that allows the marketing of the artificial sweetener, saccharin, for use in food and beverages. However, the FDA was considering allowing the sale of saccharin as an over-the-counter (OTC) drug — one that is available without a prescription — and, in June, delayed its ban until the fall of 1977 to study new test results.

Weight-conscious Americans consumed more than 3.2 million kilograms (7 million pounds) of saccharin, the only artificial sweetener presently available in the United States, in 1976. Three-quarters of it was in food and beverages. The FDA removed the substance from the "Generally Recognized As Safe" list of food additives in 1972, and recommended maximum daily levels for average adult use.

The FDA ban is based chiefly on the results of a Canadian study, which demonstrated that rats fed high doses of saccharin developed cancerous bladder tumors. Because a clause in the Food, Drug, and Cosmetic Act prohibits the use in foods or beverages of any ingredient that has been shown to cause cancer in humans or animals, the new ruling was mandated by law.

The test rats that developed cancer had received saccharin as 5 per cent of their total diet. This is equivalent to an adult human drinking an average of a thousand 355-milliliter (12-ounce) diet soft drinks daily for a lifetime. The methods available for determining whether the much smaller amounts consumed in diet drinks and food really increase the risk of cancer in humans are complex and controversial. Some experts also question the application to humans of data derived from toxicity studies in only one species of animal. Although most scientists agree that saccharin has not been shown to harm humans in over 80 years of use, an unpublished study, in June, linked its use to bladder cancer in men.

FDA reviews. In its continuing effort to evaluate OTC drugs marketed in the United States, the FDA published the reports of two of its nongovernment expert panels. One study, published in September 1976, examined some 50,000 products used for colds, coughs, allergies, and asthma. Each of the 90 active ingredients found in these preparations was rated for safety and effectiveness when used alone

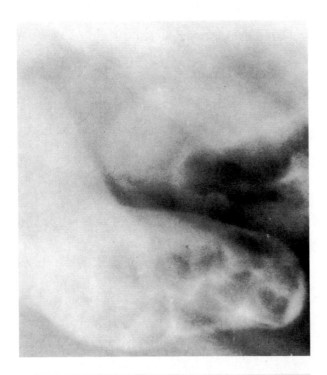

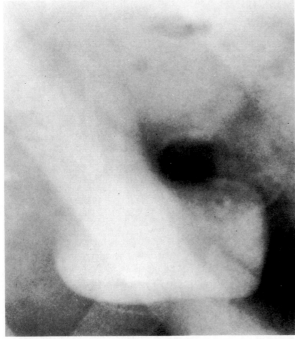

Dark spots in X ray, *top*, are gallstones in a patient's gall bladder. The stones are almost completely gone, *above*, after two years of treatment with a new drug, chenodeoxycholic acid, that dissolves the stones.

and in combination products. Although some of the drugs may relieve such symptoms as cough, runny nose, and congestion, the panel said that none of them could be shown to prevent, cure, or even shorten the duration of the common cold.

The experts also concluded that many drug combinations are unreasonable because some of the drug ingredients or amounts would be unnecessary or unsuitable for certain individuals. They called for stricter labeling and a ban on extravagant claims that are not scientifically supported. The panel also noted the need for detailed directions for the use of drugs in treating children.

The other report reviewed antibiotics used for skin wounds. The experts found that some of the antibiotics studied prevent harmful bacteria from getting into a wound, but there is not enough data to prove that these antibiotics prevent or cure an infection by slowing the growth of bacteria. Careful washing of skin wounds is still the best treatment; antibiotics should be used for no more than one week or when there are such signs of infection as redness, swelling, and pain.

New drugs introduced in the United States in 1977 included:
- Amikin (amikacin), an antibiotic that is effective against a number of disease-producing bacteria, especially certain strains that are resistant to other antibiotics. However, this compound may impair hearing and balance and cause kidney damage in some patients.
- Minipress (prazosin), a substance to reduce blood pressure in hypertensive patients. The drug dilates the blood vessels but apparently does not affect the nervous system, so it does not cause the side effects of some other commonly used antihypertensives. Prazosin can occasionally produce faintness with the first dose or after a sudden large increase in dosage.
- Imodium (loperamide), a long-acting antidiarrheal agent, acts directly on the intestinal wall. The intestine absorbs it poorly, thus preventing unwanted reactions elsewhere in the body. Chemically related to the opium-type narcotics, very high doses of loperamide can cause physical dependence in monkeys. This effect has not been seen in humans.   [Arthur H. Hayes, Jr.]

Ecology

The impact that lead pollution makes on plants continued to interest ecologists in 1976 and 1977. Many reports have pointed out that soils adjacent to roads contain increased levels of lead and other heavy metals. Extensive evidence indicates that plants growing in soils contaminated by mining and smelting activity develop metal tolerance. But until ecologists Lin Wu and Janis Antonovics of Duke University in Durham, N.C., reported on lead tolerance in the winter of 1976, there had been no studies on the metal tolerance of plants growing along roadsides.

Wu and Antonovics chose *Plantago lanceolota,* or ribwort plantain, and *Cynodon dactylon,* or Bermuda grass, for their study. Both species are widespread in North Carolina and are common along roadsides. The ecologists collected plants from sites close to a heavily traveled road in Durham and from sites farther from the road, where there was little or no lead pollution.

They transplanted five adult plants from each of the various sites to their laboratory and removed their side shoots. They grew the shoots for an additional three weeks and then removed all the mature leaves and roots from the stalk for a lead-tolerance test. They wanted to determine how well the leaves and roots would grow back in the presence of lead. From each set of cuttings, they chose two to grow in a standard nutrient solution and two to grow in a nutrient solution with 15.6 parts per million (ppm) of lead added.

After six weeks, the scientists harvested the plants, separated them into roots and shoots, and dried this material at 60°C (140°F.). Then they weighed each plant to determine how heavy, and therefore how healthy, it was. The greater the dry weight of a plant grown in the presence of lead, the greater its lead tolerance. The cuttings from adult samples of plantain taken from close to the road had a higher lead tolerance than those from sites farther away.

A number of previous studies showed that tolerance to heavy metals is a highly heritable characteristic. Wu and Antonovics investigated this phenomenon further. They grew plantain seeds

An ecologist works with automated air-sampling equipment that helps to determine how much carbon dioxide gas is released from the soil and leaves, twigs, and other debris lying on the forest floor.

Ecology
Continued

A species of aphid, when preyed upon by a larger nabid, secretes a sticky substance that repels the enemy and alerts nearby ants to attack the nabid.

collected from the roadside sites in a growth chamber for two weeks, then tested 36 seedlings from each set of seeds for lead tolerance.

For three days, the scientists measured the rate at which the longest root of the seedlings grew in nutrient solution. For an additional six days, they repeated the measurements when the plants were placed in nutrient solution containing 15.6 ppm of lead. The scientists estimated lead tolerance by measuring how well the roots grew in the presence of lead, with good growth indicating high lead tolerance. They found that seedlings grown from seeds collected from close to the roadside had a higher lead tolerance than those grown from seeds collected farther away. The data indicated that lead tolerance in the adult roadside plants is transmitted through their seeds.

The ecologists investigated lead tolerance in Bermuda grass using 10 adult plants taken from the roadside sites. They followed much the same procedure as they did with plantain.

Bermuda grass had generally a higher lead tolerance than plantain. However, Wu and Antonovics found no evidence that Bermuda grass near the road had a greater lead tolerance than samples taken farther away. While the 15.6-ppm lead solution inhibited root growth in plantain seedlings by from 55 to 91 per cent, the lead solution hindered Bermuda grass root growth only 17 to 20 per cent.

This study showed that the effects of lead pollution vary according to the species. It also provided evidence that lead tolerance evolves in plantain, but not in Bermuda grass. Since lead pollution at roadsides has only been present for about 40 years and pollution drops off rapidly away from roadsides, this study also illustrated how evolutionary change can be rapid and localized.

Bass and crayfish. In the summer of 1976, ecologists Roy A. Stein and John J. Magnuson of the University of Wisconsin in Madison reported on the influence of a predator fish, the smallmouth bass, on the crayfish *Orconectes propinquus*. They tested the response of various sizes of crayfish to smallmouth bass. They tried to determine whether crayfish respond to this predator by moving to an area of the lake or pond

bottom where it is easier to hide and by changing their behavior patterns; and if so, whether behavioral response differs with the size and sex of crayfish. The scientists also wanted to know if the response of the crayfish was correlated with how vulnerable to attack they sensed themselves to be.

Stein and Magnuson determined bottom preference and activity by placing some young crayfish in aquariums where half the bottom was covered with sand and the other half with gravel. They placed other young crayfish in tanks with half-gravel, half-pebble bottoms. Adult crayfish were tested in half sand and half pebbles.

The ecologists measured, fed, and placed five males and five females into each tank 24 hours before they began observing them. The scientists made their observations daily at 4 P.M. for six days. On the fourth day, they placed a well-fed smallmouth bass in three of six tanks containing young crayfish and in two of four tanks containing adults. They ran each experiment twice, with the same bass but different crayfish.

When the fish appeared, young crayfish and small adults moved from sand to pebbles, where they could hide more easily. Given a choice between gravel and pebble, young crayfish chose the larger gravel when the fish was present. However, large adults did not try to hide or alter their behavior when the fish was in the tank. Small crayfish suppressed behavior that might attract the attention of the fish, such as walking, climbing, and feeding. Males were more active and less intimidated than females by the fish.

The scientists also made field observations that supported these laboratory findings. Every two weeks throughout the summer, they marked off twelve 0.5- square-meter (5.4-square-foot) areas on a lake bottom and dived to examine the bottom material, the type and location of crayfish, and any predators that were present.

Predatory fish influence the behavior of the bottom-feeding crayfish, and therefore are a key factor in maintaining the balance of lake and pond ecosystems. When young or small crayfish fear attack by bass, they try to hide in areas covered by gravel or pebbles and stop eating temporarily. This helps to

269

Ecology

Continued

Using a protective glove box, a biologist at the Oak Ridge National Laboratory tends to plants growing in soil that is contaminated by plutonium. The tests showed the plants take up little of the radioactive element.

reduce the feeding pressure that crayfish may exert on smaller animals in aquatic food chains.

Energy and the average lawn. Ecologist John H. Falk of the University of California, Berkeley, reported in the winter of 1976 on a study of the various forms of energy used to maintain a suburban lawn ecosystem. An increasing percentage of the land in the United States is being converted into lawns and turfs – an estimated 6 million hectares (16.5 million acres) as of 1977. Maintaining such a large area requires a great deal of human energy.

Falk set out to develop energy-flow equations for the output and input of this system, including the activities of humans as managers and experimenters. He calculated all measurements in kilocalories per square meter per year to measure quantitatively the diverse roles of humans, animals, and grasses within the lawn system, other managed ecosystems, and natural grasslands.

Falk conducted his study on a lawn planted in 1964 in Walnut Creek, a suburb of Oakland, Calif. The lawn

covered an area of about 110 square meters (1,184 square feet) and contained a mixture of Kentucky bluegrass, creeping red fescue, and chewings fescue. Three trees – an American elm, a palmlike tree, and a Chinese toyoh – also grew in the yard. The trees intercepted solar radiation, dropped leaves, twigs, and other debris on the lawn, and absorbed nutrients and water from the soil.

This lawn was mowed with a gasoline-powered rotary lawn mower that left stubble 4 centimeters (1.6 inches) high. The grass was cut either weekly or once every two weeks, except during winter when it was mowed once a month. Prior to mowing, Falk raked and weighed all litter, including weeds, leaves, twigs, and old grass clippings. After mowing, he weighed the fresh clippings, oven-dried them at 98°C (208.4°F.) for 24 hours to remove moisture, and then reweighed them. The lawn was fertilized twice during the study, once with an all-purpose mixture that contained herbicides and insecticides, and once with manure.

Ecology
Continued

Twice a month, Falk spent the entire day observing all vertebrates — primarily birds — on the lawn. He began these observations 30 minutes before sunrise and ended them 30 minutes after sunset. He recorded the time and nature of every interaction between vertebrates and the lawn. He also recorded all human management and experiments with the lawn, such as mowing, including the date, time, and type of activity. He measured and recorded all energy sources introduced by humans, such as water and fertilizer.

Falk's measurements, observations, and calculations showed that the lawn system was extremely productive. The net annual amount of vegetation produced by lawns was 1,020 grams per square meter, slightly less than the 1,066 grams per square meter that ecologists have measured for cultivated cornfields, but greater than the 1,000 grams per square meter produced by uncultivated prairies.

Energy inputs for making the lawn grow totaled 205,000 kilocalories per year, of which providing water for irrigation and gasoline for mowing represented the largest share. The vegetation "harvested" from the lawn was almost 63,000 grams (2,222 ounces) dry weight from an energy expenditure of 271,000 kilocalories per year in mowing and raking. This net drain of 63,000 grams of clippings that could have decayed to provide nutrients was largely compensated for by adding 14,223 grams (502 ounces) of fertilizer.

In terms of grassland productivity, the food the lawn provided for suburban birds considerably exceeded that provided by natural grassland for rural birds. Lawns are ideal foraging sites for flocks of birds that feed in open areas. But in terms of energy, humans were clearly dominant in this lawn ecosystem. Energy inputs in labor, gasoline, and fertilizer amounted to 576 kilocalories per square meter per year, which equals or exceeds the energy needed to produce corn. Therefore, the study concluded that for roughly the same effort expended on suburban lawns, homeowners could just as easily raise vegetables.   [Stanley I. Auerbach]

Electronics

The microprocessor revolution continued to dominate electronics technology and the consumer market as more complex and powerful versions of these tiny computers became available in 1977. Microprocessors had required several component-laden chips working in sets, or families, to form a complete computer. This year, however, marked the debut of the single-chip microprocessor, a device that contained all the logic, controls, and memory circuits required to perform as a full-sized digital computer.

Intel Corporation of Santa Clara, Calif., the pioneer in microprocessor development, was the first to introduce a one-chip microcomputer, the 8048. This is an 8-bit system which contains a central processing unit, program memory, data memory, connections between input and output signals, and clocks and timers, all on a silicon chip that is only a little more than 0.51 centimeter (0.2 inch) on a side. Texas Instruments Incorporated introduced the extremely powerful TMS 9940, a 16-bit, one-chip microcomputer.

One popular commercial application of the microprocessor is in video games. It permits the increasingly popular games simulated on a television set to be programmed. Programmable games not only offer much more variety and complexity, but also they can be changed by inserting a prerecorded tape cassette containing a program for a game. Furthermore, the microprocessor permits more interaction between player and TV screen. For example, in the early video tennis games, the players could control only vertical motion of the paddles. Now it is also possible to move the "players" toward and away from the "net."

The first programmable game was introduced in mid-1976 by Fairchild Semiconductor's Consumer Products Group. It uses a 4-chip Fairchild microprocessor, called the F-8, with an additional chip for the memory. The model consists of a console for the top of the TV set and attached joy sticks, with which the participants control the play.

The game can be changed by digitally encoded tape cartridges called

Electronics

Videocarts, each of which offers 10 games and four mathematical quizzes. One of the first cartridges allows the viewer to draw in three colors on the screen. In addition, there is a "doodling" game in which the processor forms random, kaleidoscopic patterns on the screen.

Communication by light moved closer to application in 1977. The light beams are transmitted over optical fibers—tiny filaments of specially compounded glass. Ever since the first low-energy-loss fibers were developed in 1970, communications engineers visualized replacing bulky and costly copper cables now being used to transmit most of the world's telephone traffic with cables made of optical fibers. The fibers provide large bandwidths, which means many more telephone circuits per cable, and are lighter in weight and easier to install.

Not surprisingly, various divisions of the Bell System in the United States are among the most active in optical fiber research. During the past year, field trials were conducted at the Western Electric Company cable works near Atlanta, Ga., with cable containing 144 individual fibers. Losses encountered on the 10.9-kilometer-long line were only 6 decibels per kilometer. Signals were transmitted over the fibers at a rate of 44.7 million pulses per second, which is equivalent to more than 48,000 two-way voice circuits.

In April, the American Telephone and Telegraph Co., Bell Laboratories, Western Electric Co., and Illinois Bell began testing two optical fiber transmission systems beneath the streets of Chicago. The links will carry voice, data, and video signals over 24-fiber cables. See COMMUNICATIONS.

One of the most ambitious optical communications projects is a 140-megabit system, claimed to be the world's first high-capacity optical telephone system. Standard Telephones and Cables, a division of ITT Corporation, is installing it in suburbs north of London for a two-year field trial. The 9-kilometer (5.5-mile) system will carry up to 920 simultaneous telephone conversations or two color-TV signals.

Bowling scores are displayed automatically with a new electronic pin-detector and scoring system. Microprocessor uses reflected sound signals to spot pins, activate the pin setter, and calculate the score.

A Fingerscan system, *above*, identifies an employee who wants to enter a restricted area. In the Fingermatch system, *right*, a light pen records fingerprint data for computer coding and matching with law-enforcement files. Both systems provide fast and accurate electronic identification.

Electronics

Continued

Medical electronics technology continued to make important contributions to medicine during 1976 and 1977. In West Germany, for example, Dr. Hermann D. Funke of Bonn University's Surgical Clinic developed a cardiac pacemaker that can meet a patient's immediate needs.

Most conventional pacemakers furnish electrical pulses to a patient's heart at a fixed rate that cannot be altered when physical exertion calls for a faster heartbeat. To overcome this, Dr. Funke developed a pacemaker that is controlled by the patient's respiration rate. Through electronic circuitry, it maintains a little more than a 4 to 1 ratio between pulse frequency and the patient's respiratory rate. This means the device operates between about 60 pulses a minute, when the patient is breathing slowly at 16 times a minute, and 146 pulses a minute, when breathing is more than twice as fast.

Surgeons implant a ceramic transducer in the pleura, or lining, of the lung, where pressure varies with inhalation and exhalation. The pacemaker amplifies and otherwise processes the signals that control the pulse frequency.

The year also saw the introduction of more experimental devices for providing electronic sight to the blind. One device, the Opticron IV, was developed by Zaid Diaz, of CID Corporation in Alto Viejo, Puerto Rico. It creates images in the brain through a matrix of electrodes that is pressed against the skin of the blind patient's back. With suitable training, the patient perceives the information provided by the conversion of electronic signals to tactile sensations as a kind of sight.

The images to be "seen" are picked up by a lens worn on an eyeglass frame and transmitted over a fiber-optic cable to a vidicon amplifier tube. The tube and other electronics are housed in a portable pack about the size of a copy of *Science Year* that hangs from a belt at the waist. Signals from the vidicon are fed to the matrix, which has 3,600 electrodes mounted on the inner surface of a cloth vest which the blind person wears so that the electrodes press against his back.

Electronics

Continued

A light pen registers information from a bar-encoded label on a library book. This and similar data from the borrower's library card go to a computer for speedy checkout.

A control device that can continuously inject insulin into a diabetic's bloodstream at very precise and extremely minute rates was developed by Dr. Manfred Franetzki and his colleagues at the Siemens AG medical facility in Erlangen, West Germany. Called the Micro-Dosage System, it consists of an electronic controller and an insulin-storage device. The insulin-dosage rate is set in the electronic controller according to the patient's needs. The controller operates a miniaturized pump in the storage unit, which is about the size of a cigarette pack and is powered by a mercury-oxide battery designed to last about two years. The insulin is injected into a vein in the patient's arm through a catheter.

Free-electron laser. A high-power, tunable laser was announced by physicist John M. J. Madey and his colleagues at Stanford University in Stanford, Calif., in April 1977. In contrast to conventional lasers that stimulate coherent light from a material, the new device fires high-energy electrons into a magnetic field, producing a coherent monochromatic beam whose wavelength can be tuned by varying the initial energy of the electrons.

A 43 million electron volt (Mev) beam generated by the university's linear accelerator produced a light beam of 3.417 microns (millionths of a meter), but theoretically tunable from 10 microns to 0.1 micron. It reached a peak power of 7,000 watts in short bursts, and according to Madey is capable of generating 100 kilowatts from a Mev input of 240 Mev.

Solar cells. In May 1977, scientists at International Business Machines Corp. announced the achievement of a major increase in the efficiency of solar cells — devices that convert sunlight into electricity. The cells are formed from gallium arsenide coated with a thin layer of gallium aluminum arsenide. They convert 22 per cent of the light impinging on them to electricity, close to the theoretical maximum of 27 per cent. Conventional cells of the highest quality, normally made from silicon, now operate at only 18 per cent efficiency.   [Samuel Weber]

Energy

A Department of Energy, proposed by President James Earl Carter, Jr., was being considered by Congress as of mid-1977. Both houses had passed the proposal; it awaited final approval.

In April 1977, Carter announced that he would discourage the use of plutonium as a nuclear fuel on the grounds that plutonium's potential for making nuclear weapons posed too great a threat. Plutonium is an essential ingredient in atomic bombs. Carter's stance caused repercussions among other nations dependent upon importing plutonium, some from the United States, to power their nuclear reactors. Meanwhile, Great Britain, France, and Japan continued to move rapidly toward the plutonium-producing breeder reactors. See SCIENCE POLICY.

The new plutonium policy removed the Liquid Metal Fast Breeder Reactor (LMFBR) development project from its long-held position of supremacy in the U.S. research effort. It also set off a scramble within the nuclear industry to step up investigation of new fuels and fuel-processing techniques.

From the beginning of the atomic energy era in the early 1950s, the breeder reactor and related technologies for reprocessing spent nuclear fuel were in the forefront of nuclear research, because the earth contains a limited amount of easily recoverable uranium. The breeder's ability to produce more fissionable fuel, in the form of plutonium, than it consumed held the promise of using uranium resources to their fullest extent.

In ordinary nuclear reactors, the uranium isotope U-235 absorbs a low-energy neutron and the uranium atom splits, or fissions. As a result, these atoms release high-energy neutrons that can split more atoms, causing a chain reaction. But these high-energy neutrons must be slowed down by some material, such as water, so that they can be captured easily by the fissionable isotope U-235, rather than be absorbed by another isotope, the nonfissionable U-238. However, in a breeder, some neutrons released by the fission of U-235 are intended for capture by U-238 to produce U-239, which then

In a new burner system developed at the General Electric Co., small tubes and a large pipe collect and carry oxygen-enriched air from containers holding ultrathin membranes that allow more oxygen than nitrogen to pass through them. Since flames burn hotter in oxygen-enriched air, this system could cut fuel consumption by up to 50 per cent for some industrial processes.

decays into highly fissionable plutonium which can be separated from the spent fuel. Plutonium can also be recovered on a more limited basis from spent fuel in ordinary reactors.

However, one of the first efforts to take the weapons threat into account in designing nuclear power systems had no connection with the breeder reactor. Instead, it involved ordinary nuclear power plants. The U.S. Arms Control and Disarmament Agency released a study in June 1976, proposing a novel, tandem fuel cycle as an alternative to separating plutonium from spent fuel. In a tandem cycle, uranium would first be "burned" in conventional light-water (ordinary water) reactors. Then the spent fuel, still rich in uranium and plutonium, would be used to power a heavy-water reactor.

In heavy water, the heavy hydrogen isotope, deuterium, replaces the ordinary hydrogen atom. Heavy water is better than light water at absorbing and slowing down neutrons to efficiently promote the chain reaction. Therefore, fewer neutrons are needed to continue the reaction. Because of this greater efficiency in absorbing neutrons, it is easier to fine-tune the rate of burn in heavy-water reactors and conserve uranium.

Burning the same fuel in first light-water then heavy-water reactors would extract nearly twice as much energy from the original uranium while avoiding the nuclear weapons potential created when pure plutonium is separated from the spent fuel.

Since this initial study, university scientists, government laboratories, and federal agencies have advanced dozens of other proposals for new fuels and processing techniques and even for new types of reactors. Many of these proposals are nuclear schemes discarded in the 1950s, including ways of modifying reprocessing techniques so that plutonium is mixed with uranium and never separated in a pure form, and proposals to replace part of the light-water coolant in existing U.S. reactors with the more efficient heavy water.

Scientists also revived ideas for alternate reactor concepts, such as a high-temperature, gas-cooled model that uses a combined thorium-uranium fuel. Another is the homogeneous reactor that produces its own fuel from raw uranium or thorium. The fuel is produced only within the reactor, and therefore cannot be retrieved easily for building bombs.

Coal research. Progress toward finding new ways to use the vast U.S. coal reserves continues to be slow. In October, the Energy Research and Development Administration (ERDA) signed contracts with a consortium of oil companies to build an experimental coal liquefaction plant at Catlettsburg, Ky. The plant is designed to convert 544 metric tons (600 short tons) of coal per day into a low-sulfur fuel oil. Using a catalytic technique known as the H-coal process, finely ground coal, recycled oil, and hydrogen undergo chemical reactions in the presence of a catalyst to produce new oil, while removing sulfur, ash, and other impurities. The $180-million plant will be the largest coal-conversion facility yet built in the United States.

The technical aspects of the process are rated highly by most coal scientists, but the economic prospects are more uncertain. The synthetic oil produced by this method will probably cost nearly twice the current $13-per-barrel price of imported petroleum.

In December, ERDA in effect canceled another experimental plant that was designed to produce both synthetic gas and liquids from coal. The project, known as Coalcon, was begun in haste by the Department of the Interior after the Arab oil embargo of late 1973 and early 1974. The process involved several features that had not been adequately tested, and cost estimates rose steadily during design. As a result, several members of the industrial consortium backing the project decided not to invest money to construct the plant.

The Coalcon failure cast doubt on the technical adequacy of some other coal-research projects, particularly those designed for gasification of coal. Several such projects were launched in 1976, but experienced observers pointed out that many of them may turn out to be failures because of inadequate planning and testing.

If progress in synthetic-fuel research was mixed, the performance in coal-combustion research was better. Experiments with a new technique for burn-

Rows of heliostats, or frames holding mirrors to focus the sun's rays, at New Mexico's Sandia Laboratories are part of the world's largest solar-thermal test facility. The mirrors reflect sunlight to heat water in a boiler. This drives a steam turbine that can produce 1.7 megawatts of electricity.

An experimental wind turbine being tested in Ohio has blades that measure 18.8 meters (62.5 feet). When they turn in a wind blowing at more than 12.8 kilometers (8 miles) per hour, the turbine generates up to 100 kilowatts of electricity, enough to meet the needs of about 30 average homes.

Energy

Continued

ing coal cleanly, known as a fluidized bed process, began in Rivesville, W. Va., in November. In this type of process, coal and limestone particles are suspended in a stream of air so that the coal will burn more completely. The limestone also reacts with sulfur dioxide released during burning, capturing and chemically converting the gas into a dry material, calcium sulfate, before it can escape and pollute the air.

Renewable energy resources. In September 1976, the Brazilian government approved funding for a number of facilities to distill alcohol from sugar cane. Brazil is trying to develop energy from its immense resources of biological materials by literally growing its own fuel. The sugar or sugar-containing materials in the plants are distilled to produce ethyl alcohol. Brazilian scientists find that ethyl alcohol is a nearly ideal automobile fuel that burns cleaner than gasoline. Brazilian government researchers report that they have improved agricultural techniques and distillery designs and have successfully tested cars that operate on pure alcohol or alcohol-gasoline blends. Brazilian scientists also plan to experiment with cassava, or manioc, a root crop widely produced in the country.

The huge amount of unused land in Brazil, its abundant water supplies, and its year-round growing season make the plan feasible. More than 70 new distilleries and about 500,000 hectares (1.2-million acres) of new cultivations have been proposed.

Brazil imports 700,000 barrels of oil a day. It is cheaper for Brazil to produce alcohol than to make gasoline from its imported oil. Therefore, the country hopes that alcohol will replace 20 per cent of its gasoline consumption by the early 1980s.

Solar energy. Sandia Laboratories solar-energy researchers began experiments with a small power tower, a steam-generating boiler that is heated by reflected sunlight. The 61-meter (200-foot)-high tower is surrounded by a field of about 300 mirrors. Water heated in the boiler atop the tower produces steam which is used to drive a turbine.   [Allen L. Hammond]

Environment

Human health was threatened on a number of occasions by exposure to man-made chemicals in 1976 and 1977. The most dramatic incident involved the accidental discharge of from 1 to 5 kilograms (2.2 to 11 pounds) of the deadly 2,3,7,8-tetrachlorodibenzo-p-dioxin (TCDD) from a chemical plant in Seveso, a suburb of Milan, Italy. A safety valve burst there when a vat overheated on July 10, 1976, releasing a cloud of toxic fumes over an area of about 132 hectares (325 acres).

The town was evacuated, 50 residents were hospitalized, and more than 1,000 animals died. The most heavily affected area may not be safe to live in for more than 14 years.

TCDD is often found as a contaminant in 2,4,5-T, a pesticide that United States forces used to defoliate large jungle areas in Vietnam during the war there. It can cause kidney, liver, and lung damage. Studies by the National Academy of Sciences on its use in Vietnam revealed that it degrades slowly in the soil. The U.S. Air Force received Environmental Protection Agency (EPA) permission in April 1977 to dispose of nearly 10 million liters (2.6-million gallons) of Orange Herbicide, which contains TCDD, by burning it in an isolated area in the Pacific Ocean.

The U.S. Forest Service was prohibited on March 7, 1977, from spraying 2,4,5-T in national forests in Oregon and Washington to control shrubs—particularly the alder—which compete with Douglas fir. The project was canceled because the environmental-impact statement covering the spraying did not adequately consider the toxic effects of TCDD in the herbicide.

On April 11, 1977, the EPA effectively banned all pesticides containing Kepone by withdrawing its approval for their sale. The chemical, which had been manufactured in Virginia under a license from Allied Chemical Corporation, caused neurological disorders in workers exposed to it. The Kepone plant was closed in 1975, and legal action was begun by Virginia.

Early in 1977, Allied Chemical established an $8 million environmental fund for Virginia, thereby reducing the

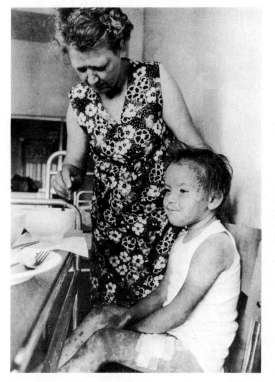

A 4-year-old resident of Seveso, Italy, suffered a skin disorder after a cloud of dioxin, a poisonous gas, was released on the town in a plant explosion, *left*. Hundreds of people were affected by the highly toxic chemical. It contaminated the town so badly that Italian authorities ordered parts of it and surrounding areas evacuated and posted warnings, *above*.

Environment

Continued

fine levied on it for the Kepone incident to $5 million.

Troubled river. Discharges of chlorinated hydrocarbons in the Ohio River in early 1977 caused concern to communities that use water from the river. In a court suit, Evansville, Ind., accused the FMC Corporation of discharging more than 60,000 kilograms (132,000 pounds) of carbon tetrachloride from its South Charleston, W. Va., plant into the Kanawha River, an Ohio River tributary. In March, the plant was also charged by the EPA with violating an agreement to reduce its carbon tetrachloride discharges into the river by 80 per cent.

Also in March, workers in the Louisville, Ky., sewage-treatment plant were sickened by fumes of hexachlorocyclopentadiene, a toxic chlorinated hydrocarbon used in pesticides and flame retardants. The municipal sewage-treatment plant had to be shut down and 400 million liters (106 million gallons) of raw sewage was dumped into the Ohio River each day until the system could be decontaminated.

Cancer compounds. Biochemists Arlene Blum and Bruce N. Ames of the University of California, Berkeley, warned in January 1977 of the possible threat of cancer from the use of Tris (tris [2,3-dibromopropyl] phosphate), commonly used as a flame retardant in children's sleepwear. The researchers based their warning on evidence that Tris is a mutagen – an agent that causes genetic changes – and there is a high correlation between mutagenicity and the development of cancer. The Consumer Product Safety Commission (CPSC) banned its use in children's wear on April 7, 1977. Ironically, the extensive use of Tris was a direct result of a CPSC standard set in 1972 to make fabrics used in children's clothing less flammable.

The Food and Drug Administration (FDA) on March 9, 1977, proposed a ban on saccharin, the only synthetic food sweetener still available to the U.S. public. The FDA based its proposal on findings from a Canadian government study that large amounts of saccharin produce bladder tumors in

When the tanker *Argo Merchant* ran aground and broke up
in heavy seas off the New England coast, *opposite page,*
its cargo of oil leaked into the North Atlantic Ocean,
left. In the North Sea, workers sprayed an oil platform
with water to prevent fire until a leaky pipe was capped.

Environment

Continued

mice. The FDA move met with strong public and congressional opposition. See DRUGS.

Ozone threat. A National Research Council (NRC) report released in September 1976 confirmed that chlorofluorocarbons used as propellants for aerosols constitute an environmental threat. The NRC concluded that continued use at the 1973 rate would eventually reduce stratospheric ozone by from 2 to 20 per cent. The ozone layer blocks some of the sun's ultraviolet radiation from reaching the earth. Excessive amounts of this radiation can cause skin cancer in man and damage plant life.

The CPSC, the FDA, and the EPA all announced in November 1976 that they will restrict the use of chlorofluorocarbons. Oregon banned their use as propellants on March 1, 1977.

The long-awaited Toxic Substances Control Act was passed by the U.S. Congress in October 1976. The act gives the EPA authority to require testing on any chemical for which inadequate information exists to establish its safe use, and to restrict or ban the use of chemicals that pose an "unreasonable risk." Chlorofluorocarbons will be among the first chemicals restricted by the EPA.

Nuclear safety questioned. The 1975 reactor safety study conducted by nuclear engineer Norman C. Rasmussen of Massachusetts Institute of Technology for the Nuclear Regulatory Commission has been subjected to increasing criticism. EPA representatives reported at a congressional hearing on June 11, 1976, that a reactor accident could cause up to 10 times the number of deaths estimated in the study. And a study sponsored by the Ford Foundation and conducted by a panel headed by Spurgeon M. Keeney of MITRE Corporation, an engineering consulting firm, found in March 1977 that the Rasmussen study may have underestimated the hazard of a reactor accident by as much as 500 times. Nonetheless, the Ford-MITRE study concluded that this risk is still lower than that presented by coal-fired generating plants.

A far greater danger, according to the Ford-MITRE study, is that the

Will Spray Keep Smog Away?

Passive measures against air pollution, such as automobile-exhaust controls, have not proved highly successful in controlling the severe photochemical smog that plagues Los Angeles and other cities. As a result, some scientists have suggested more aggressive measures to combat smog. For example, chemist Julian Heicklen of Pennsylvania State University in State College, Pa., believes it is possible to release into the air chemicals that will keep photochemical smog from forming.

Photochemical smog is formed primarily by a chain reaction between two components of automobile exhaust — unburned gasoline vapor and nitric oxide, a by-product of high-temperature combustion. In the chain reaction, certain wavelengths of sunlight convert some gasoline molecules into high energy forms known as free radicals. The free radicals react with other chemicals in the air to produce ozone and other photochemical oxidants, the most toxic and destructive components of smog. In the process, more free radicals are produced and the reaction continues.

A very small number of free radicals can thus cause a large amount of photochemical oxidants to form. But if the free radicals could be removed from the atmosphere, the chain would be broken and no smog would form. Heicklen thinks that cities should use chemicals known as free radical scavengers to intercept and react with the radicals. He believes the chemical diethylhydroxylamine (DEHA) is a good candidate for such a job.

DEHA is a colorless, volatile liquid that has no odor when it is present in low concentrations. Heicklen proposes releasing DEHA into the air over a city on mornings when atmospheric conditions indicate that pollutants will be trapped over the city. DEHA could be sprayed from helicopters or airplanes or released from trucks driving over heavily traveled highways during the morning rush hour when the important wavelengths of the sun's radiation are most intense.

Heicklen and three scientists at the Hebrew University in Jerusalem, Israel, hope to test the concept in Jerusalem. They chose that city because virtually all the photochemical smog there results from automobile traffic, weather conditions in the summer are highly reproducible, and it is a small target. Nearly 80 per cent of the traffic in Jerusalem occurs within an area of 9 square kilometers (3 square miles).

Israeli officials seem inclined to grant permission for such a test. Before they do so, however, they want assurance that the chemical is safe — that it will not create a problem more severe than the one it is meant to alleviate. As a result, Heicklen is testing DEHA to determine if it produces any ill effects in animals. So far, it has not.

But toxicologist Marvin Legator, who is at the University of Texas Medical Branch in Galveston, has found that urine from rats exposed to relatively high concentrations of DEHA contained trace amounts of a chemical that produces genetic changes in a sensitive strain of bacteria. This suggests that DEHA might be converted into a more hazardous form by the animal's metabolism. However, Legator has not observed the mutagenic chemical in urine from humans exposed to DEHA.

Still, most scientists strongly oppose Heicklen's proposal. Critics such as chemist James N. Pitts, Jr., of the University of California, Riverside, argue that it is too dangerous to release a chemical into the air without a complete knowledge of its effects on humans. They also argue that Heicklen's laboratory experiments may not provide an accurate indication of what will happen in the atmosphere.

Pitts suggests that trapping free radicals might simply delay the formation of smog until the pollutants have moved downwind of the treated area. Other critics argue that Heicklen's scheme would represent an infringement on individual rights.

If the question of genetic damage is resolved satisfactorily, it will probably still be necessary to test the chemical for at least three more years, to see if it is poisonous or might cause cancer, before the atmospheric tests could begin. If DEHA is found to be a hazard, then Heicklen must find another chemical. He already has a candidate, a derivative of DEHA that would probably not be metabolized. But testing it would delay the project so long that the scheme would probably be shelved instead.   [Thomas H. Maugh II]

Environment

Continued

plutonium that would result from nuclear fuel reprocessing could be used to make bombs. Such reprocessing is currently blocked by a Natural Resources Defense Council suit. President James Earl Carter, Jr.'s national energy plan, announced in April 1977, followed the Ford-MITRE study in recommending that nuclear fuel reprocessing be discontinued, and seemingly doomed the development of the Liquid Metal Fast Breeder Reactor, which would have created large amounts of plutonium. See ENERGY.

Oil on troubled waters. The *Argo Merchant,* a Liberian tanker, ran aground and spilled its cargo of 29-million liters (7.7 million gallons) of oil into the Atlantic Ocean off Nantucket Island in December 1976. And a blowout on a North Sea oil rig dumped more than 30 million liters (7.9 million gallons) of oil before it could be contained on April 30, 1977. Favorable weather conditions prevented the *Argo Merchant* spill from reaching land, and also saved the Norwegian coast from the North Sea spill. However, the *Argo*

Merchant oil seriously damaged the plankton, microscopic sea life that is an essential part of the food chain, in the George's Bank fishing area off New England. These two massive oil spills re-emphasized the environmental hazards of increased petroleum development and transport.

Endangered species. The Endangered Species Act of 1973 was applied to block the construction of the Tellico Dam in 1976 and 1977 because the dam threatened the existence of the snail darter, a 7-centimeter (3-inch) fish that lives in the Little Tennessee River. The act prohibits spending federal money on projects that would threaten the habitat of any plant or animal species on the endangered species list. The law may also prevent the construction of the controversial Dickey-Lincoln Dam in Maine, because the Furbish lousewort, a recently rediscovered species of wild snapdragon, might be lost. Construction of the La Farge Dam in Wisconsin may also be prevented to protect the monkshood, a rare poisonous wild flower.   [Harold R. Ward]

Genetics

Geneticists, psychologists, and statisticians provided new insights in 1976 and 1977 on how genes work and how genetic makeup affects behavior.

Two genes in one. In February 1977, molecular biologist Frederick Sanger's research group at the Medical Research Council Laboratories in Cambridge, England, published the complete sequence of nucleotides in the genetic material, deoxyribonucleic acid (DNA), of the tiny bacterial virus Phi X174. A nucleotide is a deoxyribose sugar hooked to any one of the four bases – adenine (A), guanine (G), thymine (T), or cytosine (C). The sequence of nucleotides determines, or codes for, the sequence of amino acids in proteins. The DNA in Phi X174 codes for the making of only nine proteins, but it does so with fewer nucleotides than are normally needed.

Ordinarily, each amino acid in a protein is specified by one DNA code word consisting of three adjacent nucleotides, a triplet. Adjacent amino acids are coded by adjacent, nonoverlapping nucleotide triplets. For exam-

ple, a protein of 100 amino acids would require a DNA 300 nucleotides long to code for it. Were Phi X174 to follow this rule, more than 6,000 nucleotides would be required to code for the approximately 2,000 amino acids in its nine proteins. Yet the DNA in Phi X174 has only 5,375 nucleotides.

Biologists Bart G. Barrell, Gillian M. Air, and Clyde V. Hutchison III of Sanger's group found part of the secret of Phi X174's coding efficiency. In two cases, one nucleotide sequence codes for two proteins. One case involves two proteins known as D and E, which have different amino acid sequences. All of the nucleotides coding for E are included in those coding for D. The amino acid sequence of E is not similar to any portion of D, the researchers discovered, because the nucleotide triplets are read slightly out of phase to make each protein. See BIOCHEMISTRY.

One gene in two. Immunologists Nobumichi Hozumi and Susumu Tonegawa of the Basel Institute for Immunology in Switzerland showed in October 1976 that certain genes are

Genetics

Continued

spliced together to produce antibodies. Antibodies are proteins that combine with foreign toxins or disease-causing organisms in the body and mark them for destruction by scavenger cells.

Although the body can produce thousands of types of antibodies, each antibody-producing cell can make only one type. Hozumi and Tonegawa studied antibody-producing cells from a mouse bone-marrow tumor because all the cells in such a tumor produce the same type of antibody. The DNA in these cells is first used as a template, or pattern, to align a specific sequence of nucleotides in a molecule of ribonucleic acid (RNA). This RNA, called messenger RNA, actually participates in making the protein.

In their study, Hozumi and Tonegawa made use of the fact that a molecule of RNA will stick to the DNA segment that serves as its template. The researchers ground up the mouse tumor cells, purified the messenger RNA for the antibody, purified the DNA, and cut the DNA into gene-sized fragments. They mixed the DNA fragments with the messenger RNA, which stuck to only one kind of DNA fragment. Therefore, the messenger RNA is produced from a single segment of DNA – a single gene – in the cells.

But when they mixed the antibody messenger RNA with DNA fragments purified from ground-up mouse embryo cells, each half of the messenger RNA stuck to a different piece of DNA. Therefore, the genes for the two halves of the messenger RNA in mouse embryo cells are not fused together. Hozumi and Tonegawa concluded that the separate genes are somehow spliced together into a single gene to create an antibody gene during the development of the embryo.

XYYs and crime. A widely held assumption linking a particular chromosome makeup to aggressive criminal behavior was greatly weakened by a study published in August 1976. The work, by a team of six American and six Danish researchers headed by psychologist Herman A. Witkin of the Educational Testing Service in Princeton, N.J., was carried out in Denmark be-

Drawing by Lorenz; © 1976 The New Yorker Magazine, Inc.

"I have the weirdest feeling that someone was fiddling with my genes during the night."

Viruses in Evolution

Viruses are usually thought of as infectious agents that produce diseases ranging from the common cold to cancer. But recent research indicates that some viruses may play an important role in the evolution of life because they can transmit genetic information from one animal to another, even from one species to another. In evolutionary terms, this natural information-transfer system of viruses may be more important than the fact that they can sometimes cause cancer.

The viruses in question are the ribonucleic acid (RNA) tumor viruses, which can cause diseases such as leukemia and breast cancer in mice. They are members of a larger family called retroviruses. Some retroviruses produce cancers that are fatal in a few weeks; others have more subtle effects; some appear to have no effect at all.

Genetic studies with chickens and mice, as well as studies of mouse cell cultures showing that these viruses are sometimes released from normal mouse cells, led to the theory that the viruses, or more precisely the genes from which they can be reproduced, have survived by being transmitted from generation to generation in the chromosomes of the host animal rather than by infection and multiplication within the hosts. Furthermore, these genes can escape from their original location and insert themselves in other parts of the cell, the other cells of the same animal, the cells of other animals of the same species, and even in those of genetically distant species.

Classical evolutionary theory says that species evolve through point mutations, small changes in the deoxyribonucleic acid (DNA) of individual genes that take place over long periods of time. The ability of viral genes to carry genes from one species to another might allow much greater evolutionary change in a much shorter time.

RNA tumor viruses are the only family of viruses known to transfer genes naturally from one species to another. They can integrate into cellular DNA, the chemical that carries genetic information in most forms of life, and re-emerge with some of the host's genes. For example, chicken viruses in duck cells pick up duck DNA that the viruses can then transmit to chickens.

Theoretically, RNA tumor viruses represent an efficient information-transfer system because they can recombine with cell DNA and are often so benign that they do not affect normal differentiation and development. In most cases, the transfer of genes between species will be unimportant to the recipient. But in the rare cases that are beneficial — by conferring resistance to disease, for example — the host animal may gain a significant advantage.

Although it is not yet known whether these viral genes have actually affected evolution, they do provide scientists with a new tool to study evolution. Studies of viruses from baboons, mice, and chickens indicate that species closely related to the one from which a retrovirus is isolated should also have closely related viral gene sequences. The viruses, then, can be used to estimate the evolutionary distance between species.

Scientists do this by comparing the viral RNA to cellular DNAs of the species from which the virus was isolated and to the DNAs of evolutionarily related species. Where the viral RNA is similar to a section of the cellular DNA in the species from which it was isolated, we know that the species is carrying that virus' genetic material in its own DNA. We can then compare the viral RNA with cell DNA from related species to see how closely they are related. The more similar the viral RNA is to parts of the DNA of the related species, the closer the relationship.

One such study showed that all Old World monkeys and apes have cellular DNA sequences related to the baboon virus RNA. Those primates that evolved in Africa, such as colobus monkeys and gorillas, have DNA sequences much closer to the baboon virus RNA than do their Asian counterparts — rhesus monkeys and orang-utans.

Human DNA from all races matches the baboon virus RNA roughly as closely as does that of the Asian apes and is quite different from African apes, suggesting that much of human evolution occurred outside Africa. Paleontology gives some support to this, because fossils of *Ramapithecus*, a presumed ancestor of man, have been found in Hungary, Turkey, Pakistan, and China.   [George J. Todaro]

Genes Within Genes

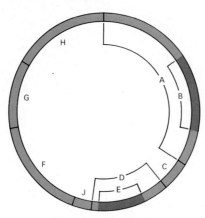

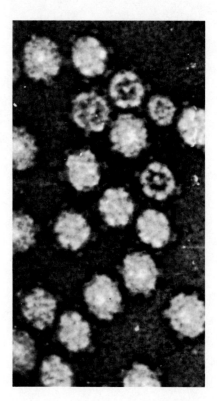

Phi X174 viruses, magnified about 500,000 times, *right,* have an unusual genetic sequence, *above.* Gene B is completely contained in gene A and gene D contains gene E. This allows the virus to make the nine proteins it needs to survive with less genetic material than would normally be needed.

Genetics

Continued

cause of the extensive social records kept there. It focused on males who have an extra Y chromosome. Normal (XY) males have 46 chromosomes including one X and one Y chromosome. XYY males have one X and two Ys, for a total of 47.

Previous studies had shown that XYY males tend to be tall – more than 184 centimeters (6 feet). A few preliminary studies had also found a higher frequency of XYYs among criminals, especially among tall criminals convicted of extremely violent crimes. This suggested that the extra Y chromosome might predispose a man to criminal aggression. Unfortunately, these preliminary results were sensationalized by lurid "born to kill" newspaper stories, and a few psychologists even proposed examining all newborn males for the presence of an extra Y in order to subject XYY boys to surveillance and perhaps decriminalization training.

Witkin and his collaborators studied all living males more than 184 centimeters tall who were born in Copenhagen between 1944 and 1947. The research-

ers visited each man, explained the study, and requested a few drops of blood for chromosome study. They identified 4,096 XYs and 12 XYYs.

Among the XY men, 9.3 per cent had been convicted of one or more criminal offenses. The rate was much higher among XYYs – 41.7 per cent. However, only one offense was an act of violence against another person. The investigators concluded that XYYs are no more prone to acts of violence than others.

The study also found that XYYs scored slightly lower on the average than XYs on a military intelligence test and had lower levels of educational attainment. Lower-than-average scores on these two measures of intellectual ability are associated with higher crime rates even among XY men. The researchers concluded that the higher XYY crime rate may be attributed to their mild intellectual impairment. Indeed, the researchers suggested that the XYYs may not commit more crimes than XYs – they may merely be more likely to be caught.   [Daniel L. Hartl]

Geoscience

Geochemistry. Major advances were made in 1976 and 1977 toward a fundamental goal of geochemistry—that of understanding the distribution of the chemical elements within the earth's interior, and the extent to which this distribution represents a state of chemical equilibrium. Perhaps the most important advance was an apparatus built by David Mao and Peter Bell of the Geophysical Laboratory at the Carnegie Institution of Washington in Washington, D.C., which can compress rocks to much higher pressures than before.

Geochemists have long tried to build an apparatus that reproduces in the laboratory the extremely high pressures and temperatures of the earth's deep interior. The goal has been to produce a pressure of 1 million atmospheres at high temperatures; as high as 2500°C (4532°F.). This would simulate conditions at a depth of 2,300 kilometers (1,380 miles), which is only about 600 kilometers (360 miles) from the boundary between the earth's lower mantle and the core.

The Mao and Bell apparatus has achieved this goal by use of a diamond anvil pressure cell based on a design originally developed at the National Bureau of Standards in 1959. In the Mao and Bell device, a sample of rock or mineral is heated by a laser beam that passes through one of the two transparent diamond blocks that are squeezing the rock. The transparency of the diamonds permits the material under study to be observed and photographed through a microscope while under high pressure. Other data, such as that from X-ray studies of crystal structure and Mossbauer spectra, can also be derived from the sample while it is under high pressure.

Chemical equilibrium. Geochemists are using the diamond anvil pressure cell to determine whether or not the earth's interior remains in a state of equilibrium—a relatively unchanging state. If the interior is in equilibrium, the chemical and physical processes we can observe at the earth's surface must represent the steady running of a natural engine. This engine is powered by radioactive heat generated by the small quantities of uranium, thorium, and potassium in rocks. If it is not in equilibrium, the processes we observe may represent the continuing evolution of physical and chemical changes started under the unstable conditions of the earth at the time it was formed.

Working with the diamond anvil pressure cell, Mao and Bell found that ferrous iron, the form in which iron oxides are usually found in minerals such as olivine, is unstable at high pressures. At pressures corresponding to depths of more than 350 kilometers (210 miles) in the earth, it becomes a mixture of metallic and ferric iron. The ferric iron atoms in such a mixture occur as ions (charged atoms) with three positive charges. The ions in ferrous iron have two positive charges. The presence of ferrous iron in the pressure cell was a surprise because in a state of chemical equilibrium metallic and ferric iron are never found together under normal atmospheric pressures. In laboratory experiments at normal pressures, metallic and ferric iron react with one another to form ferrous iron.

Scientists believe that the core of the earth consists principally of metallic iron, and they had found that the outer portions of the earth contain rather large quantities of ferric iron. This, presumably, is a consequence of the entire earth never having been in a state of chemical equilibrium. They assumed that the two forms of iron would never be found together. However, the new data shows that the two forms can coexist without turning into ferrous iron at high pressures.

Scientists also had believed that rocks thought to be typical of those in the earth's mantle had more oxidized nickel than could be in equilibrium with a metallic iron core. Nickel combines much more readily with iron to form nickel-iron alloys than it does with oxygen. In a state of equilibrium, it seemed reasonable to suppose that the earth's nickel would alloy with metallic iron in the core, and leave the oxidized rocks of the mantle nearly free of nickel. However, there has previously been little experimental work in this area. It has not been known how elements such as nickel form metallic and oxidized compounds under the tremendous pressures in the earth's core and in chemical equilibrium with the principal elements of the earth's interior—oxygen, magnesium, silicon, iron, and sulfur.

The Diamond Anvil Pressure Cell

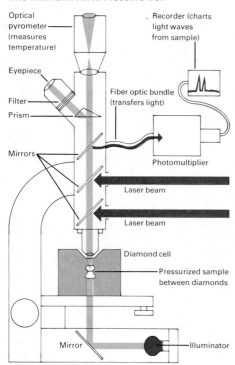

Optical pyrometer (measures temperature)

Recorder (charts light waves from sample)

Eyepiece

Filter

Prism

Fiber optic bundle (transfers light)

Mirrors

Photomultiplier

Laser beam

Laser beam

Diamond cell

Pressurized sample between diamonds

Mirror

Illuminator

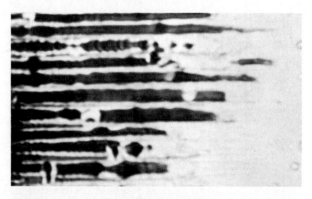

Geoscience

Continued

Diamond anvil pressure cell, *above,* squeezes rock fragments and heats them with lasers to simulate conditions deep in the earth's interior. Minerals under such conditions exhibit unusual traits. The dark area in the photomicrograph, *top right,* indicates that basaltic lava has changed into a metallic and ferric iron mixture. The dark region in the photomicrograph of olivine, *bottom right,* indicates that it has changed into spinel, another type of rock.

Bjørn Mysen and Ikuo Kushiro carried out such experiments in 1976 at the Geophysical Laboratory. They found that a considerable portion of nickel remains in the oxidized form in such a state of equilibrium rather than being entirely concentrated in the core of the earth. The geochemists' measurements indicate that there could be an even higher concentration of nickel in such mantle minerals as olivine and spinel than is actually known to be there. This suggests that the earth's composition is probably somewhat different from that of the nickel and other compounds used in the experiments. The researchers chose their experimental material to approximate the average composition of the nongaseous elements of the solar system. The earth may have a similar, but somewhat different composition.

Disequilibrium. The pressure cell experiment results showed that the core and mantle could be in a state of chemical equilibrium, at least with regard to abundant elements such as iron and oxygen. But an analysis of gases from a volcano produced evidence that

this is not the case for trace elements and their isotopes. Harmon Craig and John Lupton of the Scripps Institution of Oceanography in La Jolla, Calif., analyzed volcanic gases from Kilauea volcano in Hawaii, and from lavas that flowed onto the ocean floor. They found that the proportion of hydrogen, helium, and neon isotopes in these volcanic gases differed from the proportion of isotopes of the elements normally found in the earth's atmosphere and also differed from one another. This result suggests that these highly volatile gases coming from deep in the earth represent trapped relics of gases present when the earth was formed about 4.5-billion years ago. It also shows that these gaseous elements were never in equilibrium with the earth's atmosphere. Furthermore, the ratio of these primordial gases varied greatly from sample to sample, indicating that the gases are from several different sources within the earth.

Robert Pepin of the Lunar Science Institute in Houston and Douglas Phinney of the University of California,

Berkeley, reported additional evidence of a similar nature. These geochemists confirmed earlier findings that xenon gas from natural gas wells contains more of the isotope 129 than does xenon in the atmosphere. The decay of a relatively short-lived iodine isotope that was present when the earth was formed but which had transformed to xenon 129 more than 4.3 billion years ago produced this excess of isotope 129. Pepin and Phinney found that the xenon from well gas contained few of the heavier xenon isotopes produced by the decay of a similar "extinct" radioactive isotope of plutonium.

All these measurements show that some gases have been stored in the interior of the earth since its earliest history. They were not mixed with the atmosphere during the stages of the earth's formation.

Other isotope studies. Evidence for disequilibrium and lack of complete mixing also came from the study of isotopes in rare elements taken from volcanic rocks on the continents and under the oceans. Donald De Paolo and Gerald Wasserburg of the California Institute of Technology found disequilibrium in the isotopic composition of the rare-earth element neodymium. Christopher Brooks, David James, and Stanley Hart working at the Carnegie Institution found it in the isotopes of strontium.

It is becoming clear that although the earth is mixed and, to a great extent, in equilibrium, there are still pockets of material that have never mixed. This fact cannot be explained by the theory that the earth was entirely melted in the beginning and then cooled and congealed to its present state. It may be that the original state of the earth was more like that proposed by Victor Safronov and his colleagues at the Institute for the Physics of the Earth in Moscow in 1969. According to Safronov, the earth was formed by the coming together of many planetesimal bodies ranging up to 1,000 kilometers (600 miles) or more in diameter. Such a combination would produce an earth that was initially nonuniform both in temperature and chemical composition, and some relics of this nonuniformity could be expected to persist today. [George W. Wetherill]

Geology. A series of disastrous earthquakes shook the world in 1976, and geologists continued to study them in 1977. One of the greatest natural catastrophes in recorded history occurred east of Peking in northeast China on July 28, 1976, and killed an estimated 600,000 persons. The quake had a magnitude of about 7.7 on the Richter scale – severe, but not as bad as the 1964 quake in Alaska and the 1906 quake in San Francisco. Unsafe buildings in crowded cities and unstable, sandy ground contributed to the extremely high number of casualties in the Chinese quake.

Chinese geologists have released no data on the quake's geologic effects. But analysis of seismograms by Gordon Stewart of the California Institute of Technology indicates that the shock resulted from motion along a fault about 150 kilometers (90 miles) long. Several smaller shocks accompanied the main shock, and an earthquake of magnitude 7.2 occurred in the same area only 15 hours later.

The fault slips were part of a general deformation within the Eurasian lithospheric plate, which includes China. The lithospheric plate is one of some 20 giant plates that form the earth's outer shell. Paul Tapponier of the University of Montpellier in France and Peter Molnar of the Massachusetts Institute of Technology in Cambridge analyzed the geologically mapped faults, those recorded on scanner images from satellites, and determinations of slip directions during historic earthquakes. They concluded that such deformation represents the continued crowding of China eastward by the lithospheric plate that includes India.

The Guatemala earthquake that killed more than 22,000 persons and left 1 million homeless in February 1976 had a magnitude of 7.5. The quake was caused by the eastward motion of the lithospheric plate that includes the Caribbean Sea floor and southern Central America. This plate borders plates that include North and South America and part of the Atlantic Ocean. The Guatemala quake was accompanied by some of the most extensive faulting of the ground level in the Western Hemisphere since the 1906 San Francisco earthquake. George Plafker of the U.S.

Geologists monitoring La Soufrière volcano on Guadeloupe island measure radon-gas concentration, which is an indicator of volcanic activity.

Geoscience

Continued

Geological Survey and other geoscientists studied the Guatemala disaster. The major fault break curved eastward and northeastward across southern Guatemala. Land on the south side of the fault moved eastward an average of 1 meter (3 feet), and in some places more than 3 meters (10 feet), relative to that on the north side.

The earthquake that occurred off the southwest coast of Mindanao in the Philippine Islands in August 1976 registered a magnitude of 7.9 and killed more than 6,000 persons. Stewart, Robert Wallace, and James Taggart of the U.S. Geological Survey, who studied this quake, reported that the lithospheric plate containing the Celebes Sea floor is sliding northeastward beneath the plate containing Mindanao.

Oil deposits. Petroleum geologist Douglas Klemme of Lewis G. Weeks Associates, Westport, Conn., is among those who have evaluated the geothermal processes that concentrate oil in major oil fields. He found that the higher the underground temperature, the shorter the time needed to change

the average 1 per cent of organic matter in a typical dark-shale source rock into gas and petroleum.

The gas and liquid move through porous rock and accumulate where they are blocked by impermeable rock. An underground temperature that is too high produces gas only or destroys all the hydrocarbons. If the temperature is too low, few hydrocarbons form. The porosity of sandstone and other rocks decreases with depth because they are compacted by the increasing pressure. Only where temperatures are high enough to generate fluid hydrocarbons at depths shallow enough for high porosity can large pools of petroleum begin to accumulate.

Petroleum and plate tectonics. Geologists and geophysicists of the Exxon Corporation, Atlantic Richfield Company, and British Petroleum Company completed a study in 1976 of the geologic history of the largest oil field in North America, the Prudhoe Bay field in northern Alaska. It showed how complex the lithospheric plate interactions that control oil accumulation can

Geoscience

Continued

be. Oil and natural gas occur in sedimentary rock strata in basins and continental shelf areas close to the edges of lithospheric plates. The accumulation of oil and gas is a by-product of lithospheric plate motions.

The scientists found that the Prudhoe Bay oil is trapped in porous sandstone of the Permian Period (275 to 225 million years ago) at the highest point in a subsurface arch of rock layers called the Barrow Arch. Most of the oil was generated in marine shales of the Cretaceous Period (130 to 65 million years ago) on both sides of the arch and moved upward into the arch entering the older Permian rocks.

Irvin Tailleur, Gilbert Mull, and Warren Hamilton of the Geological Survey reported in 1977 that the part of the Arctic Ocean north of Alaska began to open early in the Cretaceous Period. As the plate containing Alaska rotated counterclockwise away from Arctic Canada, the edge of the continental crust slowly subsided below sea level. Sediments were deposited on the subsiding crust in a horizontal position, but continuing sinking of the rock layers tilted them to form a wedge, the north side of the Barrow Arch.

Shortly after the rotation began, an arc of volcanic islands, part of another lithospheric plate, collided with the south edge of Alaska near what is now the south edge of the Brooks Range. The pressures of this collision formed the Brooks mountains and produced the basin that defines the south side of the Barrow Arch. The Permian reservoir sandstones, laid down long before these events, came from mountains produced by an earlier collision between two continental masses. Complex events, mostly related to the sinking of the edge of the Pacific lithospheric plate beneath Alaska, formed modern central and southern Alaska after the Cretaceous island-arc collision.

Upraised Pacific Ocean floor. Wyn Hughes and Colin Turner of the Solomon Islands Ministry of Natural Resources reported in 1976 on comparative studies of rocks from the Solomon Islands in the southwestern Pacific Ocean and of rock samples drilled from the ocean floor to the north by the Deep Sea Drilling Project. Part of the Solomons formed as deep-ocean floor. Collision of the Pacific and Solomon Sea lithospheric plates caused large masses of oceanic crust to bow upward in the late Tertiary Period, about 5 million years ago. The exposed oceanic basement rocks are early Cretaceous basalts of the type formed where lithospheric plates are separating and molten rock oozes up between them. Deep-water limestones and subordinate cherts of the late Cretaceous Period to the Pliocene Epoch (130 to 5 million years ago) lie over these basalts. The strata are also interlayered with oceanic basalt flows that are richer in alkalis and titanium than the spreading-ridge basalts.

The Solomon Islands uplift occurred when the Pacific plate was sliding southwestward beneath the northeast side of the Solomon Island arc, but now the Solomon Sea and Coral Sea floor is sliding beneath the southwest side of the Solomon Island arc.

Evolution and the atmosphere. Geobiologist Preston E. Cloud, Jr., of the Geological Survey and the University of California, Santa Barbara, reported in 1976 and 1977 that he had reached a new understanding of how the earth's atmosphere and living organisms have evolved together. Oxygen, which is produced by photosynthesis in plants, was not part of the earth's primitive atmosphere. But fossil one-celled plants have been found in rocks formed about 2 billion years ago, and indirect evidence suggests that they may have appeared about 3.8 billion years ago, 1 billion years after the earth formed. According to Cloud, seawater absorbed the photosynthetic oxygen produced by such plants by precipitating iron oxide. By 2 billion years ago, the amount of oxygen had increased to a point where the seawater could no longer hold all of it and it slowly began to accumulate in the atmosphere.

Then limestones and red beds, sediments containing small amounts of iron oxidized atmospherically or by ground water, began to form. *Metazoa* (organisms in which differentiated cells perform different functions) appeared about 700 million years ago. But the atmosphere's oxygen content did not become high enough for animals to develop shells and muscles, which require much oxygen, until 600 million years ago. [Warren Hamilton]

Geophysics. Peter Molnar of the Massachusetts Institute of Technology (M.I.T.) in Cambridge and Kevin Burke of the State University of New York at Albany reported in 1976 that the 20 or more hot spots in the earth's mantle have shifted their positions over the last 60 million years. The hot spots are areas where molten material rises to the surface as volcanoes or otherwise affects the surface so that it is hotter there than elsewhere. Among the best-known hot spots are those under the volcanically active areas in Hawaii, Yellowstone National Park, and Iceland. See HARNESSING EARTH'S FOUNTAINS OF FIRE.

Princeton University Geophysicist Jason Morgan suggested in 1972 that these hot spots are fixed plumes and that their heat propels a convection current of molten material in the earth's mantle. These convection currents, Morgan said, might provide the force that moves 20 giant lithospheric plates that form the earth's crust.

But Molnar and Burke report that the hot spots have moved an average of 1 to 2 centimeters (0.4 to 0.8 inch) a year over the last 60 million years. This is from 20 to 40 per cent of the rate at which the giant plates move.

A somewhat different conclusion can be drawn by comparing the ages of the volcanic rocks that have erupted from the hot spots and the rock material that oozes from below the earth's crust at the mid-ocean ridges. Geochemists Sun Shen-Su of Columbia University in New York City, who is now at the University of Adelaide in Australia, and Gilbert Hanson of the State University of New York at Stony Brook made that comparison in 1976. They based their work on refinements of original studies by physicist Paul R. Gast of Columbia University, who measured the lead isotope ratios of these rocks to determine their ages.

Sun and Hanson found that the various minerals in the rocks that were once part of the molten material rising in the ocean ridges could have been mixed together within the last 200 million years. However, the material of the volcanic island rocks cannot have been

Location of the Palmdale Bulge

The land for many miles around Palmdale, Calif., has risen appreciably enough over the past decade for geologists to refer to the area as the Palmdale Bulge. Parts of the area have also recently become depressed, and some geophysicists believe that this movement may herald an earthquake in the populated area.

mixed within the last billion years. This suggests that the islands were formed by the upward thrust of volcanic material from a source below the layer in which the ocean ridge material was mixed, the molten lava forcing its way through one of the giant lithospheric plates. The ages of the rocks differed because the upward thrust of molten material occurred at different times in the earth's history as the giant plate slowly moved over the hot spot. Each island resulted from a different eruption of this molten material.

Computer simulation. Geophysicists have developed highly sophisticated computer programs to simulate conditions in the molten material that flows under the earth's giant plates so they can determine how the plates move. Because it is impossible to drill deep enough into the earth to observe the processes directly, these approximations are particularly important to the study of the earth's movements.

A group of geophysicists at M.I.T. led by M. Nafi Toksoz has concentrated on computer descriptions of subduction, the process in which the edge of one plate drops beneath the edge of another and into the earth's mantle. The tremendous pressures the plates exert against one another as they move causes subduction. Subduction, in turn, causes many volcanoes to erupt, earthquakes to occur, and mountains to rise. The M.I.T. team's computer work has shown that there are appreciable differences in the amount of heating and folding of rock strata between the subduction that takes place when two plates, each containing a continent, collide (as under the Himalayas) and when one plate containing a continent collides with one containing an ocean bottom (as under Japan).

Palmdale Bulge. Geophysicists at the U.S. Geological Survey's National Center for Earthquake Research (NCER) in Menlo Park, Calif., have been investigating an unexplained uplift of the earth along the San Andreas Fault, which runs from north of San Francisco through central California to the Mexican border. In March 1977, they announced that the swelling has occurred over a much larger area than first thought, and part of the bulge has subsequently collapsed.

Robert O. Castle, leader of the team that detected the land bulge, said that an uplift of 15 to 25 centimeters (6 to 10 inches) occurred over an area about 80 by 160 kilometers (50 by 100 miles) near Palmdale, Calif., between 1960 and 1970. The axis of this Palmdale Bulge is nearly parallel to the San Andreas Fault. The earthquakes of 1952 in Kern County and 1971 in San Fernando occurred near the bulge.

Wayne Thatcher of NCER discovered that portions of the bulge recently have been depressed by unknown geologic processes. In some cases, parts of the bulge are lower now than they were before the bulge was found in the early 1960s. Large parts of it have subsided by as much as 50 per cent. Castle said that it was still changing shape.

The NCER geologists re-examined survey data from the early 1900s and found that a similar uplift occurred along the fault between 1897 and 1914, and that it later collapsed. While there was no evidence to link this sequence of events to the earthquakes that have occurred in the area, Castle said that two major earthquakes occurred close to the affected area in 1927 and 1933, not long after the land had collapsed.

Impact cratering. The latest studies of the effects of meteorites and other objects that form impact craters on the earth, moon, and planets was the subject of a symposium held by the Lunar Science Institute of the U.S. Geological Survey in September 1976.

J. Dugan O'Keefe of the University of California, Los Angeles, and Thomas J. Ahrens of the California Institute of Technology reported on their calculations of the energy distribution in hypervelocity impacts, those greater than the speed of sound. They based their work on laboratory experiments and theoretical calculations.

O'Keefe and Ahrens found that a much higher proportion of impact energy at such high velocities—as much as 90 per cent—goes into heat than had hitherto been assumed. This helps explain why widespread melting appeared to have enveloped meteorites upon impact on the moon. It also explains the assumed slowdown of relative velocities of planetesimals (small asteroidlike bodies) before they united to form planets.   [Foster Stockwell]

Geoscience

Continued

A fossil brachiopod with a hole drilled in its shell by some unknown predator was found in eastern Quebec, Canada, by Peter Sheehan of the University of Montreal. The brachiopod lived 400 million years ago.

Paleontology. David M. Raup, a paleontologist at the University of Rochester in New York, published the most extensive and comprehensive study of invertebrate fossils ever made in January 1977. His purpose was to try to determine how rapidly animals diversified after the first multicellular creatures evolved some 600 million years ago. The question has been vigorously debated among paleontologists for many years. Raup examined the records of about 70,000 fossil species. He believes that this is somewhat less than half the estimated number of invertebrate species for which paleontologists have found fossil evidence.

Raup arranged the species according to what time they lived during each million-year interval over the 600 million years, and then counted them. He found that the Cenozoic Era, which started 65 million years ago and includes the present, had the highest number. About 648 different species of fossil invertebrates have lived in each of the million-year intervals of this era.

The Cretaceous Period, which began 130 million years ago and lasted until the Cenozoic Era, had 310 invertebrate species in each million-year interval. There were fewer during the Jurassic and Triassic periods, 130 million to 225 million years ago—only 147 species in each million-year interval. But the earlier Devonian Period, from 345 million to 405 million years ago, had more species. There were 270 in each million years during these times. Continuing back in time, there was a successive decline in the number of species for each of the earlier periods—the Silurian, Ordovician, and Cambrian.

The data at first appears to show that the number of invertebrate species has increased fairly steadily except for the decrease during the Permian (275-225 million years ago), Triassic, and Jurassic periods. However, Raup disputes the idea that there had to be a decrease during these three periods. He argues that it is easier to recover fossils that lived in more recent times. More fossil-bearing rocks remain on the earth from the Cenozoic Era than from any earlier period. So there should be more Cenozoic fossils. The other periods with large numbers of fossil-bearing rocks are the Cretaceous Period and Devoni-

an Period, the times when the fossil record shows the other two greatest number of species. Raup believes that the earth had many species of invertebrates early in its history, and that the great diversity of species has changed little since the earliest invertebrates.

Raup's data also showed changes in the preponderance of various invertebrate groups over most of the entire history of life on earth. Trilobites, shell-covered sea creatures, were the dominant invertebrate group in the earliest geologic periods. They began to decrease about 310 million years ago, and mollusks (clams, snails, and cephalopods) became more numerous. Mollusks account for more than half of all invertebrate fossil species in the Triassic Period. Protozoans, such single-celled animals as foraminifera and radiolarians, have left many fossils in the later periods of the earth's history.

Environment and diversity. Geologist Richard Bambach of the Virginia Polytechnic Institute and State University at Blacksburg studied a smaller sampling of fossil material similar to that used by Raup, grouping the individual specimens not only by age, but also by the environment in which the species lived. He concluded that species diversity is directly related to environment. Bambach assumed that the number of fossil species recovered from a fossilized community in any of the rocks of a geologic period represents a percentage of the total number of animal species of that community. He tabulated the fossil species recovered from 359 communities of different ages and divided them into three groups on the basis of the environmental conditions in which the rocks containing the fossil communities were formed.

The first group was found in rocks formed by sand and silt deposits in estuaries and intertidal zones, where many environmental factors fluctuate unpredictably, and would likely contain fossils of animals that lived under the most stress. The second group was in rocks formed by sand and silt deposits in deltas, shorelines, and shallow subtidal regions, which are variable but not as stressful environments for animal life as those of the first group. Bambach contends that the third group, animals found in rocks in deep-water areas,

A 200-million-year-old skeleton, the first complete specimen of the dinosaur *Heterodontosaurus tucki,* was found in South Africa.

Geoscience

Continued

Fossil worm burrows up to 20 centimeters (8 inches) long have been found in rocks in Zambia that date to 1 billion years ago. They are the oldest traces of multicellular animal life discovered.

where the salinity of the water is relatively stable and there is less disturbance to the underwater environment from surface storms, experienced the least stress.

Bambach's calculations showed that the diversity of species over the many ages of geologic history varied in these three environments. The number of species did not change significantly in the high-stress environments. He found the greatest change in fossils deposited in deep-water areas; the species number per community increased from about 21 in the early Paleozoic Era, beginning 600 million years ago, to 32 in the late Paleozoic Era, about 275 million years ago.

Predacious gastropods. Geologist Peter Sheehan of the University of Montreal in Canada found a group of fossil brachiopods, marine animals with two-piece shells, with circular holes bored in their shells by some ancient predator. Sheehan found the fossils in Quebec's eastern Gaspé Peninsula in rocks of the Lower Devonian Shiphead Formation. This means that the brach-

iopods lived about 400 million years ago and the holes are the best evidence of such early marine predators.

Today, predacious gastropods, snails, drill small holes in brachiopod shells and suck out tissues and juices, but it is difficult to prove that the same thing happened 400 million years ago. Some algae, sponges, and annelids also bore holes in brachiopod shells, but they do so to live there, not to eat the animals.

Sheehan concluded that the holes were bored by predators, probably snails, seeking food. Each shell had only one hole, and all the holes were carefully placed in the center of the shell.

Sheehan also noted that the borer selected only the medium-sized shells, which would contain more food than small shells and could be drilled more easily than large, thick brachiopod shells. The holes were also drilled in only one species of brachiopod, further evidence that they were the work of a particular species of predator. Sheehan suspects that the predator was a species of gastropod found in the same fossil-bearing rocks.   [Ida Thompson]

Immunology

Immunologists continued in 1976 and 1977 to learn more about how the immune system works. In some of the most interesting and significant experiments, researchers discovered and investigated what might turn out to be a new type of white blood cell, produced large numbers of specific antibodies, and used a white blood cell product to control disease.

New white blood cells? Immunologists now commonly accept that certain white blood cells called thymus-dependent lymphocytes, or T cells, provide the major immune defense against cancer cells. So it has been difficult to explain why mice from a strain born without a thymus, which produces T cells, do not have more cancer arising spontaneously than do normal mice.

A number of explanations have been offered to account for this dilemma. Of particular interest are experiments reported over the past year or two that have revealed and begun to probe the characteristics of what may be a previously unknown type of white blood cell that destroys cancer cells.

Ron B. Herberman and his colleagues at the National Cancer Institute (NCI) in Bethesda, Md., and Swedish immunologist Rolf Kiessling and his associates at the Karolinska Institute and the University of Uppsala in Sweden reported in June 1976 on these cells. They said that the cells, found in normal mice as well as in those without a thymus, look like lymphocytes but do not have the characteristics that would classify them as such. For example, they lack proteins found on the cell surface of lymphocytes.

Most of the experimental evidence for the anticancer role of these cells came from tests in which they destroyed tumor cells in the test tube. This ability led to their being called natural killer (NK) cells.

Other experiments have indicated that human blood probably also contains NK cells. For example, D. Bernard Amos and his colleagues at Duke University Medical Center in Durham, N.C., reported in March 1977 that white blood cells from human beings killed certain leukemia cells in a test

X ray of a child born with a defective immune system, *right,* reveals no sign of a thymus, an organ that produces immunologically active cells. After a new treatment with the enzyme adenosine deaminase, the child's thymus has developed (arrow), *below right,* and his immune system was vastly improved.

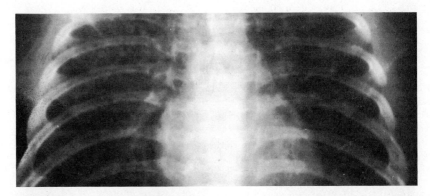

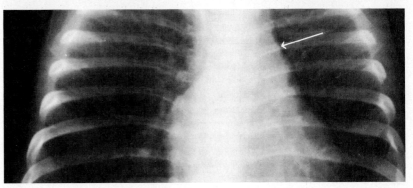

Fine Tuning the Immune System

Heather recently weathered a bout with chicken pox. That would not be very exciting for most 9-year-olds, but for Heather it was a triumph. When she was 5½ years old, Heather weighed 12 kilograms (26 pounds) and was near death because a defective immune system left her almost totally unable to fight off disease and infection. Treatment with an experimental hormone called thymosin seems to have changed all that, however, and today Heather enjoys a relatively normal life.

Biochemist Abraham White and I isolated thymosin from calf thymus glands in 1966 at Yeshiva University's Albert Einstein College of Medicine in New York City. The thymus gland is a central organ in immunity. Located between the breastbone and the heart in humans and most other mammals, it triggers the development of certain cells into white blood cells that are called thymus-dependent lymphocytes, or T cells. Thymosin is a key hormone in this development.

T cells combat viruses and several other organisms that cause disease. They also attack and destroy cancer cells. So we hoped that one day thymosin might help in treating patients suffering from immune-deficiency diseases or cancer. Although Heather is one of our more dramatic cases, doctors giving thymosin to more than 50 other patients with immune-deficiency diseases in the United States and Europe report improvement in a large majority.

Chemical and radiation treatment of cancer always lowers the number of T cells and the patient's total immune response. Thymosin treatment has raised the T-cell count in more than 75 per cent of the cancer patients receiving it either alone or along with chemical and radiation therapy. We have also seen unexpected tumor regression and other clinical improvement in some of these people, but we do not yet have good evidence linking thymosin to these changes. However, clinical trials designed to determine if thymosin helps against cancer are in progress.

Thymosin is a complex mixture of substances, each of which probably has a different function. In our lab at the University of Texas Medical Branch in Galveston, my colleagues and I isolated and have been studying a partially purified fraction of thymosin called thymosin fraction 5 since 1972. Fraction 5 has many of the effects of thymosin observed in experimental animals and in test-tube studies. For example, it increases T-cell production and triggers the maturation of T cells.

We have isolated and purified several active components from fraction 5 that produce some of thymosin's effects in animal and laboratory tests. For example, these substances will increase T-cell production. But more revealing, some of them increase the production of lymphokines, substances made only by certain subpopulations, or types, of T cells. So the active substances in fraction 5 apparently influence the development of subpopulations of T cells and, thus, specific lymphokines. If this is true and we can sort out the substances and their roles, we will have the tools to manipulate a major part of the immune system with great precision.

The medical potential is great. Each subpopulation of T cells acts, primarily through its lymphokines, in special ways. So if we could selectively increase the production of specific T cells and their lymphokines, we might be able to develop precise and potent treatments for many maladies.

For example, several diseases, including rheumatoid arthritis, may be caused by an overactive immunological response in which the immune system attacks the victim's own tissues. There is good evidence that a lack of suppressor T cells, a subpopulation that helps regulate immunological response, may cause such autoimmune diseases. Too few suppressor T cells may also be partly responsible for allergies.

Some lymphokines are already being used experimentally to treat cancer and immune-deficiency diseases. One of these, interferon, is a potent antiviral substance and is active against a variety of other harmful agents and tumor cells as well. In fact, several lymphokines are effective against cancer cells, acting by weakening their outer membranes, attacking them directly, and increasing other immune activity against them.

Much must still be learned before we can manipulate thymosin, its active components, T-cell subpopulations, and lymphokines. But we are on our way.   [Allan L. Goldstein]

tube. And when lymphocytes and other known types of white blood cells that might attack cancer cells were removed, the remaining cells continued to kill the leukemia cells.

Researchers must still answer a number of questions about NK cells. Are they really a unique type of white blood cell? Or are they merely an early stage of lymphocyte, a stage which does not allow lymphocyte identification? Also, do they really attack cancer cells naturally or do they recognize and attack them because the cancer cells chemically resemble other material foreign to the body? In addition, do NK cells attack some cancer cells – namely, lymphomas and leukemias, as has been suggested in experiments – more efficiently than other cancer cells?

Antibodies in quantity. In April 1977, Cesar Milstein and his co-workers at the MRC Laboratory of Molecular Biology in Cambridge, England, reported that they had produced large quantities of those antibodies that react with and help reject grafted tissues and organs. The scientists used a technique similar to one they reported in 1975 and 1976 with which they had produced large quantities of antibodies against sheep red blood cells. A plentiful supply of antibodies is welcome to immunologists, who usually can collect only small quantities to study.

In the initial experiments, the scientists injected mice with sheep red blood cell antigen, the portion of the cells that antibodies attack. The mice then produced cells that manufactured antibodies against the antigen.

The scientists collected some of these cells and used a process called cell fusion to join them with a type of cell that grows easily in the test tube. From the new combination cells, the scientists isolated and established a line of combination cells that constantly synthesize antibodies against the sheep red blood cell antigen.

The latest work indicates just how useful the new technique can be. For example, large quantities of graft-rejecting antibodies from a person in need of a transplant could be made and tested on the tissues of possible donors to find which donor's tissues would be least likely to be rejected. Also, some of the patient's antibodies could be used as a source to raise anti-antibodies – that is, antibodies that would immobilize the graft-rejecting antibodies. These anti-antibodies could then be used to control rejection very precisely, without the general weakening of the immune system that accompanies present methods of rejection control.

By fusing other cells in the immune system with cells that can be grown in the laboratory, scientists may be able to produce several important immunological substances in quantity, including the lymphokines. See Close-Up.

T-cell product controls disease. In April 1977, Randall S. Krakauer and his colleagues at NCI reported a series of experiments in which they dramatically decreased the incidence of a disease in mice. The disease is much like systemic lupus erythematosus (SLE), in which the human immune system produces antibodies that attack the victim's own cells.

Both diseases are probably caused by a decrease in suppressor T cells, a type of lymphocyte that controls antibody production. It may be that antibody production progresses helter-skelter when the number of suppressor T cells falls, and that some of that antibody unfortunately turns against oneself.

Krakauer and his colleagues took advantage of previous findings made by Robert R. Rich and Carl W. Pierce at the Harvard Medical School. Rich and Pierce showed that a protein called concanavalin A stimulates suppressor T cells to release a substance that inhibits various antibody responses. The NCI group collected this substance and repeatedly injected it into mice from a strain that is particularly susceptible to the SLE-like disease. The incidence of the disease in these injected mice was substantially lower than in similar mice that had not received the substance.

Thus, it appears that it is possible to by-pass a problem presumably created by a decrease in the number of an immunologically active cell by administering one of the substances the cell produces. We cannot yet say whether similar treatment will help in human SLE, because the animal experiments prevented rather than treated the condition. However, the NCI experiments certainly encourage further investigation in this area. [Jacques M. Chiller]

Medicine

Dentistry. A reduction in the efficiency of the white blood cells that attack harmful bacteria is involved in periodontosis, a relatively rare gum disease that destroys tooth-supporting bone in adolescents, according to researchers. Oral biologist Louis J. Cianciola and his associates at the State University of New York at Buffalo reported in February 1977 that white blood cells, called polymorphonuclear leucocytes, taken from victims of periodontosis, did not find and destroy harmful bacteria as well as did polymorphonuclear leucocytes from normal persons.

Bacteria were not implicated in periodontosis until 1975 because there is little dental plaque—the thin film of food particles and bacteria that harden and eventually cause decay—or inflammation associated with it. The Buffalo researchers do not know whether some chemical factor in the harmful bacteria impairs the leucocytes' functioning, or if a pre-existing malfunction in these cells allows the disease to arise.

In persons over 35, periodontal disease causes the loss of more teeth than does dental caries, or decay. Bennett Klavan, professor of periodontics at the University of Illinois Dental School in Chicago, estimated in June 1977 that 90 per cent of all adults have some degree of periodontal disease. The bacteria that cause the disease produce toxic material that attacks and inflames the gums. As the disease progresses, it gradually destroys the bone around the teeth. Eventually, the teeth must be removed because of recurring abscesses and infections.

Decay vaccine. A vaccine to prevent dental caries has been successfully tested in rhesus monkeys. A group of immunologists and microbiologists headed by Thomas Lehner at Guy's Hospital Medical and Dental Schools in London reported in November 1976 that they had manufactured the vaccine from killed *Streptococcus mutans*, the bacteria that is responsible for caries. The vaccine stimulates the body's immune system, especially both the T cells that provide cellular immunity and the antibodies that work against the bacteria. [Paul Goldhaber]

An acrylic device that corrects bad bite and accompanying lower-jaw pain is fitted roughly to a patient's palate, *right*. With the excess acrylic trimmed off, *below*, the artificial palate keeps the upper and lower teeth separated. This allows the lower jaw to reposition itself after about four months of using the palate.

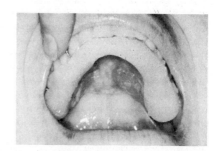

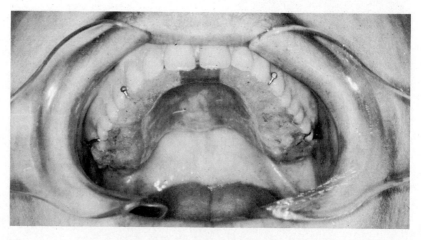

Medicine

Continued

Internal Medicine. Skin cancer has been firmly linked with a component of sunlight. Joseph Scotto, senior health services officer of the National Cancer Institute (NCI) in Bethesda, Md., reported evidence of this in October 1976. Skin cancer is the most common form of the disease in the United States, accounting for more than 300,000 cases annually.

NCI established "sunburning ultraviolet meters" in 1974 at 10 National Oceanic and Atmospheric Administration observatories in the United States to test the theory that exposure to sunlight was related to skin cancer. Ultraviolet radiation was measured because it produces erythema, reddening of the skin and ultimately sunburn, and was therefore suspected as the component of sunlight responsible for inducing malignancies. The study showed this radiation was most intense about noon and that 60 per cent of the total daily level occurred between 10 A.M. and 2 P.M. As expected, ultraviolet radiation was greatest in the summer months and increased from north to south.

The researchers compared these findings with data drawn from another study, the NCI's Third National Cancer Survey which detailed the incidence of different types of cancer in different parts of the United States. The findings for skin cancer in Minnesota, Iowa, and Texas showed that the incidence of skin cancer correlated with the magnitude of ultraviolet radiation exposure from north to south.

Fats and heart attacks. Heart researcher William Castelli, director of laboratories for the Framingham Heart Study, presented his findings on the role of cholesterol in clogging the small coronary arteries that nourish the heart muscle in a report to the American Heart Association Annual Science Writers Forum held in San Antonio, Tex., in February 1977. Coronary artery disease, which is responsible for heart attacks, is widely considered to be influenced by the cholesterol level in the blood. Castelli and his co-workers reviewed the evidence relating cholesterol levels to subsequent heart disease among more than 6,000 participants in an 18-year study of heart disease in Framingham, Mass. They found significantly higher cholesterol levels among

people who developed signs of coronary artery disease before age 50 than among those without heart disease. However, elevated cholesterol was a less reliable predictor when heart disease appeared after age 50.

Applying an idea suggested in 1951 by internist David Barr of the Cornell University Medical College in New York City, Castelli also examined the various protein components called lipoproteins that transport cholesterol in the blood. High-, low-, and very-low-density lipoproteins carry cholesterol. Barr had suggested that high-density lipoproteins (HDL) actually protect against heart disease. Castelli found that the greater the amount of HDL in blood serum in both men and women over 50, the less risk of coronary artery disease and therefore heart attack.

Apparently about half of an individual's cholesterol is carried by HDL at birth. But exposure to an American diet, with high cholesterol and saturated-fat levels, results in steadily increasing concentrations of low-density lipoproteins (LDL) and very-low-density lipoproteins (VLDL) carrying cholesterol in the blood. As a result, as blood studies show, the adult American has an average of 25 per cent of total cholesterol carried in HDL. While LDL and VLDL change according to diet, the relationship of HDL to nutrition is less clear.

Perhaps HDL levels are already determined at birth. There is evidence, however, that long-distance runners and people on diets composed primarily of vegetables, cereals, and fish tend to have higher levels of HDL and may have less heart disease. It has also been shown that clofibrate and niacin, two drugs that lower total cholesterol, increase HDL. By contrast, women taking oral contraceptives tend to have reduced HDL.

How HDL can protect the coronary circulation is not as yet firmly established. However, there is some evidence to support two possible explanations. It appears that HDL attracts and combines with more free cholesterol molecules than the other lipoproteins and forms larger molecules that are slower to enter cells and therefore more likely to return to the liver for elimination. The second hypothesis is that HDL

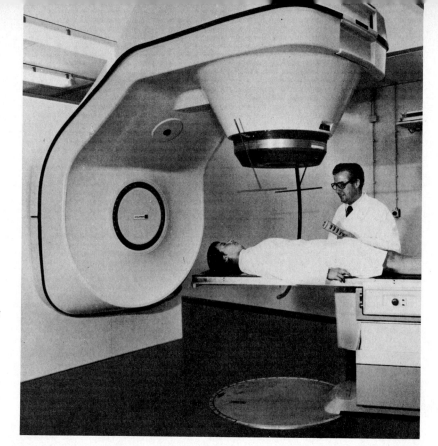

A neutron generator, *right,* is used to treat cancer. Mouth and jaw cancer had spréad to the chin of a patient, *below,* and damaged her lower jawbone, *center top.* After neutron treatment, the bone has healed, *center bottom,* and tumor is gone, *below right.*

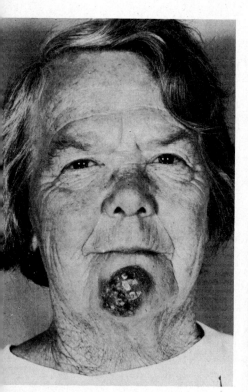

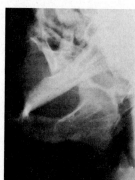

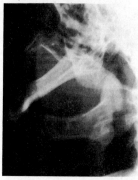

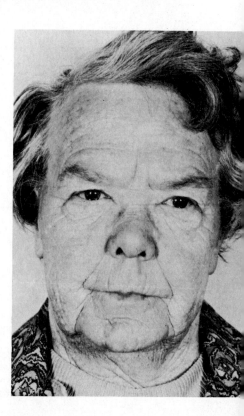

Medicine
Continued

binds preferentially to the walls of arterial smooth muscle cells, thereby preventing the adherence of LDL which can more readily introduce its cholesterol into cells.

Bone-marrow transplants. Aplastic anemia is an uncommon bone-marrow disease in which the marrow loses its ability to manufacture blood cells. The disease is almost always fatal, usually within six months of onset, and treatment has been limited to care rather than cure. But in October 1976, R.P. Gale, M.J. Klein, and a team of physicians at the University of California School of Medicine in Los Angeles described their efforts to restore the capacity of victims of this disease to produce blood cells by transplanting bone marrow from related donors.

They treated 21 patients at a hospital that followed consistent and rigorously controlled procedures. Nine patients did not have suitable related donors and received conventional treatment. The other 12 patients had relatives who met stringent criteria for genetic compatibility, which reduces the likelihood of the powerful immunological rejection that destroys grafts.

Patients were prepared for transplantations by chemotherapy and radiation treatment designed to destroy as much as possible of their own blood-cell-manufacturing system. The red and white cells and platelets necessary to sustain life were obtained from donors who were genetically matched to the recipients and gave transfusions before and during transplantation. Bone marrow was successfully implanted in the bones of eight recipients. The grafts were rejected in four, although one reimplantation was successful.

The major hazards these patients faced were inadequate oxygen supply to the blood, immunological rejection, and "graft versus host disease" in which a patient develops hepatitis, dermatitis, or diarrhea. Presumably, this problem results when graft-produced antibodies attack host organs. Almost all the patients experienced these complications; an autopsy showed that one who died of hepatitis had apparently normal bone marrow. By October 1976, seven patients had survived for from four months to three years following transplant. By comparison, despite vigorous conventional therapy, all of the nine patients given conventional treatment, who were similar in age, sex, race, and medical conditions to those who received transplants, died after an average of only 82 days.

These results demonstrate that patients with aplastic anemia who are fortunate enough to have immunologically compatible donors of tissue can achieve prolonged survival from what was an invariably fatal disease.

Normal and malignant cells. The rapid, uncontrolled, destructive cellular process called cancer is believed to result from the evolution of normal cells into malignant ones, a change probably triggered by radiation, chemicals, viruses, or unknown factors. When normal cells enter a phase of rapid growth called hyperplasia, they may produce benign tumors or cancer cells. Cancer researchers Steven S. Brem and Pietro M. Gullino of the NCI and Daniel Medina of Baylor College of Medicine in Houston reported in March 1977 on a method that may identify hyperplastic cells that have the potential to become malignant.

These researchers had previously shown that tissues from a cancerous growth arising in the ducts of the breast could induce angiogenesis, the formation of new blood vessels. The new blood vessels provide the nutrients for the cancer to grow rapidly. By contrast, normal breast tissue lacks this capacity, leading to the conclusion that angiogenesis indicates the ability to become malignant.

Further experiments by the researchers show that hyperplasia can be consistently induced in the normal breast tissue of one strain of rats by treatment with certain chemicals. When the hyperplastic tissue is transplanted to a fat pad of a genetically matched rat, it invariably produces further hyperplastic growth within 10 to 51 weeks. These tumors act and look benign and host rats survive their presence for many months. However, malignancies usually occur when fragments of these tumors are transplanted in other rats.

To see if precancerous hyperplastic tissues might display the angiogenesis that is characteristic of cancerous tissue before malignancy occurred, Brem, Gullino, and Medina transplanted the

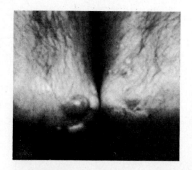

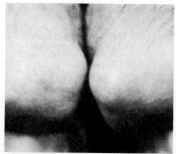

Severe xanthomas, heavy cholesterol deposits on the skin, often on the elbows, *far left,* afflict victims of a genetic disease that sharply raises the cholesterol level.
A blood-cell separator, *below,* allows doctors to replace a patient's cholesterol-rich plasma with new plasma, reducing the cholesterol level. The xanthomas then disappear, *left.*

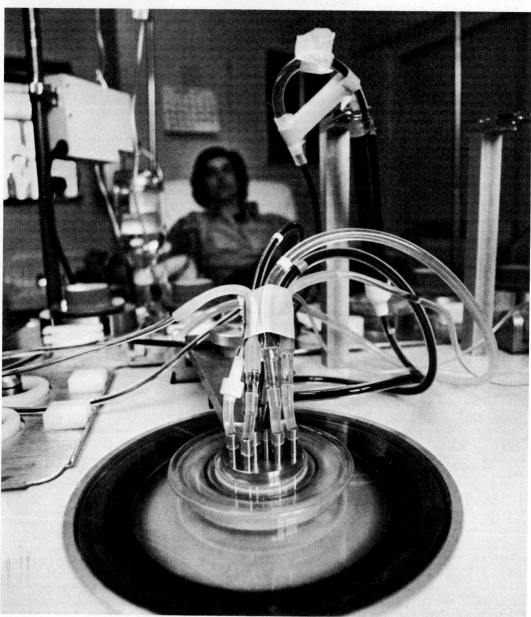

hyperplastic rat tissue to the iris of a rabbit. In three to five days, a ring of new blood vessels appeared around the tissue. Similar implantations of normal breast tissue or heat-killed hyperplastic tissue showed no such capacity.

These observations suggest a possible way to identify tissue that has the potential to become cancerous. It may be possible to test tissue biologically by measuring its capacity to induce new blood-vessel growth in an experimental animal. Identifying tissue in this way before it becomes malignant might permit earlier, more effective treatment.

Rabies vaccine. From mid-1975 through January 1976, 45 persons who had been bitten by rabid dogs or wolves in northwestern Iran were treated with a combination of hyperimmune mule serum, a vaccine supplement developed in the 1950s, and a new World Health Organization vaccine developed by microbiologist Hilary Koprowski of the University of Pennsylvania and coworkers at the Center for Rabies Research in Teheran. None of the victims died of rabies; all were alive and well a year after the vaccination except a 90-year-old man who succumbed to a heart attack five months later.

Rabies has been known for more than 23 centuries. In 1884, Louis Pasteur produced a vaccine from the rabies virus to give some protection against what had been an almost invariably lethal infection. Doctors later supplemented this vaccine with hyperimmune serum containing a high level of antibodies from a mule inoculated with the virus. This combination was highly toxic and only partially effective.

Without the new combination treatment, studies indicate that 35 per cent of the dog-bite victims in rural Iran would have died. Further evidence of its remarkable potency is that many patients did not receive the vaccine treatment until 7 to 14 days after exposure because of slow communications. The new vaccine can also apparently stimulate sufficient antibody production in noninfected humans to produce the serum needed to treat rabies victims. If sufficient donors can be found, the need for mule serum will be eliminated, and the last major obstacle to treating rabies effectively will finally be overcome. [Michael H. Alderman]

Surgery. A new skin-grafting technique that is designed to help severely burned children was described by John F. Burke, professor of surgery at Harvard University Medical School in Boston in 1976. Burke's system ensures that skin grafts will close the victim's wounds within 7 to 10 days, a much shorter time than was possible with past treatments. Rapid closure minimizes the risk of infection, the leading cause of death in child burn victims. The technique has not yet been tried on adults.

Surgeons usually use the patient's own skin to replace burned skin, but it is difficult to get enough healthy skin if more than 50 per cent of the body is burned. Areas of unburned skin can be used repeatedly for grafts, but the surgeon must wait at least 10 days to allow the donor site to heal before using it again for a skin graft. This greatly prolongs the time needed to replace the burned skin and close the wound.

In the treatment devised by Burke and his colleagues, all burned skin is removed promptly, thus avoiding the secondary infections that usually occur after the first week. Skin grafts from unburned portions of the patient's body are then applied over as much of the wound as possible, and the remainder is covered with skin taken from relatives or from cadavers. These skin grafts are taken from donors who have been tissue-typed and found to match the patient's tissue as closely as possible.

There were earlier attempts to use skin taken from relatives or cadavers. Pigskin was also used. Unfortunately, the victim's immune system rejected all these skin grafts within a week or so. This made it constantly necessary to obtain more donor skin.

Burke concluded that extending the duration of foreign skin grafts would be of great value, so he uses immunosuppression techniques similar to those used in kidney transplantation.

The sound of gallstones. Surgeons have developed two new methods to find and remove gallstones in the bile duct, a tube that leads from the gall bladder to the intestine. Heinz D. Reiss of Hannover Medical College in West Germany uses electronic amplification to find stones as small as 2 millimeters (0.07 inch) in diameter. Other surgeons are using flexible catheters to remove stones overlooked in normal operations.

Gallstones usually form in the gall bladder. Finding all the stones there is not important because the entire gall bladder with its stones can be removed. However, stones migrate into the bile duct in about 10 per cent of gallstone cases, and this vital structure cannot be removed. To get stones out of the duct, the surgeon must make a small opening in its middle because the upper and lower portions of this 8- to 10-centimeter (3- to 4-inch)-long structure are not easily accessible. Surgeons have traditionally used saline, a salt solution, to wash the stones out through the surgical opening. But overlooked stones are still a serious problem.

Reiss inserts a probe into the duct during the surgery. When it touches something solid, such as a gallstone, it vibrates. The vibrations are amplified electrically and can be heard through loudspeakers in the operating room.

Despite all these efforts, if a gallstone is overlooked and left in the bile duct the surgeon can still correct the situa-tion. In the past, the surgeon usually had little choice but to operate again if a retained stone became evident. Now the surgeon can use a new device that takes advantage of a rubber tube that is routinely left in the bile duct for from 7 to 10 days to drain bile during healing.

The surgeon inserts a catheter in the opening left by the tube and uses a fluoroscope to guide it into the bile duct. A small flexible basket, inserted into the duct through the catheter, is then used to trap and remove the stone. But more widespread use of the electrified probe will enable surgeons to detect all gallstones at the time of the operation, so that such procedures will rarely be needed. See DRUGS.

External esophagus. David B. Skinner, professor of surgery at the University of Chicago, described a new technique to help patients with blocked esophaguses at the 1976 meeting of the Society of Thoracic Surgeons in Washington, D.C. The esophagus is the tube that carries food from the mouth to the stomach. Skinner has used an artificial esophagus, originally developed in Ja-

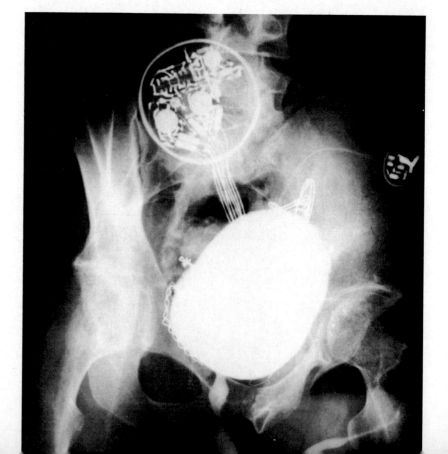

A mechanical device called a bladder pacemaker has been implanted to relieve bladder paralysis. It substitutes electronic impulses for signals that damaged nerves can no longer provide.

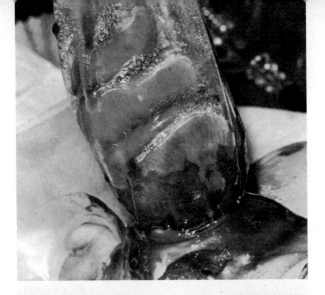

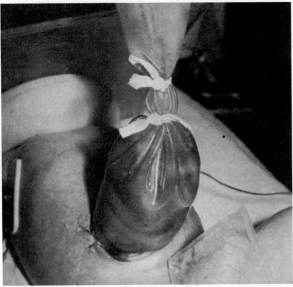

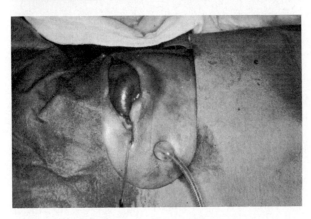

A plastic "silo" temporarily protects the intestines and other organs of infants who are born with defective abdominal walls, *top*. As the abdomen grows and stretches, the organs settle into it, *middle*. It can be closed from six to 12 days after birth, *bottom*.

pan, on six patients. The external device consists of a tube leading from an opening in the stomach, one from an opening in the throat, and a connector.

The natural esophagus can be blocked by tumors, by scarring from accidentally ingested acid or other substances, or even by the patient's own stomach acids. The block usually takes place in the chest. The earliest surgical replacement techniques used segments of either large or small intestine. But it is difficult to maintain the blood supply to these intestinal segments, and without blood, they become gangrenous. Surgeons have long sought a simpler, safer method.

The stomach tube used by Skinner is a common device designed to be inserted by minor surgery. The throat surgery is more complex because there is no dilated pouchlike structure that corresponds to the stomach. The surgeon cuts the natural esophagus, sutures the lower end shut, and sutures the upper end to the skin of the opening in the throat. After this has healed, he inserts the artificial esophagus tube into the upper esophagus. A small cap keeps the tube in place. The connector is used to link the throat and stomach tubes.

Patients must eat while standing or sitting up because the tube cannot reproduce the muscular contractions that guide food down the esophagus. However, even solid foods will pass down the tube with the aid of gravity.

Abdominal valve. A new valve and shunt tube developed by Harry H. LeVeen, professor of surgery at the State University of New York's Downstate Medical Center in Brooklyn, appears to be a significant advance in treating ascites, a fluid that accumulates in the abdomen. This condition results from a number of diseases. It is easy to insert a needle and draw off the fluid, but valuable plasma proteins and other substances are also drawn off and the ascitic fluid promptly reaccumulates.

LeVeen implants a tube under the skin that carries fluid from the abdominal cavity to one of the large veins in the neck. The key feature of the shunt tube is a one-way, pressure-activated valve that allows the ascitic fluid to flow into the neck veins but keeps blood from flowing into the abdominal cavity. Once in the bloodstream, the fluid is eliminated from the body as the blood is purified by the kidneys.

Parathyroid grafting. Surgeons Carl Feind of Columbia-Presbyterian Medical Center in New York City and Samuel A. Wells, Jr., of Duke University in Durham, N.C., described in November 1976 a new way to treat hyperactive parathyroid glands. The four parathyroid glands, located in the neck behind the much larger thyroid gland, play a major role in regulating calcium metabolism. When they misfunction, they draw calcium from the bones. The calcium then gathers in and damages the kidneys before it is excreted.

In such cases, the surgeon must remove most of the functioning parathyroid tissue, yet leave enough to avoid the severely reduced parathyroid activity that sometimes results in tetany, a disorder marked by tremors and muscle pain. Previously, surgeons removed 3½ parathyroid glands, hoping that the remaining fragment would produce the correct amount of parathyroid hormone. Repetitions of this operation were dangerous because scar tissue made it difficult to identify the vital nerves in this part of the neck.

Feind removed all the parathyroids and then implanted a small portion of one gland in a muscle on the side of the patient's neck. Although these implants were no more than small slivers, they survived because tiny blood vessels grew into them and provided them with blood. Wells later implanted 1-square-millimeter (0.01-square-inch) pieces into the forearm muscle and proved that they functioned as a gland because they showed a higher level of parathyroid hormone in the blood draining the arm with the implants. By removing all of the parathyroid tissue from the neck, these surgeons were certain of correcting the original problem — overproduction of parathyroid hormone. By implanting some tissue in the arm or neck muscle and getting it to survive, they avoided the opposite problem — tetany. And if the patient continues to secrete an oversupply of the hormone, they can remove tissue from the arm more safely than from the neck.   [Frank E. Gump]

Meteorology

The most dramatic meteorological development in 1976 and 1977 involved the weather observations from Mars obtained by Seymour L. Hess of Florida State University and his colleagues after the Viking 1 Lander touched down on July 20, 1976. These were the first direct meteorological observations on another planet that has weather similar to that on Earth.

Martian weather. A day on Mars is only 37 minutes longer than a day on Earth, so the effects of planetary rotation on the weather are similar on both planets. Furthermore, the Martian equator is tilted to the plane of its orbit at just about the same angle as the Earth's, so Mars also has seasons.

Major weather and climate differences arise from the absence of oceans and the scarcity of water on Mars. Thus, Martian weather lacks the moderating effect of large bodies of water and the complicating effects of precipitation and extensive clouds. These differences suggest that the atmospheric circulation on Mars may be much simpler than that on Earth.

Data sent back to Earth by the Viking 1 Lander during its first month on Mars provided a reasonably complete picture of summer weather on the broad slope of a basin that is named Chryse Planitia.

As the Sun rises over the calm Chryse basin, the temperature begins to rise from its predawn low of −85°C (−121°F.). The warming of the slope generates gusty winds that reach a maximum speed of about 32 kilometers (20 miles) per hour by late morning. The maximum temperature occurs during the afternoon — a comparatively warm −31°C (−24°F.). As the Sun sinks lower in the dusty Martian sky, the winds subside and the temperature falls rapidly. A light southwesterly wind, probably caused by cold air sliding down the slope, prevails during the early evening hours.

The Viking 2 Lander settled down on Utopia Planitia on September 3. Because it was summer in the northern hemisphere, temperatures at Utopia were similar to those at the first site, even though Utopia is much farther

Meteorology
Continued

The view from above a tornado-bearing storm indicates that tornadoes may form as a cloud top collapses downward, rather than by building up to enormous heights.

north. Winds were light and showed a regular pattern from one Martian day to the next. As the northern hemisphere shifted from summer to winter, and the north-south temperature contrast increased, the daily winds, temperatures, and pressures became more irregular, indicating a more variable weather pattern. See VIKING'S VIEW OF MARS.

Long-term climate changes. Because of its enormous complexity, climate research is usually slow and full of contradictory theories. Breakthroughs are rare and theories that are popular one year may fade the next.

One of the more concrete developments in 1976 was reported in December by James D. Hays of Columbia University's Lamont-Doherty Geological Observatory in Palisades, N.Y., and his colleagues. Their work confirmed a theory of climate change originally proposed by Milutin Milankovitch in the 1930s. Milankovitch suggested that the fluctuations in ice sheets over the past million years resulted from variations in the incoming solar radiation. According to his theory, the amount of

solar radiation received was affected by three types of cyclical changes in the Earth's orbital characteristics. These cycles are the 41,000-year obliquity cycle, based on changes in the angle between the Earth's equator and the plane of its orbit; the 93,000-year cycle based on the orbit's eccentricity, or departure from a circle; and the 21,000-year cycle based on precession, or the changing of the time of year when the Earth is closest to the Sun.

Hays and his associates analyzed ocean sediments laid down in the Indian Ocean during the past 450,000 years. By studying layers of the remains of minute sea creatures that are extremely sensitive to temperature changes, they found peaks in the variations of global climate with periods of 23,000, 42,000, and 100,000 years. These periods correspond closely to the periods of the changes in the Earth's orbit predicted by Milankovitch, providing strong evidence that orbital variation plays a significant role in producing climate changes. See NEW CLUES TO CHANGING CLIMATE.

Meteorology

Continued

Based solely on these orbital variations, the Northern Hemisphere appears headed for major glaciation over the next several thousand years. However, many other processes, including human activity, may enhance or override the changes predicted solely on the basis of orbital variations.

Sunspots and climate. One theory of short-term climate change, on time scales of up to several hundred years, involves changes in the amount and type of radiation emitted by the Sun, particularly the variations associated with sunspots. In June, John Eddy of the National Center for Atmospheric Research in Boulder, Colo., reported on his re-examination of observational records of sunspot activity since 1600. He focused particularly on the period between 1645 and 1715 when there were far fewer sunspots than normal. This period coincides with the "Little Ice Age" in Europe. From an analysis of carbon 14 found in tree rings, Eddy also deduced that a prolonged sunspot maximum occurred between 1100 and 1250. If these periods of maximum and minimum sunspot activity represent extremes of a solar cycle, the next maximum — and presumably the next warm-climate period — would occur around 2200. Eddy found that the number of sunspots has been increasing over the long term, and the amount of solar radiation reaching the Earth also seems to be slowly increasing. Thus, sunspot activity indicates a warming over the next 200 to 300 years.

Hail research. The National Hail Research Experiment continued in 1976, with a field program operating through June and July. The project is designed to evaluate the effect of cloud seeding on hailstorms. David Atlas, former director of the project, reported in January 1977 that seeding can either increase or decrease the amount of hail, depending on the storm's structure.

The formation of hailstones depends on temperature, content of supercooled water (water cooled below its normal freezing point), and updraft speed. Under certain conditions, seeding a storm cloud with silver iodide nuclei at freezing temperatures may increase the

Russian scientists at a meteorological station near Moscow bounce laser signals off the atmosphere to get data on clouds, temperature, atmospheric pressure, and moisture content. The instrument operates at ultraviolet, visible, or infrared wavelengths.

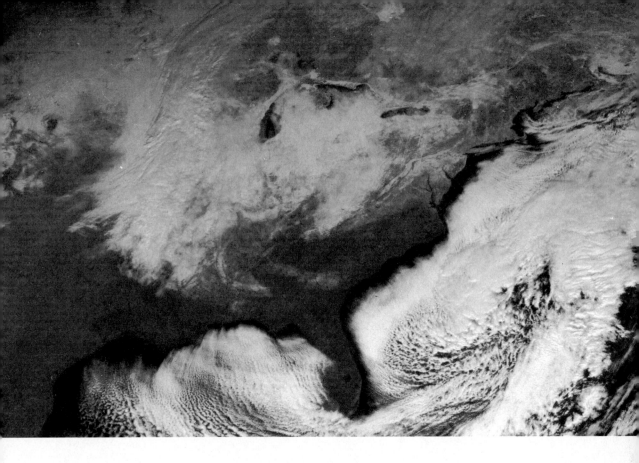

Meteorology

Continued

Satellite picture taken on January 19, 1977, reveals the severity of the winter in the United States and Canada. Snow blankets Midwestern and Northeastern states, and reaches down as far as southern Georgia and Florida. Chesapeake Bay is frozen solid, and ice covers parts of Lakes Michigan and Superior.

number of small hailstones at the expense of larger, more damaging stones. The small hailstones melt before reaching the ground. However, if the cloud contains enough supercooled water, more large hailstones will be produced.

Atlas suggested that future hail-suppression experiments classify the hail-producing storms according to their physical characteristics. It should then be possible to identify storms in which seeding may reduce the number of large hailstones, and others in which an increase is likely.

Heat-transfer research. Results began to appear from the 1975 Air Mass Transformation Experiment, which studied the rates of heat and moisture transfer from the ocean to the air. The study focused on the transformation of cold, dry, polar air to milder, moist, oceanic air as polar air masses travel along the east coast of Asia in winter. During cold periods, the energy transfer from the warm ocean to the cold air was about 3.5 times greater than during warmer periods. The larger figure was similar to the rate at

which solar radiation is absorbed at the Earth's surface during a typical summer day in middle latitudes. This amount represents a significant energy source for atmospheric motions.

Hurricane prediction. During the 1976 hurricane season, an experimental high-resolution computer model of the atmosphere helped to predict tropical storms and hurricanes in the western Atlantic and eastern Pacific. Computer models provide a three-dimensional picture of atmospheric conditions in numerical form.

This model, developed by John Hovermale and his associates at the National Meteorological Center in Washington, D.C., has a much higher horizontal resolution than previous models. The horizontal resolution is determined by the size of the grid on which the atmospheric data are represented. Hovermale's model has a mesh size of 60 kilometers (37 miles), three times smaller than the next mesh size used at the center. Also, Hovermale's model can be adjusted to follow a hurricane.   [Richard A. Anthes]

Microbiology

Biochemist Julius Adler and his colleague Sevec Szmelcman of the University of Wisconsin at Madison reported in December 1976 on experiments designed to determine how bacteria respond to chemical stimuli. An intensive investigation of the response, called chemotaxis, was begun by Adler and his co-workers about 10 years ago. They showed that the common intestinal bacterium *Escherichia coli* swims toward some chemicals and away from others. They also showed that the bacterium has at least 20 receptor sites for the chemicals, sites where the chemicals attach to the bacterium's cell membrane. The combination of chemicals with receptor sites somehow causes hairlike structures that are called flagella to rotate rapidly, and this propels the bacterium.

Membrane and motion. Adler and Szmelcman wondered exactly how the chemical-receptor combination signaled the flagella. They suspected that the chemical-receptor reaction might cause a change in the electrical voltage of the bacterium's membrane similar to the changes in nerve-cell membranes during the transmission of nerve impulses in higher forms of life.

Scientists usually measure membrane voltages by inserting microelectrodes into cells. Because bacteria are too small for this procedure, Szmelcman and Adler devised an indirect way to measure the membrane voltage. They used a charged molecule called triphenylmethylphosphonium. Molecules of this material pass freely through the membrane in both directions when the voltage across the membrane is at a level typical of the bacterium's normal resting stage. When the voltage goes above the resting value, the molecules pass into the cell almost exclusively; when the voltage goes below the resting value, the molecules pass almost exclusively out of the cell.

The scientists added a chemotactic chemical, the nutrient sugar galactose, to a suspension of the bacteria. Triphenylmethylphosphonium was taken in by the bacteria almost instantaneously, signifying an increase in the voltage across the membrane.

When the scientists added a variety of chemicals that are not chemotactic, but some of which are also foods for the bacteria, the membrane voltage was not affected.

Adler and Szmelcman further linked the voltage change to chemotaxis by using a strain of the bacterium that does not react chemotactically to galactose. Bacteria from this strain consume and metabolize the galactose just as well as the normal organism, but they do not automatically swim toward the sugar. When galactose was given to the nonchemotactic bacteria, they did not increase their intake of triphenylmethylphosphonium. This strain of bacterium did not change its membrane voltage in response to galactose, which is why it does not swim toward it.

Bacteria are often ideal for such studies because many billions of the organisms are easily confined in a test tube or flask and because they grow rapidly, some multiplying every 20 minutes. But, most important, experience has taught biologists that what they learn from studies of a process in bacterial cells can usually be applied to cells of higher organisms, including humans. Adler and his colleagues have demonstrated that bacteria have a rudimentary nervous system that depends on changes in membrane voltage, just as does the nervous system of higher organisms. The next step is to try to discover how this change in voltage controls a bacterium's flagella and thus directs motion.

One bacterium from two. Research groups in France and Hungary reported in July 1976 on the results of independent investigations of what may prove to be an important new way to manipulate bacterial genes. Both groups fused bacteria from two different strains into one, and the resulting hybrid bacteria had the combined genetic characteristics of the originals.

The new technique qualifies as genetic engineering, a highly publicized and controversial area of research, because it transfers genes from one organism into another under conditions that allow the genes to continue to function in their new "home." The technique involves fusing protoplasts of the bacteria. Protoplasts are cells which have had their outer cell wall removed, thus exposing the underlying cell membrane. Protoplasts of plant cells were first fused in 1972.

Microbiology

Continued

One of the research teams that fused the bacteria in 1976 included microbiologists Pierre Schaeffer, Brigitte Cami, and Rollin W. Hotchkiss of the University of Paris-Sud in Orsay, France. They worked with two strains of the common soil bacterium *Bacillus subtilis*. The strains differ in their requirements for growth factors, such as certain amino acids and nucleic acids that normal bacteria can synthesize.

The researchers suspended bacteria from each of the two strains separately in solutions and treated them with lysozyme, an enzyme that digests and completely removes the cell walls of some bacteria without affecting the membrane. The scientists then mixed the resulting two cultures of protoplast bacteria. Next they added a solution of polyethylene glycol, which for some unknown reason causes protoplasts to fuse. After several minutes, the scientists placed samples of the mixture on a solidified growth medium containing sufficient nutrients to allow the protoplast bacteria to produce cell walls once again and become normal bacteria.

To determine if protoplasts of the two strains had fused, the scientists placed the bacteria on a growth medium containing none of the growth factors needed by the two original strains. The only cells that could survive on such a medium would be those that resulted from a mixture of the genetic material of the two original strains, because only such hybrids could synthesize their own supply of the missing factors. Many of the bacteria survived, strongly suggesting that the protoplasts had fused and their genetic material had mixed.

The bacteria used in these experiments can exchange genes in two other ways that could account for the hybridization. In one process, a bacterium dies and deteriorates, releasing its deoxyribonucleic acid (DNA), the chemical that comprises the genes. Living bacteria then incorporate fragments of the released DNA in their own genetic makeup in a process called transformation. In the other process, genes are transferred from one cell to the other during brief cell-to-cell contact, a proc-

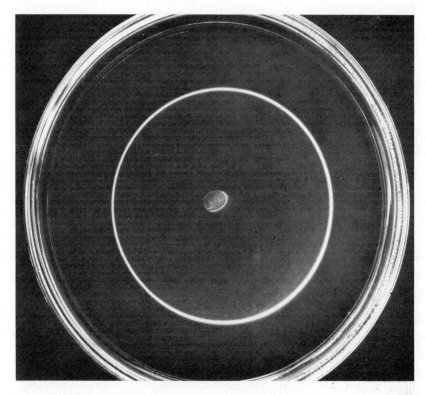

Bright ring in glass dish is composed of millions of bacteria that were originally placed in the center. The organisms have slowly moved outward because the dish contains an attractant whose concentration increases toward the edge of the dish. This response, called chemotaxis, indicates that the bacteria have a rudimentary nervous system.

ess called conjugation. The Schaeffer group designed its experiments to exclude transformation.

They added deoxyribonuclease, an enzyme that rapidly destroys DNA, to the protoplast mixture. Any DNA released into the solution by a protoplast would have been quickly destroyed by the enzyme.

Proof that conjugation did not occur was more difficult. The scientists selected a virus that attacks, enters, and kills bacteria of one of the strains used in the experiment, but becomes inactive after it attacks and enters bacteria of the other strain. The virus becomes inactive because a protein molecule called a repressor is present in the cytoplasm of these bacteria. The repressor prevents the virus from multiplying and destroying the host bacterium.

When the scientists added the virus to a solution of the hybrid bacteria, the bacteria were not killed. The only explanation for the failure of the virus to kill at least some of the bacteria is that protoplasts had fused, introducing into the cytoplasm of all the cells the repressor molecules that were originally present in only one strain.

The Hungarian research team included microbial geneticists Katalia Fodor and Lujoa Alföldi of the Institute of Genetics at the Hungarian Academy of Sciences in Szeged. The group followed essentially the same procedure as the French team did, except that they used a different species of bacteria, *Bacillus megaterium,* with different growth-factor requirements. *B. megaterium* is better for such experiments because it cannot exchange genetic material by transformation or conjugation. Consequently, the Hungarian group's results were clear cut — protoplasts of the two strains fused to form hybrid bacteria.

Bacteria fusion promises to be a major advance in research involving microbial genetics. One ultimate goal of this kind of research is the manipulation of genes to create organisms with entirely new and useful properties. An example would be an organism with the ability to produce large quantities of insulin.   [Jerald C. Ensign]

Neurology

The opening of vesicles at the nerve ending following an electrical impulse in the nerve was demonstrated for the first time in October 1976 by neuroscientist John Heuser of the University of California Medical School, San Francisco. Vesicles are small sacs containing chemicals that transmit information from one nerve cell to another.

Working with physiologist Thomas W. Reese of the National Institutes of Health, Heuser has developed a fast-freezing method that stops the activity of the nerve in a fraction of a millisecond while it is being stimulated. Then he photographed the nerve's ending through an electron microscope at magnifications greater than 100,000 times. The photographs show vesicles frozen as they were opening and emptying their contents into the synapsaptic cleft, the space between two nerves. Nerve cells communicate with one another at the synapses, where their membranes almost touch.

Heuser's pictures confirm other evidence that when an electrical impulse travels down an axon, or branch, of a transmitting nerve cell to its ending, a certain number of the tiny vesicles in this ending discharge their chemical contents. Chemical receptors in the receiving nerve's cell membrane absorb the chemical transmitter, and generate a new electrical impulse that travels down the second nerve cell to communicate with another nerve cell at the synapse.

Neuroscientists have seen or measured many of the phenomena associated with synaptic discharge. But they still do not know where and how vesicles are formed, how they get to the membrane at the synapse, and how they discharge their transmitter contents when given the appropriate signal. Some of these questions are now being answered.

Brain anatomist E. George Gray of University College in London reported new evidence in 1976 that microtubules, long fibrous proteins commonly seen in the axons of nerve cells, may be involved in delivering the vesicles. Using a new method of staining extremely thin fragments of rat and frog

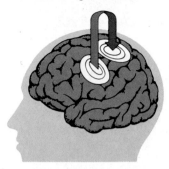

A researcher measures small magnetic fields in the brain of a test volunteer, *right*. Such new methods allow scientists to find the specific brain area, *above*, that responds to such stimuli as a mild electrical shock given to the right little finger.

Neurology

Continued

brain, Gray produced electron micrographs showing that microtubules extend into the axon terminal area and up to the membrane at the synapse. The vesicles seem to run close to the microtubules as if they were being guided by them to the synaptic sites. These findings may prove to be key elements in understanding the brain's information-transmitting mechanisms.

Huntington's chorea. A condition in rats that mimics some features of Huntington's chorea has been produced by brain specialists Joseph Coyle, Robert Schwarcz, and their associates at the Johns Hopkins University School of Medicine in Baltimore. Their findings were reported in September 1976. Previously, there had been no direct way to study this disease except by work with patients and human brain tissue.

Huntington's chorea, an inherited neurological disorder, usually strikes people between 30 and 40 years of age. The primary symptoms are continuous twitching of the limbs and mental deterioration. There is no cure at present; death usually occurs within 15 years.

The brains of patients who have died of Huntington's chorea show a marked degeneration of nerve cells in an area called the corpus striatum and have considerably lower than normal levels of two neurotransmitter chemicals that are involved in processing information in this area. The Johns Hopkins researchers injected kainic acid, a chemical known to attack brain cells in mice, directly into the striatum of rats.

After the injection, the rats behaved abnormally in ways consistent with the human disease symptoms and showed the same type of nerve-cell deterioration that occurs in humans. Such animal studies may allow detailed examination of the biology, chemistry, and physiology of cells in this tragic disease and enable scientists to test various drugs for detection and treatment.

Infantile autism. Looking for common features in autistic children, researchers Michael Rutter and Susan Folstein of London's Institute of Psychiatry, reported in February 1977 on their examination of 21 sets of twins of the same sex, one or both of whom were autistic. Autism is a rare condition in which children withdraw from reality, often right after birth.

In four of the 11 pairs of identical twins, both twins were autistic. In pairs where only one twin was autistic, the other had less severe problems, such as learning difficulties and delayed speech. In the 10 pairs of fraternal twins, only one of the nonautistic children had difficulties; the rest were normal. Because identical twins develop from a single fertilized egg and fraternal twins from two, these findings indicate a strong genetic factor in autism.

Injury at birth also seems to be a factor in some cases. Among pairs in which the nonautistic twin had learning difficulties, the autistic twin was injured during birth. This may indicate that the other, nonautistic twin may have suffered a much milder injury and that autism involves a genetic predisposition that has to be triggered by a certain level of brain injury at birth. Parental-psychological factors, once thought to be involved in the disease, seem a much less likely cause.

Bird brains. Striking size differences between male and female brains in canaries and zebra finches were reported in October 1976 by animal behaviorists Fernando Nottebohm, Arthur P. Arnold, and their colleagues at the Rockefeller University Field Research Station in Millbrook, N.Y. The differences seem related to the ability to sing.

Although female songbirds can produce calls, they cannot produce the full-fledged song that the adult male learns and sings, usually to attract mates and establish territory. The researchers report that the brain centers known to control song are larger in the males and contain more cells and fibers, reflecting the males' complex songs.

Because the patterns of adult birdsongs are usually learned, the findings may increase our knowledge of an aspect of the learning mechanism itself. Research on bird brains and birdsong may throw light on questions of human differences in sexual behavior, communication, and learning, which, in some basic elements, may be like those of birds. [George Adelman]

Nutrition

Research in 1976 and 1977 raised the possibility of a previously unknown nutrient requirement in infant diets. Kenneth C. Hayes and his colleagues at Harvard University's School of Public Health in Boston found that kittens developed sight problems when fed a diet based on the milk protein casein plus other essential nutrients. The retina of the eye degenerated because the kittens lacked taurine, a sulfur-containing substance found in muscle, lung, and other tissue.

These findings are surprising because investigators have believed that cats could metabolize enough of the sulfur-containing amino acids methionine and cystine in the casein to form sufficient taurine to prevent disease.

Scientists had also assumed that humans can convert these amino acids to taurine. Hence, the Harvard findings may be significant for human nutrition. Previous experiments showed that humans, cats, and monkeys given methionine and cystine supplements fail to show higher taurine levels in the blood. Other investigations had shown that

the human liver may not make enough of the enzyme required for this metabolic change.

Hayes noted that premature human infants have lower taurine levels in certain tissues than do normal infants. Moreover, premature infants are usually fed a formula based on milk protein, which does not contain as much taurine as does a nursing mother's milk. Hayes plans more research to determine the consequences of such a diet and the possible need to add extra taurine to it.

Iron in the diet. Concern about the low levels of iron in many diets usually focuses on the lack of enough iron-containing hemoglobin to carry oxygen to the tissues. Fatigue and limited ability to perform muscular work — common symptoms of iron-deficiency anemia — are usually blamed on this lack of hemoglobin. But Clement A. Finch and his colleagues at the University of Washington in Seattle reported in August 1976 that iron deficiency might also directly affect muscle tissue. They tested rats running on a treadmill and found that those fed an iron-deficient

Nutrition

Continued

diet stopped running sooner than those on a normal diet.

The researchers gave the anemic rats blood transfusions in order to raise their hemoglobin level to that of a control group, but their running time did not increase. Also, subsequent biochemical measurements of the anemic rats' muscle cells showed that they had fewer cytochromes, the iron-containing compounds required to use oxygen inside the cells. More important, some enzyme levels dropped during the course of the experiment. The researchers reported that a decrease in one such enzyme, α-glycerophosphate dehydrogenase, markedly reduced the anemic rats' muscle cells' ability to use oxygen to produce energy.

Iron levels in the diets of nursing infants also concerned nutritionists. Julian A. McMillan, Stephen Landaw, and Frank A. Oski of the Upstate Medical Center in Syracuse, N.Y., reported in November 1976 that breast-fed infants do not show signs of iron deficiency even though human milk is known to be low in iron.

To determine the reason for this, the researchers compared the body's ability to absorb iron from human milk and from cows' milk. They added radioactively labeled iron to samples of each type of milk and fed them to adults. They found that 21 per cent of the iron from human milk was absorbed in the intestine, but only 14 per cent of the iron from cows' milk was absorbed there. Although they tested adults, the nutritionists noted that other studies have shown that infants absorb iron at twice the rate of adults.

From these figures, they calculated that human milk, because its iron is absorbed better, would supply more iron than an infant needs, but that unfortified cows' milk would not. Hence, the practice of adding iron to infant diets based on cows' milk should be continued.

Iron fortification in the diets of elderly people was also studied. Stanley N. Gershoff and his colleagues at the Harvard School of Public Health tried to determine the effect that iron-fortified cereal products had on elderly people

Milk containing trillions of *Lactobacillus acidophilus* bacteria, which can break down lactose, was developed for the many people who inherit an inability to digest milk lactose.

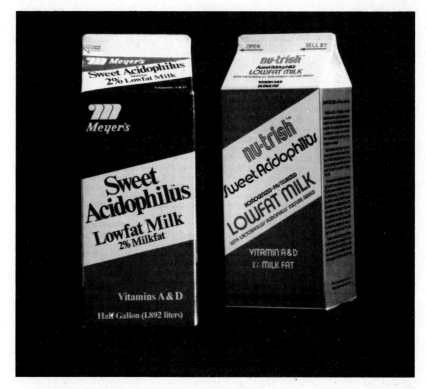

Nutrition

Cylindrical hard-boiled eggs that are made from separated and recombined yolks and whites ensure uniform proportions and size in each slice.

whose blood had moderately low iron levels but who were not anemic. They gave one group cookies, crackers, and hot cereals to which an iron salt, ferric orthophosphate, had been added. Those eating the iron-fortified cereals showed a small rise in hemoglobin levels. But those eating unfortified cereals also had higher hemoglobin levels.

The researchers concluded that the fortification did not produce the increase. They theorized that interest in the nutritional study had influenced both groups to improve their dietary habits. This explanation seems credible because all the subjects knew that their blood iron levels were less than ideal.

The study demonstrated how difficult it is to separate the individual's response, the effect of a single nutrient, and the interplay of dietary components in nutritional studies. Consequently, education and motivation may be as important in improving diet as are nutrients for fortification.

Alcohol and exercise. Joan Orlando, Wilbert S. Arnow, John Cassidy, and Ravi Prakash of Long Beach Vet-

erans Administration Hospital in California and the University of California at Irvine in July 1976 tested the effects of alcohol on people with angina pectoris, a type of heart disease in which the heart does not receive enough blood. Although some physicians suggest that patients with angina pectoris drink small amounts of alcohol to relieve some symptoms, the researchers found that as little as 60 milliliters (2 ounces) of alcohol significantly lowered ability to exercise without pain.

The researchers related these changes to the heart muscle's greater need for oxygen when alcohol is in the body. They also noted some changes in the electrical activity of the heart as measured by an electrocardiograph. They concluded that people suffering from angina pectoris could not increase the blood flow to the heart enough to supply the oxygen they needed. They drew no conclusions about the effects of alcohol on healthy people, but some nutritionists suspect that alcohol has a similar effect on people who do not have heart disease.   [Paul E. Araujo]

Oceanography

Ocean scientists reported new findings during 1976 and 1977 on the causes of global climate change, the oceans' effect on weather, the origins of new metal deposits on the sea floor, and the hazards of marine pollutants.

Oceans, climate, and weather. Scientists working in the Climate: Long-range Investigation, Mapping, and Prediction (CLIMAP) project reported in December 1976 that they had determined by examining samples cored from beneath the ocean floor that long-term changes in the earth's climate, such as the onset of ice ages, are linked to changes in the earth's orbit around the sun.

Serbian geophysicist Milutin Milankovitch suspected as early as 1930 that the earth's orbital changes triggered global climate shifts. But not until 1976 did scientists confirm the theory.

CLIMAP investigators analyzed variations in the type and amount of microorganisms in deep-sea sediment cores taken from the southern Indian Ocean. Sediment in this area built up over millions of years. Therefore, the

deeper scientists can penetrate into the ocean floor, the further they can look into the past. The Indian Ocean cores penetrated deep enough to provide an uninterrupted record of the earth's climate for the past 450,000 years, a period three times longer than scientists had previously been able to analyze.

The scientists knew that different species of radiolaria, marine microorganisms, preferred either warm or cold water. So they could identify periods when the ocean water was warm or cold by tracing the fossil layers of the different radiolaria species embedded in the core. In addition, they checked for evidence of ice in the oceans by measuring the concentration of two different types of oxygen atoms found in the shells of other marine microorganisms called plankton. Most of the lighter, ordinary oxygen atoms were locked in ice when the climate was cold. The oceans contained a high proportion of the heavier oxygen isotope, because it did not evaporate as easily. The plankton fossils contained the record of this oxygen ratio. Their analysis enabled

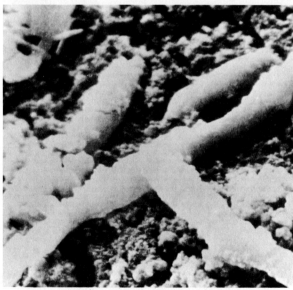

Scientists test a new water-sampling device, *top,*
that protects deep-ocean bacteria by sealing them
in a chamber that maintains the high pressures and
low temperatures existing in the ocean depths. Marine
bacteria retrieved and preserved in the chamber, *above,*
can be grown in culture and studied in laboratories.

the researchers to estimate about when changes occurred in the volume of ice in the earth's oceans.

The CLIMAP study indicated that there were three cooling cycles in the earth's climate during the past 450,000 years that lasted 23,000, 41,000, and about 100,000 years. These climate cycles were nearly identical to cycles in the tilt, wobble, and other changes in the earth's orbit around the sun.

The apparent regularity of these cycles should enable scientists to predict the earth's climate. They may also be able to determine whether future changes are part of natural cycles or the result of industrial activities. See NEW CLUES TO CHANGING CLIMATE.

Scientists in the North Pacific Experiment (NORPAX) continued their efforts to link large areas of unusually warm or cold surface waters in the northern Pacific Ocean to seasonal weather patterns over the United States. For example, NORPAX meteorologist Jerome Namias of the Scripps Institution of Oceanography used Pacific Ocean conditions to predict weather patterns three months in advance. In June 1976, he correctly forecast a warm, dry summer in the North Central States and cool, wet weather in the Pacific Northwest. Namias also correctly predicted that the winter of 1976 and 1977 in the Eastern United States would be the coldest since 1971. These successes promise improved seasonal and long-range forecasts as research progresses on the influence of the ocean on atmospheric conditions.

Exploring the sea floor. During February and March 1977, geologists tested the theory that metal deposits are formed when hot seawater circulating within the upper area of the earth's crust extracts metals from the rock and carries them up to the sea floor. Scientists made 25 dives to the Pacific Ocean's Galapagos Rift Zone in the U.S. research submersible *Alvin.* In the rift zone, two rigid sections of the sea floor are pulling apart. There, lava pushes up to form new crust.

From *Alvin's* windows, nearly 3.2 kilometers (2 miles) beneath the ocean surface, the scientists saw hot-water geysers spew from cracks in the newly hardened lava, sending plumes of chemically rich brine into the cold sur-

Oceanography

Continued

rounding water. The surface of new lava was stained bright red, orange, white, and gold by the freshly deposited minerals. The water from the geysers was nearly 9°C (16°F.) hotter than the 2°C (36°F.) temperatures usually found at these depths.

The scientists were surprised to find an abundance of marine life near the geyser vents. They saw clams nearly 0.3 meter (1 foot) long, dense populations of mussels, crabs, sea anemones, limpets, and chitons, and clusters of organisms that looked like dandelions. The scientists continued their study of the unusual marine communities during 1977.

In the summer of 1976, Deep Sea Drilling Project scientists on board the drilling ship *Glomar Challenger* found evidence in sediment cores that the continental edge rapidly collapsed after North America broke away about 150-million years ago from the single land mass that then existed to form a continent separate from Europe and Africa. They found huge underwater mudflow and avalanche deposits on both sides of the Atlantic Ocean. What were once tidal flats off the United States rapidly dropped more than 2,700 meters (9,000 feet). Off Spain, swamps ringed with coral reefs sank 1,200 meters (4,000 feet) into the Bay of Biscay.

Powerful underwater currents swept away a layer of sediment nearly 3.2 kilometers thick. These currents carried a chaotic assortment of pebbles and mud from the fairly shallow continental shelves to deep water, probably carving canyons across the sloping shelves as they moved. Large amounts of organic matter were buried quickly by this process, and it will someday be converted into oil and natural gas by natural geologic forces.

Marine pollution. Massachusetts Institute of Technology researchers reported in February 1977 on their analysis of oil spills from 100 tankers sunk by German submarines during World War II. Some 5.48 million liters (1.45 million gallons) of crude and refined oil spilled into Atlantic waters within 80 kilometers (50 miles) of shore. Researchers found it difficult to determine the extent of environmental damage,

An Antarctic iceberg, as seen by the NOAA-5 weather satellite from an altitude of 1,448 kilometers (900 miles), is almost as big as Rhode Island. It would pose a serious hazard to shipping if it entered the open waters of the South Atlantic Ocean.

Oceanography
Continued

but the oil apparently had a minimal effect on fish and waterfowl.

A five-year study released in June 1977 by the American Petroleum Institute suggests that the abundance and variety of marine life do not seem to be affected by low levels of crude oil from offshore drilling operations or from natural seepage. The petroleum institute researchers also claimed that marine organisms exposed to different types of oils apparently purge themselves of oil quickly once it is removed from their environment. The scientists also reported that continuous low-level exposure to oil does not appear to affect the growth rate and reproduction of marine organisms.

However, other studies apparently contradict these findings. Scientists at the Woods Hole Oceanographic Institution, who have been studying the long-term effects of a 1969 oil spill on local marshland, believe oil spills are damaging. Although the marshland's animal population apparently returned to normal after eight years, the researchers suspect that oil spills have long-term effects on marine life.

Their fears seemed to be borne out by preliminary results reported by U.S., Canadian, and British scientists working in the Controlled Ecosystem Pollution Experiment (CEPEX) in January 1977. The researchers were concerned not only with oil, but also with the long-term effects of other chemical pollutants. Their experiments indicate that copper, mercury, and petroleum in concentrations of only 10 parts per billion adversely affect tiny marine organisms. The scientists suspend huge plastic bags in the clear waters of a coastal inlet of Vancouver Island, Canada, to test the impact of pollutants on the lowest links of the ocean's food chain. They fill the bags by pulling them up from the bottom of the inlet so that each bag encloses entire communities of marine microorganisms. The scientists placed pollutants in some of the bags and compared these communities with those in bags free of the pollutants.

All three pollutants decreased the population of large diatoms – forms of algae – and increased that of dinoflagellates, smaller algae. The scientists added the pollutants at the beginning of each 30-day experiment and the concentration of the pollutants was diluted steadily during the test. But there was no evidence that the population balance returned to normal.

Disruption of this microscopic end of the food chain could cause changes in the fish populations that feed on them. There is already evidence that such population changes have occurred since the mid-1960s. Experiments in Scotland and the North Sea suggest that extremely low levels of pollutants can cause an explosion in the jellyfish population and a decline in the number of edible fish. Oceanographer Nicholas S. Fisher of Woods Hole suggests that these changes may have been the results of industrial pollutants.

Pollutants may also aggravate natural changes in the marine environment. During the summer and fall of 1976, anoxic, or low-oxygen, conditions in the coastal waters off New York and New Jersey trapped and killed many fish and shellfish while many other fish fled the area. In addition, oceanographers suspect that the long-term effects – egg and larvae deaths, interruption of spawning, and disruption of food chains – may be as important to these living resources as the actual number of fish killed.

Scientists are not certain why the oxygen level dropped. But they speculate that a combination of unusual weather conditions, water-circulation patterns on the continental shelf, and the decay of algae and other organic matter were partly responsible. These conditions were apparently made worse by the discharge of natural wastes into these waters. The millions of people who live in the cities and suburbs of the New York-New Jersey coastal region discharge a large part of their waste into the ocean either directly or through rivers.

Researchers at the Atlantic Oceanographic and Meteorological Laboratories in Miami, Fla., suggested that this low-oxygen incident would not have been as serious – or may not have occurred at all – had such nutrients as phosphates and nitrogen not been carried from sewage-treatment plants and down rivers to the sea, where they then aggravated natural changes in the water. [Feenan D. Jennings]

321

Physics

Atomic and Molecular Physics. Scientists at Oak Ridge National Laboratory in Tennessee demonstrated a technique in 1976 that they claim will achieve the ultimate goal of trace analysis — to detect a single atom immersed in a mixture of other atoms. Single-atom detection has been possible for radioactive atoms, but the technique described in November by G. Samuel Hurst, Munir Hasan Nayfeh, and Jack P. Young can be used to detect nonradioactive atoms.

Their experiment combined the high power and wavelength selectivity of a laser with a gas proportional counter, a sensitive detector developed originally for nuclear physics experiments. The counter was filled with a gaseous mixture of a few cesium atoms and billions of billions of argon and methane molecules. The Oak Ridge experimenters used a pumped dye laser tuned to emit light at the precise wavelength required to excite a cesium atom from its ground, or lowest-energy, state to a specified excited state. The cesium atom absorbed a photon of light to become excited, but argon and methane molecules were unaffected.

The laser action is so intense that the excited cesium atom absorbs a second laser photon before returning to its ground state. The second photon imparts enough energy to the atom to remove an outer electron, thus ionizing the atom. An electric field accelerates the free electron, which collides with the gas atoms and knocks off more electrons to produce an avalanche of electrons and ionized gas atoms. It is the electrical pulse generated by this avalanche that is detected, signaling the presence of a cesium atom.

The researchers report that their single-atom detection method is so sensitive that they can detect one cesium atom among more than 10 billion billion atoms or molecules of another kind. They believe that the technique may provide a highly sensitive means for detecting a wide range of chemical pollutants in the environment.

Light labels energy states. Physicist Arthur L. Schawlow and his co-workers at Stanford University in California

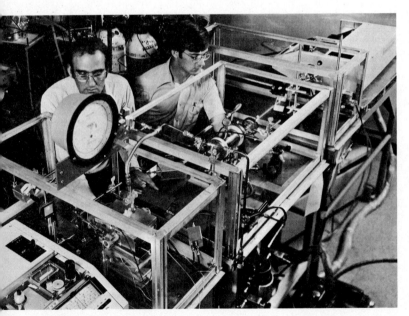

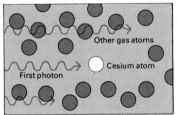

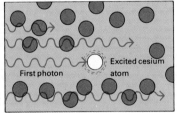

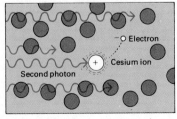

Scientists demonstrate a laser device, *above,* that can detect a single atom of an element such as cesium in a gas of 10 billion billion other atoms. A precisely tuned laser emits photons that excite, then ionize, only cesium, *right.* The energetic electron from the ion will trigger a sensitive detector, revealing the single cesium atom's presence.

demonstrated a technique in 1976 that simplifies the process of identifying a molecule's various energy states. The simplification consists of labeling those lines in a spectrum that result from transitions to the various high-energy states from one, and only one, of the molecule's many low-energy states.

Scientists study molecular structure by observing the spectra produced by light absorbed or emitted by a molecule when it changes energy states. Each state represents a different combination of rotational, vibrational, or orbital energy for the molecule. Because there are many possible combinations, there are many energy states, some of which are nearly the same. The closely spaced energy states produce a maze of overlapping spectral lines that are difficult for scientists to unravel.

The Stanford scientists studied molecular sodium that was vaporized in an oven and irradiated by two lasers. The first laser labels certain lower states; the second detects the consequence of that labeling process. The labeling, or pump, laser produces an intense pulse of circularly polarized radiation at a wavelength tuned to one of the lower-to-upper energy state transitions of the sodium molecule.

Light consists of electrical and magnetic vibrations in a plane that is at right angles to the direction in which the light is moving. In plane-polarized light, the plane is fixed. When the plane rotates uniformly about the beam direction, however, the light is called circularly polarized. It can rotate either to the right or to the left. In fact, plane-polarized light can be viewed as composed of an equal mixture of left-circularly polarized and right-circularly polarized light beams.

The lower state of the sodium molecule is generally made up of a collection of substates, all of which have the same energy in the absence of a magnetic field. However, the states differ in their reaction to polarized light. Some absorb right-circularly polarized light, others absorb left-circularly polarized light. Thus, the circularly polarized pump laser light selectively excites only a portion of the molecules in the lower, or labeled, state. Most of the molecules that remain in the labeled state respond to light of the opposite polarization.

A second probe laser that emits a very broad range of wavelengths exploits this property. Its plane-polarized light faces a filter that will block it if the plane of polarization is unchanged after it passes through the sodium cell. However, if a portion of the probe laser light changes its polarization in traversing the sodium cell, it will pass through the filter and be detected. This polarization change occurs at those probe laser wavelengths that correspond to transitions from the labeled state to higher-energy states because the labeled state no longer responds equally to the left- and right-circularly polarized parts of the light. In this way, spectral lines of a common origin are labeled, and analysis of the molecule's spectrum is greatly simplified.

Light pushes light. Observing the direct influence of one light beam on another is still far beyond researchers' observational capability. But recent experiments by physicists Andrew C. Tam and William Happer of Columbia University in New York City achieved the next best feat in 1976. They demonstrated the attractive and repulsive interaction of two light beams through an intermediary reaction with atoms of sodium vapor.

The researchers aimed two circularly polarized beams from a dye laser on a collision course at an angle of one-tenth of a degree between the beams. The two beams were to intersect inside a cell containing sodium vapor. The wavelength of the laser beams was chosen to be on the low-wavelength edge of one of the yellow absorption lines of sodium that give sodium vapor street lamps their characteristic yellow glow.

An unpolarized laser beam at this wavelength can interact with any of the sodium atoms. However, a polarized beam will be selectively absorbed by only one part of the atomic population. The circularly polarized light beam will therefore surround itself with a sheath of atoms more susceptible to one polarization than to the other. The asymmetric environment causes the beams to bend as they would in a medium whose optical density changes. The repulsion observed was great enough to cause the two beams to diverge rather than to cross inside the sodium cell.   [Karl G. Kessler]

Elementary Particles. Quarks, the tiny objects that physicists believe make up the protons and neutrons of atomic nuclei, dominated particle physics during 1976 and 1977. No sooner had experimentalists firmly established the existence of a fourth quark than other researchers reported evidence that there may possibly be more. Then, in April 1977, came news that a free quark may have been found after more than 10 years of searching. A free quark is one that is not bound to another quark inside a particle.

Despite the uncertainty about the total number of quarks, the quark theory of subatomic matter was firmly established during the year. In this theory, all matter is composed of two kinds of tiny objects—quarks and leptons, which include the familiar electron and its electrically neutral companion, the electron neutrino. Leptons and quarks differ most in that quarks are bound together by a little-understood strong force that no experiment has yet been able to overcome, while leptons are immune to this force. A better-understood force, called weak electromagnetism, can transform one kind of quark or lepton into another. Quarks and leptons also differ in their electrical charge. Quarks appear to have either $+2/3$ or $-1/3$ the fundamental unit of charge, whereas leptons are either neutral or have a -1 charge.

The two quarks and two leptons that make up ordinary atomic matter have at least two heavier relatives. For the leptons, these are the electrically charged muon and its companion the muon neutrino. The additional quarks are called strange and charmed quarks respectively, a reference to the exotic behavior of the short-lived subatomic particles that are built up out of them. None of these particles is stable; that is, all of them transform into ordinary quarks or leptons in less than one-millionth of a second after being formed in collision experiments.

The different kinds of quarks or leptons are known technically as flavors. They follow the universal rule that each particle of matter is matched by a companion antiparticle that is equal in mass but opposite in electrical charge.

Charmed, they're sure. The final proof that there are at least four quarks, including the charmed quark, to match the four known leptons, came in July 1976 at the Stanford Linear Accelerator Center (SLAC) in Stanford, Calif. SLAC was one of two labs at which evidence for charmed quarks first appeared 18 months earlier with the discovery of the psi particles.

A research team led by Burton Richter of SLAC and Gerson Goldhaber of the University of California's Lawrence Berkeley Laboratory (LBL) found the electrically neutral companion of the D^+ meson, the charmed particle they had discovered several weeks earlier. And a team led by Wonyong Lee of Columbia University in New York City, working at the Fermi National Accelerator Laboratory (Fermilab) near Chicago, found the first charmed antibaryon, a particle similar to the antiproton but containing one charmed antiquark. These composite particles fit so neatly in the four-quark theory of theorist Sheldon L. Glashow and his co-workers at Harvard University in Cambridge, Mass., that there is little doubt that the theory is correct.

A group led by Eric H. S. Burhop of University College in London found in photographic films at Fermilab what are probably charmed particles produced by collisions of high-energy neutrinos with nuclei. By examining the films under a microscope, the researchers were able to trace the paths of exotic particles that travel only a few tenths of a millimeter before they change into more common particles. Although the kind of particle involved was not precisely identified, these examples are consistent with the prediction that charmed quarks last only about one-hundred-trillionth of a second before changing into strange quarks.

More than four? Other tantalizing experiments were of potentially great significance, provided the most likely interpretation of them holds up. The first evidence that four leptons may not be the whole story came from a subgroup of the SLAC-LBL team, headed by Martin Perl of SLAC. Studying the annihilation of electrons by their antimatter companions, positrons, Perl found several cases where two ordinary muons or electrons emerge from the annihilation, accompanied by other particles. The most likely explanation is

Tracks made by charged particles in a sensitive emulsion similar to the material used in common photographic film offer visible evidence for charmed particles. A neutrino from a beam strikes a nucleus and produces ordinary particles plus a charmed particle that breaks apart at a distance that is about as great as the thickness of a few sheets of paper.

Neutrino (not visible)

Other particle tracks

Beam direction

Charmed particle track

Charmed particle breaks up

0.0002 meter

Physics

Continued

that these leptons originate from the spontaneous breakup of a much heavier lepton, one more than 4,000 times as heavy as an electron. But the new lepton's life is too short for researchers to observe it directly, so there remains some room for doubt. This discovery has been confirmed by researchers at DESY, the electron synchrotron laboratory near Hamburg, West Germany.

Additional evidence came in March 1977 from work at Fermilab by a team using electronic detectors to study collisions of high-energy neutrinos. They found examples in which three ordinary leptons emerge. It seems likely that one of the leptons originated from the transformation of the incoming neutrino, with the other two originating from a process similar to that described by Perl. But analysis of the energies of the emerging leptons suggests that yet another, heavier, short-lived particle exists. This may well be the sixth lepton necessary, according to theory, to keep the total number even. However, all of this evidence is too indirect to be conclusive.

Hark, a quark? William M. Fairbank and his co-workers at Stanford University reported in April 1977 on a 10-year search for a free quark, one that is unattached to other quarks but loosely bound in ordinary matter. They studied tiny balls of the rare metal niobium, cooled to temperatures near absolute zero. Under these conditions, it becomes barely possible to measure the electrical charge on a ball to an accuracy of a fraction of the electron charge.

The object of the search was to find a ball carrying not a whole number of electron charges, but one whose charge differs from a whole number by one-third. The Stanford researchers found this proved true for two balls out of a total of eight studied. But just two examples of such a fantastically difficult measurement are hardly enough to erase all doubt that free quarks exist.

If the Fairbank team's result is correct, it may mean that quark bonds can be broken if enough energy is provided. On the other hand, their free quarks might be remnants of the "big bang" that created the universe. Perhaps

Physicist William M. Fairbank, left, and graduate student George S. LaRue adjust the equipment they used at Stanford University to detect a fractional electrical charge—evidence for the existence of quarks.

Physics

Continued

when the quarks found partners to form protons and neutrons, a few missed the "mating game" and have lived as "bachelors" ever since.

Other news of the weak. Researchers at the Swiss Institute for Nuclear Research in Basel found preliminary evidence that a muon can transform itself into an electron simply by giving off a gamma ray. This happens to about 1 in 1 billion; the others transform themselves into their companion neutrinos. With such a rare process, there is some chance of a false signal, and the final result had not been published by mid-1977. But current theories of weak electromagnetism insist that such a process should occur this often.

Another experiment that bears on the same theory deals with the passage through bismuth vapor of plane polarized laser light – light in which the direction of vibration is fixed in a single plane. The theory states that weak electromagnetism should add slightly to the force that holds electrons in their atomic orbits, and this force makes the atoms absorb plane polarized light in such a

way that the plane rotates slightly. But two research teams – one at Oxford University in England and the other at the University of Washington in Seattle – reported in December 1976 that they found no rotation when they passed laser light through bismuth vapor. If the result holds up, it could be interpreted as evidence that new quarks and leptons are polarized oppositely to the four known ones, thus canceling out the preference.

European particle physicists gained an important new tool in December 1976 when the 400-billion-electron-volt super-proton synchrotron (SPS) began operation at the European Center for Nuclear Research near Geneva. The SPS, which lies on both sides of the French-Swiss border, is better equipped with experimental apparatus than is the Fermilab 475-billion-electron-volt proton synchrotron. The SPS may thus provide Europe, which has spent considerably more than the United States on particle research in recent years, a way to win leadership in this field of physics.   [Robert H. March]

Nuclear Physics. A new reaction in which oxygen 18 changes into neon 18 by converting two neutrons into protons in the oxygen nucleus was reported in January 1977 by Robert L. Burman and his collaborators at the Meson Physics Facility of the Los Alamos Scientific Laboratory in New Mexico.

The reaction, called pion double-charge-exchange, is a powerful probe for studying nuclear structure because it simply changes the charge of two nucleons (protons or neutrons) within the target nucleus without otherwise changing its structure. The new reaction opens up a field that theorists have studied extensively – the relationship between nuclei with the same total number of nucleons but different numbers of protons and neutrons.

Burman's group produced a beam of positively charged (positive) pions – particles one-seventh as massive as nucleons – that collided with a target of oxygen. They used magnets to separate negative pions from the other collision debris and guide them into a detector. Their measurements of the negative pions' energies agreed with predictions for the reaction in which a positive pion collides with oxygen 18 to produce a negative pion and neon 18.

The researchers reported that they can also create a negative pion beam to perform the inverse reaction, which converts two protons into two neutrons in a target nucleus and changes the negative pion into a positive pion. Or, in a single-charge-exchange reaction, either a positive or negative pion beam collides with a stable nucleus and changes only one of its nucleons into the other type. In this case, neutral pions are created. However, a neutral pion does not respond to a magnetic field and must be detected by measuring the gamma radiation produced when it decays.

In addition to probing detailed structure, such pionic reactions will permit researchers to create proton-rich nuclei. Thus, pionic reactions complement the use of fission reactors and heavy-ion reactors in which neutron-rich nuclei can be created. The reactions promise to expand the arsenal of

"You remember Newton's law of gravity don't you Benson?"

Physics

Continued

tailored radioisotopes available to physicists for use in medical diagnosis, clinical medicine, and industry.

Element 107. Georgii N. Flerov of the Joint Institute for Nuclear Research in Dubna, Russia, announced in early 1977 that his research team had synthesized element 107, an atom that has 107 protons and 154 neutrons in its nucleus. The Russian researchers fired energetic chromium 54 nuclei at a thin layer of bismuth 209 on a rapidly rotating cylinder. They reported that glancing collisions produced element 107 nuclei, which spontaneously broke up into lighter nuclei in about two milliseconds. United States nuclear scientists who specialize in making heavy elements believe that spontaneous breakup alone is not enough proof that element 107 actually existed.

Weak magnetism. Gerald Garvey of the Argonne National Laboratory near Chicago reported in February 1977 on a series of measurements made at Princeton University in New Jersey. The work supports theories relating the electromagnetic force and the weak force. Physicists had known that the weak force counterpart to the electrical part of electromagnetism is responsible for beta decay, a process in which a nucleus changes into a different nucleus by emitting an electron or its antiparticle, a positron, and a neutrino. Garvey's group probed the less understood weak magnetism, the weak force counterpart of the magnetic part of electromagnetism.

The Princeton researchers studied beta decay in nuclei that have eight nucleons: radioactive lithium 8 (3 protons, 5 neutrons), which changes into beryllium 8 (4 protons, 4 neutrons) by emitting an electron and a neutrino; and radioactive boron 8 (5 protons, 3 neutrons), which changes into beryllium 8 by emitting a positron and a neutrino. In both cases, the beryllium 8 quickly breaks up into two alpha particles (helium nuclei).

The experimenters measured the rates at which electrons and positrons emerged at certain angles relative to the two alpha particles. They also measured the rate at which photons of light emerged when some beryllium nuclei changed from a high-energy excited state to a lower-energy excited state.

This high-energy state for beryllium corresponds to the ground state, or lowest-energy state, for lithium and boron. The first measurement provided information about weak magnetism. The investigators related it to the photon measurement, which provided them with information about electromagnetism.

Garvey concluded that weak magnetism and the magnetic part of electromagnetism are alike. His finding supports the work of Steven Weinberg of Harvard University in Cambridge, Mass., who has proposed a theory to view the weak force and the electromagnetic force as different aspects of the same force. It therefore puts the extensive theoretical work toward a unified theory on a much sounder experimental footing than was previously possible. See PHYSICS (Elementary Particles).

Magnetic multipoles. Nuclear scientists used electron beams to probe the magnetic properties of target nuclei resulting from the collective motions of their protons and neutrons. In the simplest magnetic structure of the target nuclei, a magnetic dipole, all the protons and neutrons oscillate against one another. In the next most complex, the quadrupole, the nucleons' collective motion causes the nucleus to oscillate between a football shape and a doorknob shape. In the octopole, a pear-shaped nucleus oscillates end-for-end. Higher multipoles have more complex oscillations and shapes.

William Bertozzi and his colleagues at the Massachusetts Institute of Technology in Cambridge studied the lowest seven multipoles in a variety of nuclei. In February 1977, Dutch physicist Helmut de Vries reported that the technique can be used on bismuth 209 and niobium 93 to study the 502-pole magnetic structure. Such measurements pose stringent tests for theorists who develop models of nuclei.

De Vries found that nuclei with many more neutrons than protons do not have excess neutrons at their surface, as physicists previously believed. Instead, the neutrons are uniformly distributed throughout the nucleus. Scientists do not yet understand why this is so.   [D. Allan Bromley]

See also CORES IN COLLISION.

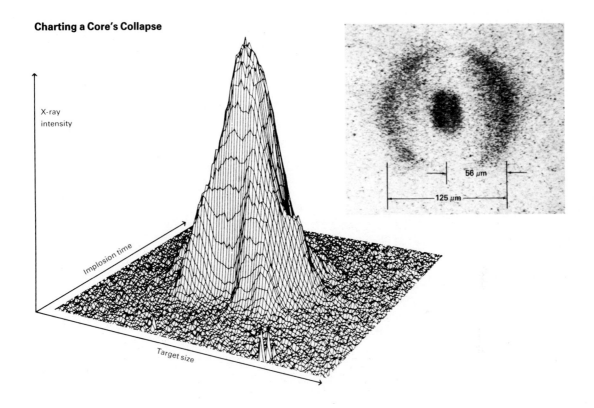

Physics

Continued

Normal X-ray image of
a fusion fuel pellet's
implosion, *above, right,*
shows only the initial shell
and the compressed
core. Scientists use
a new pinhole camera
technique to record an
implosion from start to
finish and to make a
step-by-step plot, *above.*

Plasma Physics. Researchers reported several impressive results in thermonuclear fusion experiments during 1976 and 1977. Among these were significant advances in laser fusion and magnetic confinement, the two major approaches to duplicating on earth the sun's energy-producing reaction.

In both approaches, the goal is to heat a fuel mixture of the hydrogen isotopes deuterium and tritium to more than 100 million°C so that the heated ions fuse to form helium and give off usable energy in the process. For fusion to occur, the fuel density (expressed in ions per cubic centimeter) times the length of time the fuel is confined (in seconds) must be greater than 10^{14}.

Laser fusion. Physicists Sidney Singer and Gene McCall of the Los Alamos Scientific Laboratory in New Mexico reported in March 1977 that laser fusion was possible with a laser wavelength longer than previously used.

To achieve laser fusion in a practicable reactor, scientists must direct an intense laser beam or beams with a power of about 100 terawatts (100 trillion watts) at a small deuterium-tritium pellet. The laser pulse must last less than about 10 nanoseconds (10 billionths of a second). It vaporizes the pellet surface, producing a "rocket" action that drives the surface inward like a piston, thereby creating the density, temperature, and confinement time needed for fusion to occur.

Because short-wavelength laser light penetrates deep into the pellet skin, theorists had predicted it would be absorbed efficiently. As a result, most experimenters have used the short-wavelength neodymium-glass laser, whose wavelength is 1.06 microns (a micron is a millionth of a meter), even though it converts into laser energy only about a thousandth of the electrical energy going into it. The carbon dioxide (CO_2) laser converts electrical energy to laser energy far better, but most scientists expected it to be 100 times less efficient in delivering laser energy to the target because of its longer wavelength — 10.6 microns.

The Los Alamos researchers found that this was not the case. They directed two CO_2 laser beams at a glass sphere, or microballoon, approximately 200 microns in diameter that was filled with deuterium-tritium fuel. The laser power was 0.4 terawatt. The researchers measured a fusion energy release that indicated CO_2 gas lasers can be used to compress pellet targets despite their relatively long wavelength. Scientists speculate that the electrical fields in the CO_2 laser light create turbulent motion at the surface of the fuel pellet. This accounts for the better-than-predicted laser energy absorption.

Meanwhile, scientists continued to make strides with neodymium-glass lasers. For example, John Emmett and his colleagues at the University of California's Lawrence Livermore Laboratory reported in October 1976 that they had achieved a record number of 1-billion thermonuclear reactions using a single 2-terawatt pulse. The experimenters set a record value for the most thermonuclear energy released in a laser-fusion experiment. About 1 per cent of the thermal energy in the fuel came out during fusion.

Other researchers studied the use of relativistic electron or ion beams to compress fusion fuel pellets. Moving at nearly the speed of light, the particles in such beams transfer energy to the small fuel pellet efficiently, but they are difficult to focus sharply.

Gerald Yonas and his research group at Sandia Laboratories in Albuquerque, N. Mex., have built and operated an electron-beam machine that produces a beam with a power of 8 terawatts that lasts for 24 nanoseconds.

The Sandia researchers reported in May 1977 that the intense electron beams can be guided along a plasma channel formed on the surface of a small tungsten wire. This new technique will allow several electron beams to be focused by wire tracks that lead into a common target.

In June, the Sandia scientists reported producing the first fusion neutrons from a deuterium-tritium pellet compressed with electron beams.

Intense ion beams have several potential advantages for achieving fusion in pellet targets. The ions tend to deposit their energy in a thinner surface layer, thereby creating more efficient compression.

Gerald Cooperstein and his colleagues at the Naval Research Laboratory in Washington, D.C., reported in October 1976 that the Gamble II experimental device had produced 0.2-terawatt ion beams lasting 10 nanoseconds. The beams consisted of hydrogen ions (protons).

Russian physicists were also very active in laser and beam fusion. During a visit to the United States in 1976, physicist Leonid I. Rudakov of the Kurchatov Institute in Moscow described a new technique in which an electron beam is first passed through a metal foil that converts the electron energy into an intense burst of X rays. The X rays behave like very-short-wavelength laser light and compress a pellet target very efficiently.

Magnetic fusion. Ronald Parker and Bruno Coppi, who lead a research team at the Francis Bitter National Magnet Laboratory at the Massachusetts Institute of Technology (M.I.T.) in Cambridge, reported setting several records for the confinement of toroidal, or doughnut-shaped, plasmas in tokamak machines.

In November 1976, they reported the highest magnetic field yet achieved in a tokamak—about 170,000 times the earth's magnetic field—and fuel densities of 10^{15} ions per cubic centimeter for the plasma. In addition, they produced a plasma for which the product of density and confinement time was 2×10^{13}, the largest achieved in a fusion experiment. They showed that the plasma confinement time increased in proportion to the fuel density.

Researchers at the Princeton Plasma Physics Laboratory in New Jersey, led by Harold Furth and Wolfgang Stodiek, confined plasmas for a record 0.07 second in the Princeton Large Torus tokamak. The results achieved at the M.I.T. and Princeton facilities indicate that tokamaks can be built large enough to meet the conditions needed for thermonuclear fusion.

John Clarke and his team of tokamak fusion-reactor designers at the Oak Ridge National Laboratory in Tennessee planned an experimental fusion reactor that achieves high fuel densities by directing intense beams of neutral atoms at the tokamak plasma in order to heat it.   [Dale M. Meade]

Physics
Continued

Solid-State Physics. California researchers reported in January 1977 that certain metallic salts behave partly like a solid and partly like a liquid when heated. The salts may be used to make solid-state batteries.

Physicists James B. Boyce and Bernardo F. Huberman of the Xerox Corporation Research Center in Palo Alto, Calif., studied such metallic salts as copper iodide, silver iodide, and crystals of beta alumina that had absorbed metallic sodium like a sort of atomic sponge. These materials, called superionic conductors, are of technological interest because, when heated to a high temperature, their metal ions become very mobile and diffuse electrochemically when voltage is applied. Solid-state batteries made of superionic conductors would contain the electrolyte in the solid instead of using a separate liquid electrolyte as in a car battery, or a gel electrolyte as in a dry cell.

Boyce and Huberman found that the metallic copper ions in the superionic conductor copper iodide actually form a kind of liquid when the crystal is heated to above 400° C (752° F.). But the cage of iodine ions remains rigid so that the crystal remains solid.

Japanese researchers showed in 1952 that the copper ions in copper iodide become disordered when the salt is heated to more than 400°C, and no longer form a periodic arrangement, or sublattice, in the lattice of iodine ions. However, it was not clear then whether the high ionic conductivity observed at high temperatures was caused by melting of the copper sublattice or was a slow diffusion in which the ions moved into vacant sites in the crystal lattice. Boyce and Huberman used nuclear magnetic resonance to study the copper nuclei and showed clearly that the copper ions behaved like atoms in a liquid. Hence, they had actually melted while the overall material remained solid.

Synchrotron radiation. Solid-state physicists joined chemists and biologists in using synchrotron radiation, a powerful tool to probe the atomic structure of matter. This is the intense, very broadband electromagnetic radiation that electrons emit as they circulate

A magnetic bubble circuit that was etched by exposure to X rays is much sharper in relief than any made with conventional manufacturing methods.

A scientist at Bell Telephone Laboratories holds new semiconductor material—a monolayer crystal that was made atomic layer by atomic layer. On the background screen is an electron microscope image of the structure magnified about 2 million times.

Physics

Continued

between the poles of a series of magnets in the vacuum chamber of a high-energy accelerator. When the electrons reach nearly the speed of light, they emit radiation that ranges from the infrared to the X-ray region of the electromagnetic spectrum.

Several accelerators known as electron storage rings, which were designed to study elementary particles, were also used in 1976 and 1977 as powerful sources of synchrotron ultraviolet radiation and X rays. One such machine is the Stanford Positron Electron Accelerating Ring (SPEAR) at the Stanford Linear Accelerator Center in California. There, researchers set up the Stanford Synchrotron Radiation Project (SSRP) to study the structural properties of various materials.

Stanford scientists Steve Cramer, Tom Eccles, Frank Kutzler, Keith Hodgson, and Sebastian Doniach, working at SSRP, studied complex metallo-organic compounds such as porphyrin compounds, which are important because they are similar to the active regions of hemoglobin and myo-globin, protein molecules that carry and store oxygen in the blood.

The Stanford researchers reported in December 1976 that they had used synchrotron X rays to measure more precisely the distance between the biologically active inorganic atoms and their neighboring organic atoms in the porphyrin compounds. They used a spectroscopic technique called extended X-ray absorption fine structure spectroscopy (EXAFS).

EXAFS was perfected by Farrel Lytle, Edward Stern, and Dale Sayers of the Boeing Company and the University of Washington in Seattle. They reported in 1970 that as X-ray wavelengths are tuned through an absorption edge, or sharp increase in X-ray absorption characteristic of one of the atoms of a material, the amount of X rays absorbed by the material increases and decreases in an oscillating fashion. The Seattle team interpreted these oscillations in terms of the spacing between the absorbing atoms (such as copper atoms in a copper bromide crystal) and neighbor (bromine) atoms.

Physics

Continued

Days, or even weeks, of scanning with a conventional X-ray machine are needed to get enough X rays to provide a usable spectrum of the sample. Brian Kincaid of Stanford and Peter Eisenberger of Bell Laboratories demonstrated in 1975 that extremely good spectra can be obtained in a matter of minutes when X rays from a high-energy storage ring are used.

Microelectronics revolution ahead? Researchers at the IBM Research Center in Yorktown Heights, N.Y., reported in February 1977 on another use of synchrotron radiation. Physicists Eberhard Spiller, Ralph Feder, and John C. Topalian showed that integrated-circuit elements can be made with much greater sharpness and higher relief using synchrotron radiation than using electron beams. This development could lead to even more compact integrated circuits, the tiny silicon chips that contain the equivalent of thousands of transistors and other electronic circuit elements.

To make an integrated circuit, technicians use a scanning electron beam to cut a circuit mask, a kind of stencil. The mask is then placed over a film of plastic called a resist on the silicon chip. Scanning the mask with a second electron beam weakens the exposed resist material, which can then be etched away with a solvent.

The IBM team exposed a masked resist to intense synchrotron X rays to form a pattern whose elements were about three times as high as they were wide — much better relief than electron beams produce. There are many technical problems to overcome before X-ray circuit etching can become a production-line technique. One is the need for extremely sharp registry of the successive masks when superimposing different patterns during the fabrication process of a complete microcircuit. If circuit elements less than 1,000 angstroms wide can be made, 100 times more circuit elements could be packed on a single chip using this new technology than is possible now. Thus, synchrotron radiation could be the basis for yet another revolution in microcomputer power.   [Sebastian Doniach]

Psychology

Scientists reported in 1977 that an extensive systematic investigation strongly supported the belief that heavy smokers are addicted to the nicotine in cigarette tobacco. Even though much evidence connects heavy smoking with reduced life expectancy and indicates that even nonsmokers can be harmed by cigarette smoke in the air, many heavy smokers find it extremely difficult to stop or substantially reduce their smoking. But until now, data supporting the assumption of addiction have been inconclusive and inconsistent.

Stanley Schachter and five co-workers at Columbia University in New York City reported on their investigation of this problem in March 1977. In their first experiment, they gave cigarettes that differed in nicotine content to long-time heavy and light smokers. The heavy smokers tended to keep their nicotine intake at a fairly constant level by smoking more when they were given low-nicotine cigarettes. The light smokers tended to smoke the same number of cigarettes no matter what the nicotine content.

Follow-up studies demonstrated that heavy smokers adjust their rate of smoking to keep the amount of metabolized nicotine at a fairly constant level. Analysis of urine samples showed that less nicotine is metabolized during social events and under conditions of stress. These are two situations where smoking usually increases. They also found that smoking behavior is determined more by nicotine metabolism than by the degree of stress.

Predicting academic achievement. Psychologists and education specialists use various tests to try to predict a child's achievement in school, particularly in learning reading and arithmetic. Harold W. Stevenson and his associates at the University of Michigan in Ann Arbor and the Hennepin County Mental Health Center in Minneapolis, Minn., took an extensive new look at this procedure in August 1976.

Stevenson and his associates initially tested children at an earlier age than in other such studies — during the summer before the children entered kindergarten. They used an unusually large group of tests that contained psychometric and cognitive tasks. The psychometric tasks included tests of intelligence and achievement, and of perceptual, motor, and language skills. The cognitive tasks tested simple and complex forms of learning and memory.

Kindergarten teachers rated each child on how good they were in learning, memory, vocabulary, and following instructions. Later achievement was measured at the end of the first, second, and third grades.

The original test group contained 133 boys and 122 girls who were pre-registered in 17 kindergarten classes in four Minneapolis elementary schools. When the study began, their average age was 5.4 years and their average score on the Peabody Picture Vocabulary Test, an intelligence test that uses pictures instead of words, was 104.8. When the study ended after the third grade, 84 boys and 69 girls of the original group were still participating.

As is usually the case, the best predictor of achievement in the third grade was the child's achievement in the first and second grades. But many of the prekindergarten tests also accurately predicted third-grade reading and arithmetic achievement. The group of tests that best predicted later reading achievement measured the child's familiarity with letters, the ability to form associations between what is seen and what is heard, and the ability to group common objects in categories – for example, knowing that both cats and dogs are animals.

The tests that best predicted later achievement in arithmetic were similar to those for reading, plus tests that measured analytical ability. Ratings by kindergarten teachers were less accurate than the original test scores in predicting future achievement.

The use of preschool tests to predict later school achievement raises the possibility of accurately identifying children who will have learning difficulties much earlier than was previously thought possible. The researchers said their data also indicate that special educational programs to increase school achievement should emphasize teaching the information and perception skills that are needed for learning and remembering.

Coping with crowding. Human reactions to crowded surroundings are complex and may depend on factors other than crowding, Ellen Langer and Susan Saegert of the City University of New York reported in March 1977. Studies of animals generally have shown marked harmful changes in both behavior and health as crowding increases. Data on human reactions to crowding are less consistent, however.

Langer and Saegert conducted their crowding studies in two New York City grocery stores with 80 women between the ages of 25 and 45. Each woman was given the same 50-item shopping list and asked to spend 30 minutes trying to find as many items as possible. Then she was asked to write down which brand and size of each item she would buy in order to spend the least money.

Half the women shopped when the store was relatively uncrowded and the other half when it was at least twice as crowded. But Langer and Saegert also examined the effect of a second condition – information. Half of the women in each situation – crowded and uncrowded – were told that the store might be crowded and that crowding sometimes made people feel anxious.

Increased crowding interfered with overall performance. On the average, the women in the crowded store did not record information for as many items on the shopping list as the other group. But in both crowded and uncrowded situations, those women who had been told how they might react to crowding did better than those who were uninformed. Langer and Saegert suggest that people who know that they may react negatively to crowding control their reactions better and keep them from interfering with performance.

Submerged in sex. T. R. Halliday and H. P. A. Sweatman of the Animal Behavior Research Group at Oxford University in England reported in August 1976 on their study of competing drives in the smooth newt. The smooth newt is a small air-breathing salamander that conducts its courting and mating underwater near the bottom of a pond. The male newt plays the more active role and, because courting can extend for a relatively long time, he often has to leave the female he is courting and surface for air. When

IQ and Heredity: A Taint Of Fraud?

The field of psychology was shaken in late 1976 when charges of scientific fraud were leveled against Sir Cyril L. Burt, an eminent English educational psychologist who died in 1971 at the age of 88. The charges center on his work on the inheritability of intelligence. This work strongly depended on his studies of identical twins raised in separate homes.

Ever since tests were invented to measure intelligence, investigators have argued whether the variations in human intelligence are determined primarily by the individual's genetic makeup (the hereditarian view); primarily by his social and cultural environment (the environmentalist view); or by some combination of the two.

Burt's data indicated that at least 75 per cent of the variability in intelligence can be attributed to heredity. As an adviser to the British government in the 1930s and 1940s, he was able to see his views translated into public policy with the setting up of a three-tier system of education in which a test given at the age of 11 determined for all time whether a child could go to college.

Burt's data were also cited by Arthur R. Jensen of the University of California, Berkeley, in a furiously controversial article in the *Harvard Educational Review* in 1969. The article suggested that differences in intelligence quotient (IQ) scores among races were genetically based, and therefore could not be corrected by special educational programs for minorities.

Leon J. Kamin of Princeton University in New Jersey was the first to question Burt's data publicly. Burt had published several papers on three studies comparing IQs in separated identical twins, adding more pairs of twins each time. Kamin noticed a curious feature of these papers — the correlation between the IQ scores of the separated twins was given as 0.771 in all three studies. For even one such correlation to remain unchanged through three different sample sizes is improbable, but other correlations given by Burt in the same papers also remained unchanged.

Meanwhile Jensen, a great admirer of Burt's work, was coming to a similar conclusion. In bringing together all the results Burt had published over his ca-

reer, Jensen noticed many more of the same curious consistencies in correlations that Kamin had spotted.

At this point, no one was prepared to accuse Burt of fraud. But two years later, Oliver Gillie, medical correspondent of the London *Sunday Times*, struck a further blow to Burt's credibility. Many of Burt's most important papers were coauthored by two women named Margaret Howard and J. Conway. But, in an article published on Oct. 24, 1976, Gillie reported that there was no record of Howard or Conway at London University, the address given on the papers, and that 18 of Burt's acquaintances did not remember them.

The question of Howard's and Conway's existence has not yet been resolved. A Manchester University professor claims to remember Howard from the 1930s. On the other hand, there is evidence that several papers signed only by Howard or Conway in the 1950s and 1960s may have been written by Burt under pseudonyms.

The question of their existence is important because Burt in his later years was probably too old and deaf to administer intelligence tests himself. Yet his papers of this period include data from new pairs of twins. If Howard and Conway did not exist, it is reasonable to wonder if the new twins did either.

Present evidence suggests fraud, but is not conclusive. The only sure evidence of error is the unchanging correlations, a curious mistake for a statistician as skilled as Burt to make if he were trying to fake the results. An alternative is that the aging Burt did not bother to recalculate the correlations in his later papers, but simply carried over his previous calculations.

Burt's IQ data are now unusable. Although his data are not crucial to the arguments of the hereditarians, they may now have to reduce their estimate of the heritability of intelligence from about 80 per cent down to around 60 per cent, based on other studies, according to one estimate. The longer-term impact of the Burt affair may lie in the questions it raises about the general level of scholarship in a field that allowed centrally important, but patently flawed, data to remain unchallenged for so long.   [Nicholas Wade]

335

"Watch this. Every time I pull this string, he
runs into the other room and gets the other guy."

Psychology

Continued

the male returns to the bottom of the pond, he may not be able to find the female he was courting. His plight is a classic example of conflict between two strong drives, only one of which can be satisfied at a time.

Halliday and Sweatman studied the conditions that determine the relative strength of the male newt's competing drives. The female newt responds to a courting male by moving away, remaining still, or moving toward the male. The researchers controlled the apparent state of responsiveness by placing an anesthetized female in a harness that they could move. Then they measured the length of time that a male newt stayed submerged for different degrees of female responsiveness.

The continued presence of a stationary female kept a male newt submerged about 30 per cent longer than he would stay when not breeding. The male stayed down for about the same length of time when the female was moved toward him at a rate that indicated high responsiveness. The male newt stayed submerged longest when the fe-

male was moved toward him at a rate indicating only mild responsiveness.

In the first situation, the researchers observed courtship displays but no matings. In the second, the male completed up to three mating sequences before surfacing to breathe. In the third situation, there were prolonged and vigorous courtship displays, but no matings.

Halliday and Sweatman concluded that the male newt's sexual behavior is influenced both by the female's behavior and his own internal state. If the female is highly responsive and they mate, the mating urge is reduced and the male will go to the surface to breathe. If the female is not responsive and mating does not occur, the mating drive remains high, and the male will surface to breathe only when his oxygen supply is so low that breathing becomes more important than mating. Mild responsiveness on the part of the female increases the male's desire to mate and the need for oxygen will have to be even more severe than when he gets no response before it overrides the mating urge.   [Sally E. Sperling]

Public Health

A public health poster warns that rabies-free Great Britain is in danger of a European epidemic crossing the English Channel.

More than 40 cases of a new penicillin-resistant form of gonorrhea had been identified in the United States by June 1977. The new organisms are apparently spreading slowly, and health measures, including prompt checks on the patient's sexual contacts, have so far effectively controlled them.

Microbiologist Stanley Falkow of the University of Washington in Seattle had predicted in June 1975 that a new strain of the bacterium *Neisseria gonorrhoeae* would appear in the United States and that it would be resistant to penicillin.

His prediction was confirmed in February 1976 when the first American case of the new gonorrhea was identified in St. Marys County, Maryland. The infection was traced to a prostitute in the Philippines. Legalized prostitution flourishes in Manila and most of the prostitutes regularly take low doses of penicillin to prevent venereal disease. Unfortunately, this tends to foster the growth of resistant organisms; the more common strains that are susceptible to penicillin are thus eradicated, but any penicillin-resistant organisms are left free to multiply.

Penicillin resistance in gonorrhea is not new. In 1958, 200,000 units of penicillin cured any gonorrhea infection, but by 1977, 4.8 million units were needed. This reflects increasing resistance to the drug. However, the new strain poses an entirely different problem. These microorganisms produce the enzyme penicillinase which renders the antibiotic completely ineffective. As a result, no amount of penicillin can control the infection.

Fortunately, a drug called spectinomycin is effective against the new form of gonorrhea. The effective agent should not induce a false sense of security, however. The gonorrhea bacteria probably will eventually overcome even this drug. Furthermore, this new agent is seven times as expensive as penicillin, making it too costly for people in many countries. As a result, the disease may continue to spread.

Researchers hope to develop a vaccine that will provide permanent immunity to gonorrhea or an agent that will inhibit penicillinase, thereby re-establishing the effectiveness of penicillin. But public vigilance remains the best defense against gonorrhea, which can cause arthritis; brain, heart and kidney damage; and, ultimately, death.

Duodenal ulcers down. A suspected overall decline in the incidence of peptic ulcers was substantiated for at least one urban area by surgeon Meredith P. Smith of the University of Washington School of Medicine in March 1977. He found that there had been a decline in ulcer cases and ulcer surgery in Seattle.

Doctors have long recognized that a relatively fixed percentage of ulcers break through the wall of the duodenum, a part of the small intestine just below the stomach. This perforation almost always results in hospitalization and usually in corrective surgery. Thus, although it is difficult to determine accurately the exact incidence of new cases of ulcer, the number of perforations probably accurately reflects the incidence of ulcer disease.

Smith examined the records of two metropolitan hospitals, two smaller suburban hospitals, and a Veterans Administration Hospital in Seattle covering the 10 years from 1965 to 1975. During the first five years, 183 perforated ulcers were recorded. There were 117 – 35 per cent less – during the second five years. There were a total of 625 ulcer operations in the first 5-year period and only 395 in the second.

What caused this dramatic change? Surgery might decline if there were increasingly vigorous and successful nonsurgical treatment of ulcers, even if there were no decline in incidence. However, there is no evidence of this in Seattle.

Smith could find no completely satisfactory explanation for the decline in ulcer disease. However, his findings tend to question the commonly held belief that modern urban stress and ulcers are related, because there is no evidence that stress is diminishing in the Seattle metropolitan area.

Perhaps ulcer disease is more common at the beginning of the urbanization process as towns grow into cities or when people move into urban areas, and occurs less frequently as time passes. An apparently higher rate of the disease among people who recently moved from a rural environment to the city of Seattle tends to support this hypothesis.

Requiem For a Germ

Smallpox, once the most devastating pestilence on earth, has been pushed to the brink of extinction by a 10-year World Health Organization (WHO) campaign to eliminate the disease. By mid-1977, the disease that had killed, blinded, and scarred countless millions had been driven from all parts of the world, except a remote area of Africa. The only smallpox outbreaks as of May 1977 were in Somalia and in Ethiopia, which had only four cases caused by virus imported from Somalia.

Most scientists believe that smallpox virus can only live in and be passed along by humans. Therefore, if public health workers stamp out these last outbreaks, smallpox should completely disappear. And this will be the first time that a disease has been eradicated.

Even before recorded history, smallpox was plaguing the human race. It was apparently widespread in China and India between 2000 and 1000 B.C., and scientists have found what appear to be smallpox scars on Egyptian mummies dating from 1160 B.C.

The disease probably spread from Asia to Africa and Europe. Spanish conquistadors in 1520 brought it to the New World, where it killed some 3.5-million persons in Mexico alone.

Smallpox virus ordinarily spreads from person to person in the tiny droplets expelled by sneezing and coughing. Its victims develop a high fever, aches, and a rash that usually starts on the face and spreads over the entire body. Some are left blind; others are hideously disfigured by the rash. Scabs form over the pus-filled rash blisters and fall off, leaving the skin pitted with scars.

The Asian form of smallpox virus, *variola major,* kills from 20 to 40 per cent of its victims. The African form, *variola minor,* kills about 1 per cent.

There is no cure for smallpox, but it can be prevented. Hundreds of years ago, people in Asia recognized that smallpox survivors became immune to further attacks. So they began infecting healthy individuals with material from the rash of smallpox patients to create a mild case of the disease that would immunize against more severe attacks.

This practice became common in Europe during the 1600s and 1700s, but it had drawbacks. A few who were deliberately infected died. The remainder would often transmit their infections to others, causing new epidemics. Some experts believe this crude attempt at immunization actually increased the incidence of smallpox.

In 1796, Edward Jenner, an English country doctor, developed the most effective weapon against smallpox. He knew that dairymaids who caught cowpox, a mild virus disease that causes a few skin sores, became immune to smallpox. Jenner inoculated an 8-year-old boy by putting scrapings from a dairymaid's cowpox sores in two cuts on his arm. Almost seven weeks later, Jenner exposed the boy to smallpox; he did not catch the disease.

Jenner confidently predicted that vaccination would result in "the annihilation of smallpox," but that has been long in coming. Smallpox vaccine and vaccination techniques were improved during the late 1800s and early 1900s. In the 1940s, vaccination almost completely eliminated smallpox from Europe and North America. But the disease persisted in the developing world. As recently as 1967, there were an estimated 10 to 15 million cases, mainly in India, Africa south of the Sahara, Indonesia, and Brazil.

In 1967, WHO started a global campaign to wipe out smallpox within a decade. Health workers went into city slums and remote villages, armed with high-quality freeze-dried vaccine, injector guns that could vaccinate more than 1,000 persons an hour, and special two-pronged needles that were simple to handle and conserved vaccine.

WHO planned to vaccinate most people in each affected country. But health workers discovered by chance that locating outbreaks and vaccinating only people in that area worked better. They searched for smallpox victims, isolated them, and vaccinated everyone who had come into contact with them. This strategy eliminated the disease from vast areas, even though only 6 or 7 per cent of the population was actually vaccinated.

The campaign barely missed its goal of total eradication by the end of 1976. Only 953 smallpox cases were reported in 1976, down from 19,278 in 1975. And WHO officials are hopeful that they can eliminate the disease by the end of 1977.　　[Philip M. Boffey]

Public Health

Continued

Safety and the pill. Biostatistician Anrudh K. Jain of the Population Council at Rockefeller University in New York City reported in March 1977 on the safety of oral contraceptives. His report suggests that older women might consider other forms of contraception. And women who smoke, especially those over 40 years old, should use the pill with great caution.

Jain reviewed many studies of contraceptive safety. His work focused particularly on the pill, perhaps the most satisfactory and practical technique to control fertility. It has been used by millions of women since the early 1960s.

Jain's report noted a significant relationship between smoking and the pill. Both smoking and oral contraceptive use individually carry increased risk of blood clots and heart attacks. But when a woman who smokes also uses the pill, there is a far greater risk.

The study also concluded that using the pill is less hazardous for women under 29 than becoming pregnant and giving birth. Among nonsmoking women from 30 to 44 years old, taking the pill to avoid pregnancy is still safer than undergoing pregnancy and childbirth. It is actually more hazardous however, for those who smoke. Women 40 to 44 years old who use the pill and also smoke heavily have an annual death rate linked to these factors of 83 per 100,000. This compares with a rate of 22 per 100,000 from pregnancy and childbirth for nonusers overall and a rate of 1 per 100,000 for those using an intrauterine device for contraception.

Applying these observations to individual cases requires caution. The risk of heart attack, for example, is also directly related to other health factors such as diabetes, high blood pressure, and obesity. The role of these factors in those who use the pill was not considered in the studies Jain reviewed. In addition, the estimated risk of pregnancy and childbirth is based on data from the United States and Great Britain. In other societies, the risks of pregnancy and childbirth may be much higher and, even for smokers over 40, the pill may be far safer than no contraception at all. [Michael H. Alderman]

Science Policy

The Democratic election victory in November 1976 greatly affected United States science policy. After James Earl Carter, Jr., replaced Gerald R. Ford as President in January 1977, there was a change of leadership in many federal research agencies. Carter appointed a new presidential science adviser to help plan and coordinate the government's massive investment in science and technology. He also moved swiftly to stamp his mark on energy policy and proposed radical changes in the nation's nuclear program.

Controversial issues. Moves to regulate controversial genetic-engineering experiments began at the state and local government level and eventually reached the U.S. Congress. These were the first major attempts to legislate controls on an area of basic research, and they also involved an unprecedented degree of public participation in scientific decision making. See GENES: HANDLE WITH CARE; SCIENCE AND THE DECISION MAKERS.

Laetrile, a substance extracted from apricot pits that proponents claim can cure or in other ways benefit cancer patients, was the center of another controversy. The medical establishment and the Food and Drug Administration contend that Laetrile is ineffective against cancer. Nevertheless, Laetrile supporters mounted a public campaign and won significant victories. Nine states had legalized the distribution of the substance by June and legalization was expected in several more states.

The Presidents' scientists. President Ford in August 1976 appointed H. Guyford Stever director of the newly established White House Office of Science and Technology Policy (OSTP). The director also serves as the President's science adviser. However, Stever, a former director of the National Science Foundation (NSF), resigned in January 1977 when Ford left office.

Carter nominated Frank Press as his science adviser on March 18, and the choice was widely endorsed by the scientific community. The Senate confirmed the nomination, and Press was sworn in on June 1. Press, a geophysicist from Massachusetts Institute of

Frank Press, *top,* was chosen as President Carter's science adviser. James R. Schlesinger, *above,* assumed the post of energy adviser.

Technology in Cambridge, had served on several important government advisory committees, including the National Science Board and the President's Science Advisory Committee. He is an expert on earthquake prediction and has worked on the problems of monitoring seismic waves produced by underground nuclear explosions.

Scientist administrators. Carter appointed several other distinguished scientists to high government posts. He chose Harold Brown, a physicist and former president of the California Institute of Technology in Pasadena, as secretary of defense. He named Donald Kennedy, professor of human biology at Stanford University, as head of the Food and Drug Administration and retained biochemist Donald S. Fredrickson as director of the National Institutes of Health (NIH). Psychologist Richard C. Atkinson also remained as director of the NSF.

James C. Fletcher, who had headed the National Aeronautics and Space Administration (NASA) since 1971, resigned on May 1, 1977, and Carter named Robert A. Frosch, deputy director for applied oceanography at the Woods Hole Oceanographic Institution in Massachusetts, to replace him.

Carter chose James R. Schlesinger, former chairman of the Atomic Energy Commission and secretary of defense from 1973 to 1975, as his energy adviser. Robert C. Seamans, Jr., resigned in January as head of the Energy Research and Development Administration (ERDA) and was replaced by deputy administrator Robert W. Fri.

Science funding. In the budget he submitted for fiscal year 1978 (Oct. 1, 1977, to Sept. 30, 1978), Ford proposed levels of research and development (R & D) funding that would outpace the projected rate of inflation. He slated $26.3 billion for R & D programs and $1.6 billion for research facilities. This was an 8 per cent increase over estimated R & D spending in fiscal 1977.

The largest share of the R & D funds was earmarked for the Department of Defense, which was to receive an 11 per cent budget increase from $11.1 billion to $12.23 billion. The ERDA budget increased from $3.6 billion to $4.1 billion, while the NASA budget remained about the same at $3.8 billion.

In contrast, Ford proposed virtually no increase in fiscal 1978 for the NIH. In particular, he leveled off the budget for the National Cancer Institute. However, observers expected Congress to add substantially to the proposed NIH budget later in the year.

Ford's budget contained provisions for three major research areas that had been recommended by Stever and the OSTP – significantly greater support for basic research and applied agricultural science and the launching of new earthquake-prediction programs.

Ford proposed that basic research funds be increased by 9 per cent to help make up for a long period of decline in funding. According to an analysis published in January 1977 by the Office of Management and Budget, inflation had eroded basic research support by about 20 per cent since the mid-1960s. Ford singled out the NSF for an 11 per cent increase in funds, and NSF Director Atkinson said the money would be used primarily to upgrade scientific equipment in colleges and universities.

Agricultural research will include a new program of research grants, administered by the Department of Agriculture, for studies in such areas as photosynthesis and nitrogen fixation. Ford proposed $28 million for this program in fiscal 1978.

The earthquake-prediction program will be divided between the NSF and the U.S. Geological Survey. Ford proposed that the program should receive $23 million in fiscal 1977, $54 million in fiscal 1978, and $70 million in fiscal 1979. In addition to developing within 10 years reliable methods of predicting earthquakes, the program will help communities prepare plans for responding to earthquake warnings.

Ford also proposed a new start on the Large Space Telescope, an optical telescope to be launched by the space shuttle in 1983. He endorsed the Jupiter unmanned mission, in which twin spacecraft launched in 1982 would orbit Jupiter and send probes into the planet's atmosphere. However, there was doubt whether Congress would approve these programs.

Carter's budget changes. Carter let most of Ford's R & D proposals stand, but he made major changes in energy-related areas. The revisions came in

Demonstrators at a National Academy of Sciences forum on recombinant DNA, held in Washington, D.C., in March, called for an end to genetic engineering experiments.

Science Policy

Continued

three installments and affected the nuclear program particularly.

In February, Carter proposed that Ford's requests for several long-range energy R & D projects be reduced and the savings spent on efforts likely to produce faster results. He proposed that research programs on conservation, fossil fuels, and solar heating and cooling be increased substantially.

The funds would be taken mainly from three programs. Carter recommended postponing construction of a long-term project, the 10-megawatt solar-thermal generating plant at Barstow, Calif., to save $55 million in fiscal 1978. He recommended that Ford's request of $513 million for nuclear fusion research be trimmed by $80 million. However, Carter would allow work to go ahead on a major new facility, the Tokamak Fusion Test Reactor at Princeton University. Finally, in April he recommended that the Liquid Metal Fast Breeder Reactor (LMFBR) program should receive only $483 million in fiscal 1978, some $370 million less than Ford had proposed.

New nuclear policy. On April 7, Carter announced that the United States would put off its plans for reprocessing spent uranium reactor fuel to recover plutonium for use as another nuclear fuel (see ENERGY). Carter hoped this action would help prevent the spread of nuclear weapons. Plutonium is a key ingredient in one type of atomic bomb, and he wanted to discourage nations that do not have nuclear weapons from acquiring plants capable of extracting plutonium from spent reactor fuel. So he proposed that the United States should set an example by turning its back on the recycling of plutonium and sharply downgrading efforts to build a plutonium-producing breeder reactor.

Carter also wanted to dissuade other industrialized nations from selling reprocessing plants to nations that do not have nuclear weapons. But West Germany refused to cancel its plans to sell a reprocessing facility to Brazil. Other countries also refused to give up their own plans to recycle plutonium and to develop commercial plutonium-producing breeders. However, Canada, France, Great Britain, Italy, Japan, the United States, and West Germany agreed in May to participate in a joint study of alternative nuclear fuel-cycle technologies and policies.

Carter's energy program was unveiled on April 20. Its major feature is a carrot-and-stick approach to encourage Americans to conserve energy. Carter would allow fuel prices to rise and would place high taxes on gas-guzzling automobiles. If yearly conservation targets could not be met, he would place an extra tax on gasoline. These taxes would be returned to consumers in the form of income tax credits for home insulation and for installing solar heating. Purchasers of efficient automobiles that use relatively small amounts of gasoline would also receive rebates. In addition, the plan included incentives for industrial plants to convert from oil and gas to coal.

The energy program encountered strong opposition in Congress, and observers expected that debate would continue throughout 1977.

Carter's plan for energy research and development emphasized short-term energy projects. It also spelled out proposals for the LMFBR program. The main casualty was a demonstration LMFBR scheduled to be built on the Clinch River in Tennessee by the early 1980s. Carter recommended that construction be postponed indefinitely. He allowed work to go ahead on a test reactor nearing completion in the state of Washington, but he reduced the entire LMFBR program to a low-priority effort.

Department of Energy. On March 1, Carter proposed legislation to establish a Department of Energy that would include the functions of ERDA, the Federal Energy Administration, the Federal Power Commission, and parts of the Department of the Interior. Congress was considering the legislation as of mid-1977, and Schlesinger, chief architect of the energy plan, was expected to be named head of the new energy department.

Congress also undertook a major reorganization of its science-related committees in 1977 to streamline operations and remove overlapping jurisdictions. In February, the Senate merged the Committee on Aeronautical and Space Sciences with the Commerce Committee to form a Committee on Commerce, Science, and Transportation, chaired by Senator Warren G. Magnuson (D., Wash.). The Senate established a subcommittee with authority over science policy and the space program and appointed Senator Adlai E. Stevenson III (D., Ill.) chairman. It placed jurisdiction over health and welfare in a Human Resources Committee, chaired by Senator Harrison A. Williams, Jr. (D., N.J.) with Senator Edward M. Kennedy (D., Mass.) as chairman of its health subcommittee.

A new Committee on Energy and Natural Resources under Senator Henry M. Jackson (D., Wash.) received authority over energy policy. Environmental jurisdiction was consolidated in a new Committee on Environment and Public Works headed by Senator Jennings Randolph (D., W. Va.).

Congress also scrapped the once-powerful Joint Committee on Atomic Energy and placed authority for nuclear development and regulation in separate committees. In the Senate, authority was divided between the committees headed by senators Jackson and Randolph. In the House, the Committee on Science and Technology got authority over nuclear energy research and development, while jurisdiction over nuclear regulation went to the Committee on Interior and Insular Affairs.

Science court. Scientists in 1976 and 1977 debated a proposal to establish a science "court," an idea conceived by physicist Arthur Kantrowitz, chairman of the Avco-Everett Research Laboratory in Everett, Mass. The science court, consisting of experts, would hear conflicting testimony and try to resolve disputes about scientific facts in public policy issues, such as nuclear safety and the safety of food additives.

Scientists, lawyers, and representatives of public-interest groups discussed the proposal in September 1976 at the Xerox International Center for Training and Management Development near Leesburg, Va. The group recommended that the concept be tested. However, some experts doubted that scientific judgments could be separated from value judgments in complex political disputes.   [Colin J. Norman]

Space Exploration

The two Viking landings on Mars dominated the United States space effort in 1976. Viking 1 reached orbit around Mars on June 19, and its landing craft touched down on the planet on July 20. Viking 2 reached Mars orbit on August 7 and sent its lander down on September 3. See VIKING'S VIEW OF MARS.

Man in space. The Soviet Union launched its Soyuz 21 spacecraft on July 6, 1976, carrying cosmonauts Boris V. Volynov and Vitaly M. Zholobov. Two days later, the cosmonauts boarded a new Russian space station, Salyut 5, which was put into orbit on June 22. They spent 49 days aboard the station and carried out a variety of tests, including studies of their sense of balance and of how well plants grow in weightless conditions. They also experimented with manufacturing techniques and with making multispectral photographs of the earth. They returned to earth on August 24.

Western observers had suspected that the mission was an attempt to surpass the 84-day man-in-space record set by the final U.S. Skylab crew in 1974. Some suggested that the flight was cut short because the crew suffered from what a Soviet newspaper described as "a state of sensory deprivation." In addition, the spacecraft apparently had technical difficulties.

The second attempt to use the Salyut 5 station failed. Cosmonauts Vyacheslav D. Zudov and Valery I. Rozhdestvensky tried to rendezvous their Soyuz 23 capsule, launched on October 14, with the station. But they had to return to earth two days later because a defect in the spacecraft's docking equipment prevented coupling with the station.

Cosmonauts Viktor V. Gorbatko and Yuri N. Glazkov went aloft in Soyuz 24 on Feb. 7, 1977, and boarded Salyut 5 on February 8. They returned to earth on February 25. Both crewmen had reportedly trained for spacewalks, although they apparently never left their craft. The Soviet news agency Tass indicated that no more crews were scheduled to visit Salyut 5.

Space shuttle debut. The U.S. space shuttle took to the air for the first

A U.S. space shuttle riding piggyback on a Boeing 747 jet made its first manned test flight on June 18.

Space Exploration

Continued

Astronauts try out a space "lifeboat" for carrying a person from a disabled shuttle to a rescue spacecraft. A crew member without a space suit will zip into a urethane bag containing life-support equipment so that a space-suited astronaut can carry him to safety.

time on February 18, attached to the top of a Boeing 747 jet. No one rode in the shuttle during this and four later tests, which were designed to check the vehicle's stability and aerodynamics as well as the safety of the piggyback arrangement atop the jet. President Gerald R. Ford dubbed the shuttle used in the tests "Enterprise" after the craft on the television show "Star Trek."

The first of several piggyback tests with humans aboard the shuttle took place on June 18. Astronauts Fred W. Haise, Jr., and Charles G. Fullerton were aboard. In later flights, they will alternate with Joe H. Engle and Richard H. Truly as commanders and pilots. The shuttle was scheduled to land on its own for the first time in August, when it will be carried aloft by the 747, but set loose and guided down to a landing by the crew. The first orbital shuttle flight is scheduled for 1979.

The National Aeronautics and Space Administration (NASA) in July 1976 began a yearlong "casting call" for people to operate the shuttle on its orbital missions. NASA placed special emphasis on recruiting women and members of minority groups.

In April and May 1977, the space agency tested 10 women at the Ames Research Center in California to collect data on how well they will withstand the physical stress of space flight. The women were confined to bed for several weeks to cause their muscles to weaken much as they would in the weightlessness of space. Then scientists studied their responses during tests that simulated the physical stresses caused when a spacecraft re-enters earth's atmosphere. The data collected from the test will aid NASA scientists in establishing physiological criteria for selecting future women astronauts.

NASA also sent out 10,000 requests for proposals involving small research packages that require no attention in space to be part of a shuttle payload known as the Long Duration Exposure Facility, a cylinder 9 meters (30 feet) long and 4.3 meters (14 feet) in diameter, containing racks that can accommodate up to 80 small research projects per flight.

Beyond the shuttle. NASA in the summer of 1976 conducted its second major study on the possibility of build-

ing self-sustaining colonies in space. It also awarded several substantial contracts for studies on building large structures in space.

International space research. European scientists continued to develop the Spacelab manned laboratory module that will be carried up to orbit in the space shuttle's cargo bay. Ground tests were completed in Europe to evaluate Spacelab crew procedures, and related tests were conducted aboard aircraft in the United States.

An unmanned Russian spacecraft, Luna 24, landed on the moon's Sea of Crises on Aug. 18, 1976. It drilled a 160-centimeter (63-inch) core sample from beneath the lunar surface and returned it to earth. In March 1977, Russian officials presented U.S. scientists with 3 grams (0.11 ounce) of material from the core.

NASA launched two Indonesian communications satellites into geosynchronous orbits, where they remain above the same points on earth, on July 8, 1976, and March 10, 1977. The satellites, Palapa 1 and 2, are of particular communications value to Indonesia, which has about 140 million inhabitants spread among more than 13,600 small islands.

The Japanese National Space Development Agency passed a major milestone on February 23, when it first launched a satellite into geosynchronous orbit. Two of the three rocket stages that lifted the satellite, Kiku 2, into space were based on U.S. designs. Only four days earlier, the University of Tokyo had launched a satellite, Tansei 3, into a lower, nongeosynchronous orbit to check an improved version of its own smaller rocket.

NASA launched a complex satellite named Geos on April 20. Built by the European Space Agency, it was intended to be the first purely scientific geosynchronous satellite. But because of a malfunction in the upper stage of its Thor-Delta rocket, it had to be placed in a lower orbit than is necessary to remain stationary. Scientists activated all of the instruments on the satellite, however, and the probe was expected to provide data on the earth's magnetic field for the remainder of the four-year International Magnetospheric Study, which began in 1976.

Space Exploration

Continued

Weather-watching from space. The U.S. National Oceanic and Atmospheric Administration launched its latest Improved Tiros Operational Satellite into a pole-crossing, circular orbit on July 29, 1976. It was equipped with instruments to provide visible and infrared images of cloud cover, snow, ice, and the sea surface and to provide sophisticated data on temperature and moisture in the atmosphere.

More than 100 scientists and technicians, representing 23 government and academic groups, took part in a month-long test in March 1977 to prepare for the Seasat-A oceanographic satellite launch in 1978. Seasat will provide regular data on ocean wave heights and spacing, currents, winds, temperatures, and other conditions that will be helpful for the crews of ships and transoceanic flights. The scientists used planes, radar stations, buoys, and instrument towers to collect ocean and weather information to be used for evaluating Seasat instruments and for making comparisons with information transmitted from the satellite.

Voyager mission. Preparations were underway during the year for the Voyager mission, the only U.S. interplanetary mission scheduled for 1977. Two Voyager spacecraft may visit as many as four planets and at least a dozen moons in an epic journey that may last for as long as 12 years.

The first probe, scheduled for launch on August 20, will pass Jupiter in July 1979 and Saturn in August 1981. If all goes well, it may become the first spacecraft ever sent to distant Uranus, arriving there early in 1986. There is a slim chance that it will go on to Neptune in 1989. Equipped with self-repairing computers and other aids to keep it active for a long time, the craft will undertake the longest mission ever attempted by NASA.

Because of the relative position of the earth to Jupiter at time of launch, the second Voyager probe, scheduled to lift off on September 1, will reach its targets first. It will fly past Jupiter and many of its moons in March 1979, then go on to the planet Saturn in November 1980.   [Jonathan Eberhart]

Technology

Many technologists and energy experts in May 1977 focused their attention on a new type of electric motor that engineer-inventor Cravens Wanlass, of Tustin, Calif., claims will save significant amounts of energy. The Wanlass motor can be built into new equipment or installed in existing home appliances and industrial machinery. Tests by the Southern California Edison Company indicated the Wanlass motor uses at least 10 per cent less electricity than conventional electric motors.

Electric motors convert electrical energy into mechanical energy to do work, such as running household appliances. The motors operate on the electromagnetic principle; in most of them, copper wires are wrapped around a stationary core that is usually made of iron. An electrical current flows through the wires and a magnetic field builds up, creating the torque, or turning power, needed to drive the motor.

The copper wires in the Wanlass motor are wound differently for greater efficiency. Because conventional motors must be designed to meet peak loads,

they do not operate at full efficiency while idling. For example, a washing machine designed for a momentary peak power load as it switches to a new wash or rinse cycle must operate at less than peak efficiency most of the time. Full technical details remained secret, but Wanlass claims that his motor achieves energy savings through a controlled-torque system that matches current to demand more efficiently.

Smoke detectors became more common in homes in the United States. In some localities, building codes required them in new houses and existing apartment buildings. The units can be installed easily and get their electrical power from either a wall outlet or battery. They cost from $30 to $75.

Smoke detector units fall into two categories — photoelectric and ionization. Photoelectric models contain a small light that remains on continuously. Any smoke entering the unit reflects the light to a photocell, setting off an audible alarm. In the ionization type, an extremely weak electrical current is created by the ionized molecules in a

Technology

Continued

Battery-powered smoke detector is easy to install. Such detectors, and others wired to building electrical boxes so as not to require battery checks, became popular home fire alarms in 1977.

minuscule amount of radioactive material. If smoke interrupts the flow of ionized molecules, the resulting drop in current will sound an alarm.

Instant photography. The Eastman Kodak Company in mid-1976 ended Polaroid Corporation's monopoly on instant pictures by introducing its own instant models, the EK-4 and EK-6. Kodak also introduced self-developing film much like that sold by Polaroid. Kodak's camera and film produce a rectangular print, rather than the square Polaroid shots.

However, a more startling development came from Polaroid. In April 1977, the company announced its long-awaited instant-movie camera, Polavision. The camera uses Super 8 film that comes packed in a small cassette. After the film has been exposed, the cassette may be inserted into a tabletop projector with a built-in screen. There, in only 1½ minutes, the film is automatically rewound, developed, and wound onto a take-up reel.

Digital watches. Makers of digital watches used semiconductor technolo-

gy to improve models with both liquid-crystal display (LCD) and light-emitting diode (LED) features. Gruen Industries, Incorporated; Time Computer, Incorporated, maker of Pulsar; and Litronix, Incorporated, introduced watches with LEDs that become visible with a mere flick of the wrist instead of by the push of a button. An internal "switch" – operated by either a ball of mercury moving in a tube or by a metal ball and magnet – turns on the LED.

Uranus Electronics, Incorporated, of Port Chester, N.Y., in September 1976 began marketing a solar-rechargeable digital watch that can also serve as a calculator. Instead of the conventional nickel-cadmium batteries, the Uranus watch uses thinner silver oxide batteries that the manufacturer claims last longer. The batteries recharge photoelectrically when the dial of the watch is exposed to sunlight or incandescent light. Normal usage should be sufficient to keep the batteries charged, unless the calculator is used extensively. In that case, the watch can be re-

An elliptical sprocket for bicycles allows the rider to get more thrust from each turn of the pedals, thus improving efficiency by up to 12 per cent.

Technology

Continued

An automatic-injection syringe continuously administers precise drug dosages by means of an electronic pump in a compact, portable case.

charged by attaching it to a battery charger plugged into a wall outlet.

The wearer pushes a button once for an hour and minute readout; twice, to learn the month and day; and three times to show seconds being counted off. To convert the watch to a calculator, the wearer pushes another button that clears all digits. Then 12 buttons around the dial can be used to add, subtract, multiply, and divide.

Waste and water. Water conservation became an issue in 1976 and 1977, especially in the drought-plagued West. New products appeared on the market that were designed to reduce or eliminate the 19 to 26 liters (5 to 7 gallons) of water required to flush a toilet. Welscot Enterprises in Colorado marketed a two-step device that permits the user to regulate the amount of water used for flushing. The device can be set to release a small amount of water for liquid wastes and a large amount for solid wastes.

Several versions of the composting toilet, developed in Sweden, appeared in U.S. bathrooms. All composting toilets work on the same basic principle—human and kitchen wastes settle to the bottom of the toilet or a large tank where, over a long period of time, bacteria break the waste down to produce compost that can be used as fertilizer. Odors are carried away by vents, assisted in some models by fans or blowers. These toilets use no water. The homeowner removes a small amount of the compost a few times a year.

Warm walls. The Kalwall Corporation of New Hampshire developed a "solar battery" that can be installed in the south wall of a new or old house. Translucent tubes inside the wall and behind a translucent outside panel admit sunlight that warms water inside the tubes. Fans or vents carry the air that has been warmed by the tubes to the house's living area. At night, an insulating curtain can be pulled across the outer wall to seal off heat losses. Kalwall installed the first system during the 1976-1977 winter in a rural New Hampshire home. Kalwall claimed the system resulted in at least a 15 to 20 per cent saving in fuel.   [Edward Moran]

Transportation

Transportation research continued to slow in 1976 and 1977 because of cutbacks in funding. The developments reflected a shift away from the belief that new technology can contribute significantly to a short-term solution to U.S. transportation problems.

In the early 1970s, the federal government greatly increased funding for transportation research and development in the hope of developing a variety of new transportation technologies. Research ranged from work on short, automated guideway systems for commercial centers to ultrahigh-speed vehicles supported on cushions of air or by magnetic levitation for long-haul passenger transportation. However, funding for high-speed trains was quietly withdrawn from the fiscal 1976 federal budget. Funding for short-haul guideway systems was cut back and redirected to feasibility studies rather than demonstration programs.

Experts now place greater emphasis on upgrading existing bus and rail systems to provide improved passenger service. Attention was also given to using smaller vehicles such as taxis and vans for public transit.

Energy saving. Decision makers in 1976 abandoned the once-common belief that greatly increased use of mass transit should be the primary approach in attempting to solve energy-shortage problems. The scattering of potential riders across wide city and suburban areas and the unwieldy mix of individual destinations pushed the cost of enlarging the nation's urban public-transit systems to the point where any dent they might make in energy consumption would cost more in money and time than the nation could afford. Nor would decreased use of passenger cars compensate for the expenditure. At a 1976 meeting of the Transportation Research Board, an association of transportation researchers, participants even argued that an established, modern, fixed-rail system such as San Francisco's Bay Area Rapid Transit System actually uses more energy than an improved highway system permitting an increased flow of private automobile traffic.

Other developments reinforced the view that the nation must continue to rely on the expensive and wasteful private passenger car for most short-haul travel. Data prepared for the Office of Technology Assessment (OTA) show that doubling the mileage traveled by public-transit vehicles each day would generate an increase in ridership of only 20 to 40 per cent. The OTA study concluded that buses and rapid-transit vehicles are energy-efficient, low-pollution means of travel only if they carry a large number of people on each trip. But the study found that increasing service usually resulted in fewer riders on each trip. Thus, if doubling transit mileage generated only a 20 to 40 per cent increase in patronage, this would increase energy consumption and air pollution per passenger mile.

Railroads. The Federal Railroad Administration (FRA) in 1976 concentrated on improving existing technology and equipment. This policy represented a change from the early 1970s when the FRA supported the development of new technology for high-speed transportation systems that were supported, propelled, and guided above a track by air pressure or magnetic force.

In cooperation with the Association of American Railroads, the FRA in November 1976 opened its new Facility for Accelerated Service Testing at the Department of Transportation test center near Pueblo, Colo. This facility includes a 7.7-kilometer (4.8-mile) loop track on which equipment can be operated continuously under controlled conditions. Track and equipment are subjected to 10 times the normal traffic accumulation, mileage, and load applications, so one year's testing simulates 10 years of actual use.

Despite the new emphasis on existing rail technology, the railroads still had problems. Three Amtrak passenger trains derailed in late 1976, and all were pulled by SPD40 diesel engines built by a United States manufacturer under a 1971 contract.

Until an investigation could be completed, Amtrak ordered all trains hauled by the SPD40 to reduce speed to 65 kilometers (40 miles) per hour on

"It's an import."

Transportation

Continued

curves. One theory being investigated is that the accidents occurred because the heavy engine, weighing 180,000 kilograms (397,000 pounds), was not suited to the poorly maintained U.S. track system. Some investigators believe that the SPD40 might push the tracks apart when it takes turns at high speed, derailing the cars that follow.

Supplemental transit. Many corporations began using 10- and 12-passenger vans during 1976 to solve employee transportation problems created by inconvenient public transit and the growing expense of owning and operating private cars. Using such conventional vehicles for public transportation is called paratransit.

The system gives regular passengers door-to-door service and minimizes energy costs by using lightweight vehicles. It avoids the inconveniences of public-transit systems, which must keep rigid schedules and are often some distance from the rider's destination. Most corporations subsidize employees who participate in such programs by paying part of the fare.

Safety car. Calspan Corporation of Buffalo, N.Y., and Chrysler Corporation of Detroit announced in November 1976 that they had designed a safer, more economical, small family car to meet the needs of the mid-1980s. They began building 17 test cars in January 1977, under a grant from the National Highway Traffic Safety Administration. Results from the test cars will help establish future safety standards.

The safety car is a four-door, five-passenger model that uses the latest in protection systems and structural science. It is designed to protect passengers in front and rear collisions at speeds of up to 80 kilometers (50 miles) per hour, and side impacts at up to 72 kilometers (45 miles) per hour. An inflatable front-seat-belt restraint system, which comes into place automatically when the doors are closed, protects passengers. Other features are soft front and rear ends that compress on impact, see-through head restraints, a breakaway steering column, high tail lights, and tires that can run 80 kilometers after a flat.   [James R. Wargo]

Zoology

Zoologists reported in March 1977 that weaver ants use a complicated communications system to recruit ants to invade and control new territory. Bert Hölldolber and Edward O. Wilson of Harvard University in Cambridge, Mass., discovered that weaver ants use two previously unknown glands in communication. They concluded that chemical communication in ants may be more extensive and complex than scientists previously believed.

Weaver ants of the genus *Oecophylla* are found in tropical Africa, southeastern Asia, and Melanesia. They live in trees where they build nests out of leaves bound together with silk from the cocoons that surround larvae—immature ants. The adults weave the nests by holding larvae in their jaws and shuttling them back and forth between the edges of leaves. In this way, the ants also build bridges to new areas where they can capture prey for food.

Weaver ants are important in agriculture because they live in citrus and coconut groves where they prey on crop pests. Several species of weaver ants

prey on the bug *Amblypelta* which feeds on young coconuts, causing them to fall before they are ripe. The Chinese reported using these ants to control citrus-grove pests as early as the 1100s. About 1900, weaver ants were discovered preying on the *Amblypelta* pest on coconut palms in the Solomon Islands.

For their study, Hölldolber and Wilson confined colonies of weaver ants in the laboratory in potted citrus and fig trees that were surrounded by mineral oil. In each colony, worker ants restlessly patrolled the available space surrounding the nest. When a large branch was placed close enough to the nest tree to be detected by the ants, they gathered on the twig nearest the new branch. To cross the gap between the twig and the branch, they formed a living bridge with their bodies. Next they used the larvae to lay a trail of silk across the gap, creating a bridge that workers used to move into the new territory.

To guide fellow workers to the new territory, the ants laid a trail back to the nest with liquid expelled from their

Zoology

Continued

A beetle that lives in the dunes of South Africa's Namib Desert stands on its head to collect moisture from fog that condenses on its body and trickles down into its mouth.

rectal gland. Other ants poured from the nest area to follow this scent cue into the new territory. The researchers refer to this as long-range recruitment.

Hölldolber and Wilson discovered that an ant must turn its rectal gland inside out to lay a trail. The ant lowers its abdomen, rotates the end segment downward, and pushes out the gland. It then drags the gland lightly over the surface. The gland apparently rests on a sled of long bristles that come from the upper edge of the gland opening.

The scientists also discovered that for short-range recruitment the ants secrete another odor through an opening called the sternal gland. This odor stimulates nearby ants to cluster so that they can capture insects that are too large for one ant to handle alone. This second odor thus helps the colony in foraging for food.

Breeding zoo animals. Researchers at several zoological establishments reported success in 1977 in breeding animals that are seldom bred in captivity.

At the Reptile Breeding Foundation in Ontario, Canada, Thomas A. Huff succeeded in breeding the rare and endangered Cuban boa, *Epicrates angulifer*. Although most snakes lay eggs from which the baby snakes hatch, the Cuban boa gives birth to live young. Huff reported that the snake he worked with, less than 6 years old, gave birth to at least two baby snakes. One baby boa survived and shed its skin after nine days. Oddly, it appeared more nervous and became more aggressive than other snakes of the same species captured in the wild.

Pamela S. Davis and Guy A. Greenwell reported the hatching of the first North Island brown kiwi, *Apteryx australis mantelli*, a nonflying bird, at the National Zoological Park in Washington, D.C. The zoo received a male and a female kiwi in 1968 as a gift from the New Zealand government. The female laid several eggs, but none was fertile. The zoo requested more kiwis from New Zealand, which sent two more males. However, subsequent eggs were also infertile.

The researchers then renovated the kiwis' enclosure, installing New Zealand plants and ground cover, new lights, and making other changes that made the area more like the kiwis'

native habitat. The kiwis eagerly brought the pine needles that were provided into the nest box. After the female laid another egg, the male diligently sat on it. This egg eventually hatched successfully.

Greenwell, with Sheryl Gilbert, also reported success in breeding the Borean great argus pheasant. Other successes were reported by R. J. Wheater, who bred the gentoo penguin at the Edinburgh Zoo in Scotland; and by M. L. Lubbock, who bred the white-winged wood duck in Gloucester, England. In each of these cases, breeding was stimulated by changing the zoo environment to make it more like the animal's natural habitat.

Moving day. Precautions can minimize the trauma connected with moving animals to new quarters, a West German researcher reported in 1977. Introducing wild animals to the confined quarters of a zoo is difficult and the animals need time to get used to their new surroundings. By the time visitors are allowed to see them, the animals usually appear to be quite familiar with their quarters.

But what happens when a zoo is enlarged to provide new accommodations for its residents? Uta Hick of the Zoologischer Garten (Zoological Garden) in Cologne, West Germany, described how her zoo prepared to move 123 primates to new quarters.

To prepare the animals for moving day, zoo workers placed the transportation crates in the animals' cages several weeks before the scheduled move. The crates were regularly baited with food so that the animals would become accustomed to the crates and could be easily lured into them on moving day.

Although moving went smoothly for the most part, even these precautions could not eliminate all problems. Two previously friendly male douc langurs, Boris and his son Mischa, Asiatic monkeys similar to the baboon, began to quarrel in their crate during the move and never resumed their former closeness. Boris harassed Mischa in the new quarters and refused to let him rejoin the rest of the group for several weeks. A baby langur died after the move, and several previously friendly animals became much more aggressive toward one another.

Venoms
For Science

Herpetologist Jack
Kilmon demonstrates
milking technique that
persuades a snake
to donate its venom
for medical research.

Even as a youngster, Jack Kilmon of Baltimore was fascinated by snakes. He wanted to pursue his life's work in this field of biology. But opportunities were largely limited to museums and zoos. So Kilmon turned his interest in collecting snakes into a science-serving business. He makes his living selling venoms for research. Many of the venoms that poisonous snakes, insects, and sea creatures spew are now being enlisted in the cause of biomedical research.

Only two firms in the United States produce venoms for medical research—the Miami Serpentarium and Kilmon's Biologicals Unlimited.

Kilmon, his wife, and a small staff take care of several thousand snakes, insects, and any other poisonous creature he can find. He has to be sure all get an adequate diet so they will produce venom on schedule. Snakes that do not feed on live prey are force-fed with the aid of a caulking gun filled with hamburger and nutrients.

Kilmon stimulates the poison glands of insects electrically so that they will eject their venom into containers. He collects venom from such creatures as jellyfish by removing the stinging cells and extracting their contents. Snakes are "milked" by inducing them to bite through a plastic membrane covering a test tube. The venom drips from their fangs and is captured inside.

Kilmon freezes the venoms at −100.1°C (−150°F.), then vacuum-dries them to produce a powder. It takes hundreds of cobra milkings to produce 1 gram (.035 ounce) of cobra-venom powder, which sells for $40 a gram. A gram of venom from glands under the skin of a Central American species of frog requires thousands of frog donations and costs about $10,000.

Research with venoms is still in its early stages. But researchers believe the complex and powerful chemicals that venoms contain can make significant contributions to understanding and curing major diseases, from heart attack to cancer.

The venom of every creature is different, but most are classified as either hemotoxins, which affect blood and circulation, or neurotoxins, which affect the nervous system. The hemotoxic components of venoms can either cause blood to clot, or prevent it from clotting by breaking down the walls of blood cells to cause massive hemorrhaging. Neurotoxins can block the transmission of nerve impulses to muscle cells, causing numbness and paralysis.

Advances in biochemistry and sophisticated tools for analyzing various compounds have made venom research an exciting new field. Venoms are now used mainly as a tool to investigate diseases of the circulatory, immune, and nervous systems. But, eventually, researchers may come up with a whole new family of venom-based drugs. Only two such drugs, both made with cobra venom, have been licensed for use in the United States. One is the painkiller Cobroxin, which some researchers say is more effective than morphine; the other, Nyloxin, is used mainly for arthritis pain. But researchers at Colorado State University in Fort Collins are trying to develop drugs from rattlesnake venom for treating high blood pressure, preventing blood clots, and limiting the damage from heart attacks.

In Boca Raton, Fla., physician Murray Sanders of the Sanders Research Institute is experimenting with a combination of venoms from the cobra and an African snake, the krait, to treat victims of a central nervous system disease called amyotrophic lateral sclerosis, also known as Lou Gehrig's disease because it killed the great baseball player. Sanders thinks that as-yet-unknown chemicals destroy nerve cells that activate muscle cells. He believes the venom might attach to these nerve cells in the spinal cord, thereby blocking the action of the chemicals.

So far, only a few snake venoms have been analyzed in detail. And many other creatures are potential contributors to research, including lizards, spiders, scorpions, jellyfish, and various poisonous fish and insects. Bee venom has already been used to treat arthritis.

Kilmon regards venoms as the hub of a wheel whose spokes radiate out to all branches of biological science—neurology, cardiology, hematology, molecular biology, and enzymology. At best, venoms can offer new drugs for previously untreatable diseases. At the least, they are a valuable research tool whose possibilities are just beginning to be explored.   [Constance Holden]

Besides preparing the animals for the move, zoo workers also took special precautions to prepare the new quarters for the animals. They put sponge mats on the floors so that particularly nervous animals would not injure themselves if they fell.

Sheets of black foil were layered on front of the glass cages to discourage the animals from hurling themselves against the glass. After a few days, small holes were cut in the black foil. These were gradually enlarged until the foil was entirely removed shortly before visitors were allowed to enter the new building.

The animals also learned about their new surroundings from one another. When one animal moved through a new passageway, others followed him. Soon all members of a family group were familiar with their new home.

Raising zoo babies. Charles J. Hardin of the Toledo Zoo in Ohio reported in early 1977 that he and his associates had hand-raised a baby giant anteater. Many captive animals raise their young in much the same way as they would in the wild, but the artificial zoo environment causes some species to be poor parents. A prime example is the giant anteater, *Myrmecophaga tridactyla*.

One of the major problems faced by zookeepers is giving anteaters an environment like their natural habitat. Zoos cannot provide them with colonies of ants at dinnertime, nor even substitute other insects. Instead, they are usually fed milk, eggs, and meat. Because of this, anteaters living in captivity are under more stress than animals whose natural diet consists of grasses or raw meat. As a consequence, newborn baby anteaters are often discarded by their parents.

Hardin found an infant anteater crying loudly at the edge of a pool in its enclosure, while its parents casually slept about 5.5 meters (18 feet) away. He removed the infant, cleaned it, placed it in an incubator, and gave it a rolled bath towel, which it clung to.

The baby was fed with an ordinary baby bottle filled with cat milk. While one assistant held the bottle and the infant anteater's long nose, another attendant held the baby on his lap, using one hand to restrain its powerful forelegs and the other to open its mouth and steady its head while it drank. Great care was needed to restrain the animal, yet allow it sufficient freedom to feed. All worked well, and soon the anteater outgrew its incubator and had to be placed in a large enclosure. When it finally began to investigate a dish filled with milk, it was on its own.

This kind of hand-raising has become especially important to zookeepers because many of the animals they deal with are rare and close to extinction. By perfecting methods and diets for hand-raising captive baby animals, zookeepers hope they can preserve such animals for future generations.

Tail-wagging language. Three Spanish researchers reported in early summer of 1977 that fallow deer use a series of tail signals to communicate with one another. Fernando Alvarez, Francisco Braza, and Alberto Norzagaray reached their conclusions after studying many different animals at the Estacion Biologica de Donana in Seville, Spain.

The tail of the fallow deer has a white underside that enlarges the white anal area when the tail is raised. The deer can hold its tail hanging, swaying, or in various erect positions.

The researchers discovered that each tail position imparts specific information to other deer. When the situation is calm, the deer's tail hangs still, apparently telling other deer that there is no danger nearby. A partially raised tail signals caution, and when danger is definite, the deer raises its tail fully. These motions markedly change the amount of white on the deer's rear end, sending signals to be interpreted by other deer of the same species.

The Spanish biologists proposed that deer use a gradient of tail signals, from the calm situation signaled with a lowered, gently swinging tail to the danger signal of an erect tail. Although they have not fully analyzed the complexity of the tail movements, the biologists believe the signals are likely to prove more intricate than reported so far. Signals of this sort can be of great aid in survival for animals that live in groups. The more specific each signal is, the more information they can impart. Studies of other species of deer may show whether they also use this type of tail communication. [William J. Bell]

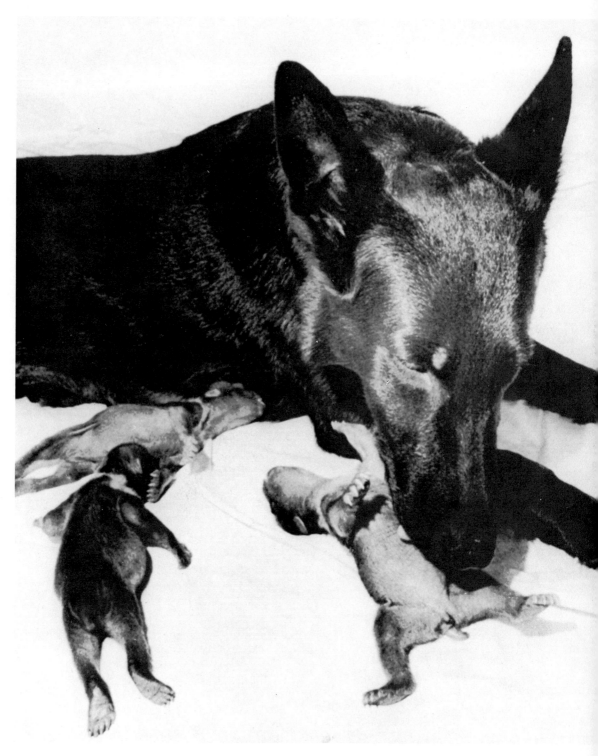

Deva, a German shepherd that belongs to an employee
of the Philadelphia Zoo, nurses three newborn Syrian
bear cubs who were rejected by their mother at the zoo.

People
In Science

The rewards of scientific endeavor range from winning a Nobel
prize to successfully completing a project for a science fair.
This section, which recognizes outstanding scientists, also
recognizes the students who may someday follow in their footsteps.

Rita Levi-Montalcini

By Joan Arehart-Treichel

Personable, intelligent, and insightful, this famed Italian neurobiologist leads a search to understand better how nerves grow

It is 9 A.M. and an elegantly dressed woman walks in the warm morning sunshine from her Rome apartment to a nearby bus stop. Her stately manner and graceful movements are emphasized by her carefully manicured hands decked with fine gold rings. Even passers-by who are not yet fully awake take a second look.

Is this fashionable Italian woman bound for one of Rome's fine shops? No, she is an internationally acclaimed neurobiologist on her way to work as the director of Italy's Laboratory of Cell Biology, which is housed in a modern six-story building in downtown Rome.

Rita Levi-Montalcini (pronounced *Lay-vee Mahnt-al-cheenee*) is a scientist of many accomplishments. In addition to heading a staff of 80 scientists at one of Italy's largest scientific institutions, she holds a professorship at Washington University in St. Louis. She is the codiscoverer of nerve growth factor (NGF), a protein in the body that makes certain nerves grow and develop, and of sympathetic immunity, a process by which antibodies made against NGF prevent certain nerves from maturing. These discoveries have set the stage for better understanding the growth of the entire nervous system.

"She made very important discoveries because she had enormous insight into a problem of basic cell biology," says biologist Salvador Luria of the Massachusetts Institute of Technology (M.I.T.) in Cambridge. Luria is a former classmate of Levi-Montalcini's and co-winner of the Nobel prize for physiology or medicine in 1969. "Her work is significant in all developmental neurobiology," he continues.

Insightful and extremely intelligent, Levi-Montalcini has devoted her life to research. Bubbling with conversation and energy, she is more vibrant and alive at 68 than most people are in their 20s. She has a youthfulness that somehow survived the long hours in the lab. Her rich laughter and lilting voice express a keen sense of humor. Tiny laugh lines crinkle the corners of her eyes as she listens to a colleague tell a joke at lunch. "Anybody could be happy with her," says a co-worker who is also a good friend.

There is no doubt, however, that Levi-Montalcini's brain, not her personality, brought her to scientific prominence. "I come from an intellectual Jewish family in Turin, Italy," she explains. "My father obtained a doctorate in engineering from the University of Turin. Most of his brothers—he was one of 18 children—became lawyers, and one of his sisters was a mathematician. With such a background, I naturally aspired to higher education and having a profession myself."

Desire was not enough, however. Although her father loved Rita, her twin sister Paola, and her older sister Nina, he was a staunch Victorian who was convinced that education and careers would conflict with his daughters' roles as wives and mothers. But Levi-Montalcini recalls, "I had no intention of being a housewife." When she completed finishing school at 18, she considered herself "the most miserable young woman in Italy." Finally, at 20, she begged her father to let her continue her education. He reluctantly agreed.

Parental permission was only the first step, however. Levi-Montalcini's finishing-school education had not included Greek, mathematics, and many other classical subjects that are prerequisites to a European university education. In order to catch up with her contemporaries, she completed the Italian equivalent of a high school education and two years of college in the United States in six months. Then, in 1931, at 21, she started medical school at the University of Turin. At last, she was to realize her dream—a profession of her own.

From the start, Levi-Montalcini was especially interested in neurology. She became a student of histologist Giuseppe Levi, no relation but a leading biologist and teacher. "He was very strong, talented, and authoritarian—like my father," she recalls, her large blue eyes warm with affection. "In a sense, I exchanged father figures, especially since my own father died from a heart attack in 1932." She made top grades in medical school. "It wasn't difficult because I was so interested," she explains with typical candor.

Levi-Montalcini graduated as a physician specializing in neurology and psychiatry in 1936, and began work as an assistant in neurology at

The author:
Joan Arehart-Treichel
is medical editor of
Science News magazine.

the University of Turin. She was equally drawn to patient care and to research and teaching. But her professional aspirations were thwarted two years later, before she could make a choice.

Dictator Benito Mussolini launched an anti-Semitic campaign in Italy. Jews were prohibited from attending Italian universities or holding academic positions in order to "avoid contaminating the Aryans," as Levi-Montalcini puts it. Indeed, she lost her Turin post because of her Jewish background. She accepted a position at the Neurological Center in Brussels, Belgium, in 1939 but returned to Turin before Germany invaded Belgium in May 1940.

All of wartime Europe soon became dangerous for Jews. Levi-Montalcini wanted to emigrate to the United States, but she loved her family too much to leave. She transformed her bedroom in Turin into a tiny lab in order to conduct experiments on the nervous system of chick embryos. As anti-Semitic pressure grew, Jews were forbidden to use university books, but devoted friends smuggled those that she needed to her. Her former professor, Levi, who had also gone to Belgium and returned to Turin, now became Levi-Montalcini's assistant in her makeshift laboratory.

"My bedroom became the meeting center for old pupils and friends of Giuseppe Levi. We worshiped in him not only the great scientist but also the valiant and undaunted anti-Fascist," she recalled in her autobiographical report "NGF: An Uncharted Route," which she presented at an M.I.T. neurosciences symposium in October 1973.

After Italy declared war on the Allies in 1940, Levi-Montalcini's life as well as her research were endangered. The Allies started bombing Turin and other Italian cities, so she and her family fled to a small farm not far from Turin. She resumed her neuroembryological research there under even starker conditions than in her Turin bedroom. She had little water, sporadic electrical power, and few eggs. But she came up with an ingenious idea. Why not extract the embryos from the available eggs, then cook what was left of the eggs for food?

"My brother didn't like that compromise very much," she says with the only bit of levity she can muster in recalling that period. "I wonder how I could have devoted myself with such burning enthusiasm to the study of a small neuroembryological problem when all the values I cherished were being crushed and the triumphant advances of the Germans all over Europe seemed to herald the end of Western civilization," she wrote. "I believe that I inherited from my father an unusually efficient defense mechanism that was to be of great help during those years. This was strengthened by my association with Levi, who was then in his 70s."

But even graver days lay ahead for the family. Levi-Montalcini's research came to a standstill as Mussolini fell from power in July 1943 and the Germans took control of Italy. Within two months, hordes of German soldiers had moved into Italy, and Jewish-Italians were persecuted and sent to concentration camps.

In her office in the Laboratory of Cell Biology in Rome, Levi-Montalcini and Pier Carlo Marchisio of the University of Turin discuss a new research project.

The Montalcini family went underground. They changed their name to that of a southern Italian, Roman Catholic family, and fled to Florence. There, Levi-Montalcini and her twin Paola became experts at making false identification cards for themselves, their family, and friends. They lived with a hospitable landlady, listening secretly to news broadcasts from London and fearing discovery by the Nazis.

When the Allies captured Florence from the Germans in August 1944, the family's agony ended. Levi-Montalcini offered her services to the Allies and worked as a physician in a refugee camp, where people were dying of diphtheria, typhus, and starvation. "Working among such rampant infectious diseases was really more dangerous than hiding from the Nazis," she says. Yet she survived, and when the German forces in Italy surrendered to the Allies in May 1945, Levi-Montalcini was free to resume her research.

Once again, she became an assistant to Levi at the University of Turin, where she studied the growth and differentiation of nerve cells. Then biologist Viktor Hamburger of Washington University offered her a position on his staff so that they could collaborate on such experiments. She accepted and went to the United States in 1947. Thus began a long association with American science that she still maintains along with her work in Rome. Even more important, her collaboration with Hamburger led to the happiest and most productive years of her professional life.

"Viktor directed the department with rules that were quite different from those of Levi," she reported at the M.I.T. symposium. "Accustomed as I was to the thundering voice of Levi and to his explosive way of expressing dissent . . . I was struck by the kindness and subtle dry humor of Viktor, who would never hurt other people's feelings or show his disagreement with more than a few gentle remarks and a firm glance of disapproval. Working with him on the same problem that had absorbed so much of my time and thoughts in my secluded laboratory in Turin was sheer pleasure."

Their collaboration in what to Levi-Montalcini was a "garden of Eden" went so well, in fact, that she postponed her return to Italy indefinitely. Then, in the fall of 1948, Hamburger showed her a short article by one of his former students on an experiment to test the ability of developing nerve fibers to grow into tumors. The article reported that if a mouse tumor known as sarcoma 180 was implanted in a chick embryo, sensory nerves from the embryo would grow into the tumor. Sensory nerves are those that transmit impulses from eyes, ears, or other sensory organs to a nerve center.

When Levi-Montalcini investigated this curious behavior, she discovered something even more remarkable—sympathetic nerves branched much more profusely into the tumor than did sensory nerves. Sympathetic nerves are those that transmit involuntary impulses, such as changes in the heartbeat or in the secretion of glands. One bright spring day in 1951, Levi-Montalcini formed a profound and exciting suspicion about her experimental results—the developing embryonic nerves might be attracted by some growth factor in the tumors. This possibility seemed all the more plausible because the nerves invaded tumors far quicker than they invaded embryonic tissues. They even forced their way into blood vessels in the tumors, as if in search of some nourishing chemical.

Impatient to test her theory, Levi-Montalcini rejected the usual time-consuming embryological experiments in which she would have grafted fragments of tumors into embryos and watched their developing nerves in order to prove the existence of such a chemical. Instead, she decided to use tissue-culture tests, the technique of growing, in the test tube, cells and tissues from living organisms. She contacted Herta Meyer, a friend who was in charge of the tissue-culture unit at the University of Rio de Janeiro Institute of Biophysics in Brazil. Carlos Chagas, director of the institute, invited her to join him and Meyer.

Levi-Montalcini arrived in Rio in October 1952 with two live mice in her handbag, each bearing a sarcoma 180. As soon as she reached her friend's lab, she dissected the tumors into small pieces, cultured the pieces in a medium of chicken blood and embryonic extract, then incubated sensory and sympathetic ganglia, or groups of nerve cells that form a nerve center, from chick embryos close to the pieces of tumor in this medium. The presumed growth factor, which had given a tantalizing hint of its existence in St. Louis, now clearly revealed itself. After 12 hours of incubation, nerve fibers reached eagerly from the ganglia toward nearby tumor pieces, then spread rapidly around them like rays of the sun. This halo effect was virtually indisputable proof that some chemical in the tumors promoted nerve growth.

Levi-Montalcini returned to St. Louis brimming with excitement and eager to isolate the molecule responsible for stimulating nerve growth. The search became a team effort with biochemist Stanley Cohen, who had just joined the Washington University Biology Department. It was Cohen who broke a sarcoma 180 into various chemical fractions until he finally isolated the growth chemical. Cohen, Levi-Montalcini, and Hamburger called the molecule "nerve growth stimulating factor" when they published their findings in 1954.

But even more provocative insights into the nature, origins, and actions of NGF were to emerge during the next five years, thanks to close collaboration between Levi-Montalcini and Cohen. For instance, Cohen identified the factor as a protein. He used snake venom to purify tumor NGF further and found, purely by chance, that the venom was another important source of NGF. In fact, the NGF it contained was 1,000 times more concentrated than that in the tumors. Cohen reasoned that a mouse's equivalent of a snake's venom gland is its salivary glands, so they tested tissues from mouse salivary glands and found even more NGF in them than in snake venom.

Meanwhile, Levi-Montalcini determined that NGF from venom produced the same effects on chick-embryo sensory and sympathetic

Biochemist Costantino Cozzari discusses his research project with Levi-Montalcini during a leisurely stroll on the Washington University campus in St. Louis.

nerves as did tumor NGF. Nerves grew out in a dense halo of fibers from sensory and sympathetic nerve tissue in culture. Cohen next developed an antiserum, a preparation of antibodies designed to neutralize NGF. When Levi-Montalcini injected the antiserum into newborn mice, she found that the animals' sympathetic ganglia were almost completely destroyed. This effect, which was called sympathetic immunity, was the greatest proof yet that NGF is indispensable to the growth of sympathetic nerves.

Unfortunately, Levi-Montalcini and Cohen's fruitful partnership ended soon after their discovery of sympathetic immunity. Cohen moved to Vanderbilt University in Nashville, Tenn., and began other research. "Stan's departure from his small and not-too-clean office on the first floor of Rebstock Hall, where he had labored for six years to unveil the mysterious nature of NGF, signaled the end of the most romantic and picturesque phase of this adventure," Levi-Montalcini wrote with both affection and sadness. "He used to spend entire days and most evenings there, meditating with eyes half-closed, smoking his pipe, and playing the flute ... while Smog, his gentle, dirty dog, slept peacefully at his feet."

Levi-Montalcini finds time to laugh, *right,* between the serious moments of science, such as when she studies a series of photomicrographs of part of the human eye with neurobiologist Boyd K. Hartman, *below.*

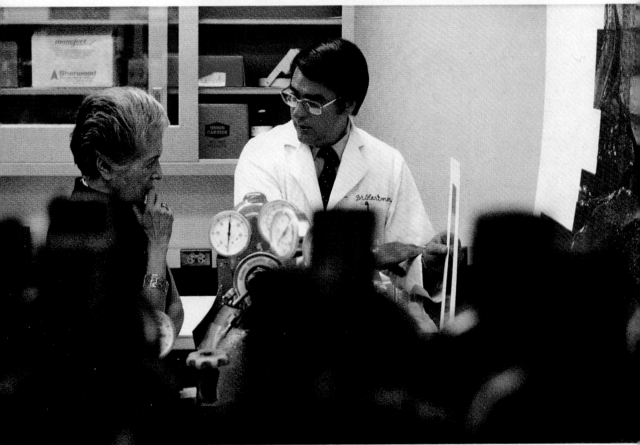

Cohen remembers Smog's impact on Levi-Montalcini. "She preferred that Smog not come into the tissue-culture lab," he says, "because each time he shook, he contaminated all the cultures." Both scientists stress that the discoveries would have been impossible without the other's work. "Neither of us could have done it on our own," says Cohen, and Levi-Montalcini heartily agrees.

Cohen's departure marked a turning point in Levi-Montalcini's career. They were partners—he the biochemist, she the neuroembryologist—and they worked together in the lab. But as NGF research grew, she assumed the role of senior partner, or group leader, and spent more time guiding the work of others and less time doing research herself. Her next NGF collaborator was Pietro Angeletti, a fellow Italian. He brought to the NGF quest youthful imagination and a strong scientific drive. Under Levi-Montalcini's overall supervision, he directed a growing research team from 1960 until 1972.

At first, Angeletti and Levi-Montalcini spent most of their time at Washington University. They worked comparatively little in Rome, where she had set up a small biology laboratory in 1961 in order to be closer to her mother. But the Rome lab—first funded by the U.S. National Science Foundation, then by the Italian government—eventually blossomed into the Laboratory of Cell Biology, Italy's second largest biological research institute. She became its director in 1969, primarily on the basis of her NGF discoveries. During the next three years, she and Angeletti spent about six months a year at Washington University and the other six months at the Rome lab—their feet in two scientific camps, as it were. Yet their research remained focused on one target—to learn more about NGF's actions on nerve cells.

Over the years, crucial insights have emerged from the work. For example, there are trace amounts of NGF present in all tissues and body fluids, such as human saliva, milk, and urine. And in 1971, Ruth Hogue Angeletti, who was then Pietro's wife, and Ralph Bradshaw, who is now a professor of biochemistry at Washington University's Medical School, finally identified the sequence of amino acid building blocks that make up a molecule of the protein NGF.

What do such findings about NGF mean to the science of neurobiology? NGF is the first and only chemical taken from mammals or other vertebrates that promotes the growth of nerves. Thus, it is a powerful tool for exploring the growth of brain and other nerve tissues, surely the most fantastic and complex biological system known. NGF's discovery also set the stage for the identification of other growth factors that stimulate other tissues in the body. For instance, after Cohen moved to Vanderbilt, he isolated a chemical from mouse salivary glands that promotes the growth of skin cells. Several other growth-promoting chemicals have since been found.

Yet Levi-Montalcini is the first to admit that what is known about NGF and other growth factors is just a beginning. "So much more needs to be learned," she exclaims. "For example, why are there so many different kinds of these factors? Exactly how do they promote the growth and development of various cells and tissues? How many more factors remain to be identified? Should they be classified as hormones or as a separate category of tissue-stimulating agents?" Research has already provided some partial answers. Extensive studies on mouse salivary gland NGF by Levi-Montalcini and Angeletti suggest that NGF works quite differently from classical hormones.

Other far-reaching questions about NGF concern Levi-Montalcini and her colleagues. How can these chemicals be exploited to help humanity, perhaps as drugs to treat various diseases? For example, in addition to making skin cells grow, epidermal growth factor inhibits acid secretion in the stomach. So it might be effective in treating stomach ulcers. A growth factor might be discovered that makes nerves in the central nervous system grow, just as NGF makes peripheral nerves such as sympathetic and sensory nerves grow. Such a factor might prove valuable in regenerating damaged adult central nerves, which usually cannot repair themselves. This would be of enormous benefit to victims of spinal cord paralysis or brain damage.

Her discoveries and the promise they hold have won international recognition for Levi-Montalcini. She is a member of the United States National Academy of Sciences and the Papal Academy of Rome. Levi-Montalcini's election to the Papal Academy is especially impressive because she is the first woman member as well as a non-Catholic.

Fellow scientists praise her scientific contributions. Biologist Renato Dulbecco, co-winner of the 1975 Nobel prize for physiology or medicine and Levi-Montalcini's classmate in medical school, considers her research "an important discovery." Some scientists believe her work merits a Nobel prize. "Why not?" says Adriano Buzzati-Traverso, senior scientific adviser to the United Nations Environmental Program and a long-time friend of Levi-Montalcini. "Yes, she would qualify for a Nobel." Pietro Calissano, who works on NGF with Levi-Montalcini in Rome, is even more explicit in his praise. "I believe that her discoveries are enough to command a Nobel prize. The reason is simple. The Nobel prize is given whenever a breakthrough occurs in a field, and NGF can certainly be visualized as a breakthrough in the field of neurobiology."

Colleagues also hold Levi-Montalcini in high personal regard. "The most impressive thing about her is her enthusiasm," says Calissano. "She has the enthusiasm of a 25-year-old. Yet at her age, she could well be satisfied with her achievements and honors." "She is a lady who knows what she wants," says Buzzati-Traverso. "She is also strict about certain moral principles, which is very nice. For example, she takes a very dim view of the petty politics that are so much a part of Italian university life."

"She has a fascinating personality and tremendous poise," says Dulbecco, now at the Imperial Cancer Laboratory in London. "A friend of ours calls her 'La Regina,' which means 'The Queen' in Italian."

Levi-Montalcini looks at a collection of slides with Viktor Hamburger who, when he was chairman of Washington University's Biology Department in 1947, invited her to join his department.

Levi-Montalcini hardly lives like a queen, however. In Rome, for instance, she arises at 5 A.M. and works on research in her apartment study until 9 A.M., then goes to the lab. She usually works there until 6 P.M., orchestrating research on numerous biological subjects and reviewing and signing papers. She dislikes such routine administrative work but took on the lab directorship because it was a challenge and enabled her to monitor new directions in NGF research.

How does Levi-Montalcini see herself as a supervisor? "Tough, sharp, and not always pleasant to be around," is her reply. She wonders whether her staff likes her and suspects that some may resent or envy her position and power. However, Calissano believes that she is not only highly respected but very much liked. "She is the best boss I could ever imagine," he says. "I have seen her give her staff both moral and financial help when they were in need. She is also democratic, which is unusual in Italy, where bosses normally maintain a hierarchical attitude." Lenore Friedman, Levi-Montalcini's secretary at Washington University for 14 years, has similar praise: "She is a marvelous person to work with. She is considerate, appreciative, and warm. I remember once I was working late and had a lot of letters to get out. She stopped to help me fold and insert them into envelopes, which is extraordinary. I have worked for a lot of people, but I have never really enjoyed a job as much as this one."

Levi-Montalcini returned to Washington University in May 1977 for the last time as professor. In June, she became professor emeritus and will no longer teach. Friedman says that Levi-Montalcini liked the contact with young students, and they in turn liked her.

Back in Rome, Rita
and her twin sister
Paola chat amid the
flowering plants
in their apartment.

Levi-Montalcini walks
slowly through a quiet
park in Rome, lost
in thought about the
uncharted future of
her biological research.

Ruth Hogue Angeletti, who is now doing research with NGF at the University of Pennsylvania School of Medicine, remembers Levi-Montalcini as a good lecturer, but holds a different view of her as a teacher. "She is very demanding, and this makes it difficult for some students to get along with her," says Ruth Angeletti. "Young researchers, especially those from non-Latin environments, have been upset by her direct emotional outbursts when she thought they were saying something wrong." But she adds that Levi-Montalcini's spontaneous emotional nature can also affect people in positive ways, "She can be unduly generous."

Yet, if her compassion for employees and her demands on students are exceptional qualities, her devotion to twin sister Paola is perhaps even more extreme. They share an apartment on Rome's Viala di Villa Massimo and often travel together. Paola has an artist's temperament and is not as robust as Rita. Paola encourages Rita to come home for lunch and to take a short nap before she returns to work.

Levi-Montalcini is proud of Paola's accomplishments as an artist. Paola's paintings and engravings are well known in Europe and are displayed in museums in Rome, Turin, and Modena in Italy; in Tel Aviv-Yafo in Israel; and in private collections. Levi-Montalcini is also fond of her sister Nina, who lives in Turin, and she was very close to her brother, Gino, who died in 1974. Another important person in her life is Maria, her devoted maid for 10 years. Maria has transformed the Rome apartment into a tropical paradise with lush plants and gaily colored flowers.

When not at work in her study, Rita reads or listens to music. But she is also "a great cook of Italian specialties," says Cohen. She takes pride in a dessert that she created and that Maria now prepares—a mold of ice cream swimming in rich liqueurs and garnished with kumquats grown on their apartment balcony.

As often as not, however, Levi-Montalcini is called from the table by a telephone call before she has finished her meal. This time, it is a young American mathematician she helped to place in Pisa with one of Italy's finest mathematicians. "Helping young people is my hobby," she smiles. She has helped some 50 young Italian scientists find positions in the United States.

Although Rita Levi-Montalcini has retired from active teaching at Washington University, she has not abandoned American science. From 1978 to 1980, she will spend four months a year as a John E. Fogarty International Center Scholar in Residence at the National Institutes of Health in Bethesda, Md. She will write, study, and occasionally lecture. But her main focus there, as in Rome, will continue to be her first love, NGF research.

Says Rita Levi-Montalcini, who "watched with awe and wonder the birth of this 'miracle' molecule from the sinister womb of malignant tissues...[NGF research] keeps us wondering where it is heading, and whether its uncharted route has, indeed, an ending."

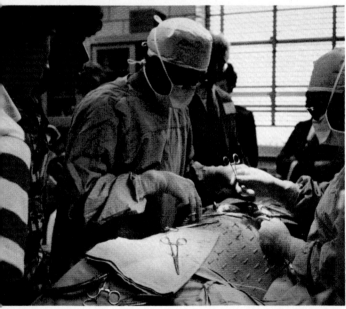

HORMONE THROUGH THE
UTILIZATION OF
IN VIVO AND IN VITRO
INVESTIGATIVE TECHNIQUES

The International Science And Engineering Fair

By William J. Cromie

**Years of study and experimentation laid
the groundwork for many of the prizes won
by young science students in Cleveland**

Lori Rhodes, 17, thinks she has discovered a new antibiotic. A senior at East Noble High School in Kendallville, Ind., Lori has worked on isolating and testing the substance since her freshman year. In 1977, she wanted to prove the value of her work by displaying it at the Northeastern Indiana Regional Science Fair in Fort Wayne. Top winners of this and 230 other regional fairs would compete at the 28th International Science and Engineering Fair (ISEF) in Cleveland. A large drug company in Indianapolis had asked Lori for a sample of the antibiotic to test. So she felt confident she could win the regional science fair and take top honors at the prestigious ISEF, even though she would be competing against the top high school science students from 50 states, Puerto Rico, Canada, Japan, and Sweden.

Other students were confident they would win at the ISEF. Among them was Richard H. Ebright, 17, a senior at Muhlenberg High School in South Temple, Pa. Officials at the United States Department of Agriculture and other experts believe he has discovered a previously unknown insect hormone. One of Richard's classmates, Paul M. Embree, 17, also hoped to capture a first-place award for a device he invented to improve AM radio reception. But first, Paul and Richard would vie against each other and about 400 other 6th- to

Facts About The Fair

Every spring, students who are interested in science show their expertise by exhibiting their work in science fairs. Their projects range from insect collections to research on ways to detect cancer.

Elementary school students show their exhibits at school and local fairs. High school students also exhibit their work first at school and local fairs, but the winners go on to larger city, regional, and state fairs. The top of the pyramid is the International Science and Engineering Fair (ISEF), held during the second week in May.

The ISEF, sponsored by Science Service of Washington, D.C., grew out of smaller science fairs that began about 1930. The first ISEF was held in 1950. It featured the work of 30 finalists chosen from some 15,000 students who competed at 13 regional fairs. The ISEF grew rapidly until the competition now involves more than 1 million students from high schools in the United States and other nations.

The quality and the complexity of the projects have increased with the number of participants. Many of the entries at the first ISEF were simple collections and models. Typical projects in recent years have included "A Holographic Study of the Sporangiophore of Phycomyces" and "A Myo-Electric Prosthetic Terminal Device."

Only students who have not reached their 21st birthday by May 1 are eligible to participate. Each finalist must be accompanied by an adult. Expenses are paid by the sponsoring local or regional fair.

The ISEF is one of the few contests in which the judges outnumber the contestants. Some 500 scientists, engineers, and physicians judge about 400 exhibits.

Science Service awards first, second, third, and fourth prizes in 11 categories that range from behavioral science to zoology. About 30 private, professional, and government organizations also present awards.

Students who want to compete in science fairs can obtain information from their science teachers, science fair directors, or by writing to Science Service, 1719 N Street NW, Washington, D.C. 20036.   [W.J.C.]

The author:
William J. Cromie is a free-lance science writer and executive director of the Council for the Advancement of Science Writing.

12th-grade students at the Reading-Berks Science and Engineering Fair in Reading, Pa.

At the regional fairs, Paul, Richard, and Lori faced a battery of scientists, engineers, and teachers who judged their projects. The judges study the displays and question the students in detail about their work. In each case, the judges want to know if the student thought of an interesting problem, then proceeded to solve it by using creative scientific methods. They try to determine how much help the student obtained from parents, teachers, and others, and whether the problem chosen was too easy or too difficult considering the ability and resources of the student.

Lori explained to the regional judges that she had participated in science fairs since she was in the third grade. Her mother teaches biology at East Noble High School, where her father is principal. In the seventh grade, she became interested in microbiology after watching older students in her mother's class doing experiments such as determining what kinds of fungi grew in the school's locker rooms.

In the ninth grade, Lori collected soil samples from woods near her home and samples from a compost heap of grass cuttings taken from the high school football field. In these samples, she found a number of microorganisms, such as bacteria and molds, which produce antibiotics against the fungi that cause athlete's foot. She displayed her work in a science fair at a local college and became one of 20 winners sent to the regional science fair. There, she captured a fourth place in the microbiology division.

For the next three years, Lori continued her work on this project. She isolated antibiotics, tested them for effectiveness, and tried to determine the properties of one she believed to be a possible new drug to combat fungi infections. In her junior year, Lori won third place in microbiology at the 27th ISEF in Denver. She also won special awards that were presented by the American Society of Agronomy, the U.S. Air Force, and the U.S. Army.

In her senior year, she found that her drug killed 10 kinds of fungi in laboratory experiments. These included fungi that cause, in various parts of the body, conditions similar to athlete's foot, such as gilchrist's disease, which affects the lungs, mouth, and lymphatic vessels, as well as attacking the skin.

"The big problem with antibiotics," Lori explained, "is that many of them can kill the person along with the germs. My research became very exciting when I found that the substance I was working on seemed to be as effective as the known antibiotics, but without potential harmful side effects."

The 1977 competition was to be the peak of Lori's 10 years of competition in science fairs. "Ever since the third grade, I wanted to go to the ISEF in my senior year," she said. "That's what I devoted myself to in high school. After winning in 1976, I never considered the possibility that I wouldn't be going in my senior year." Indeed, her parents and friends expected her to win at the regional fair. They sat in tense anticipation as fair officials read the names of the two top winners. The absence of Lori's name produced shocked silence. "I was dreadfully disheartened and cried for a couple of weeks," she said.

But earlier, Lori had entered another prestigious competition for high school science students. The Westinghouse Science Talent Search gives 40 winners the opportunity to share $67,500 in scholarships and awards, and to take part in a five-day expense-paid trip to Washington, D.C. Three of the 1977 winners were Lori Rhodes, Richard Ebright, and Paul Embree. One week before the ISEF, Lori also won one of five 2-week trips to London awarded at the National Junior Science and Humanities Symposium in West Point, N.Y. "This helped to soothe the pain of my disappointment," she said.

Meanwhile, Richard won the grand championship (first place) and Paul won the championship (second place) at the Reading regional fair. They traveled to Cleveland together and began setting up their exhibits along with 414 other regional winners, who began arriving with their parents and teacher-sponsors on Sunday, May 8.

Judging began on the following Wednesday in Cleveland's spacious Convention Center. Prohibited from entering the exhibit area during

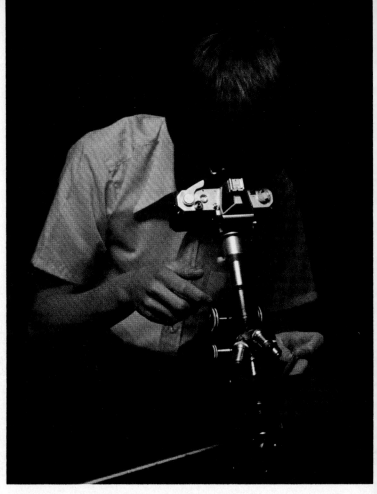

Richard Ebright catalogs a butterfly he has captured for his experiments, *below.* Using a camera attached to a microscope, *right,* he photographs one of his earlier slide specimens.

Paul Embree tests radio circuitry he designed to eliminate distortions in AM-signal reception.

the morning, students rested or took tours while 523 judges examined the 416 displays. Nearly half the judges would determine who would win the ISEF monetary awards in 11 categories ranging from the behavioral sciences through medicine to zoology. The other scientists, engineers, and teachers would decide who would receive the special awards to be given by government agencies such as the Energy Research and Development Administration, professional organizations such as the Mathematical Society of America, and private companies such as General Motors Corporation.

One reason for the large number of judges was the need for experts to judge work in a tremendous variety of fields. There were projects on such subjects as the behavior of ants; the effect of caffeine on fruit flies; detection of cancerous substances in peanuts; and how to build solar-powered devices, wind-powered generators, and laser communications equipment. A number of students constructed robots, microcomputers, programmable hand calculators, and particle accelerators. One participant even built a remote-controlled unit that could be used for exploring other planets.

Judges inspected the projects in the morning and did some preliminary rating. In the afternoon, they talked to the students. Paul stayed

Lori Rhodes shows her father a culture she tested in her search for an antibiotic that would combat fungi infections.

cool during the interviews, answering each question carefully. Richard bubbled with enthusiasm and could not resist telling the story of his scientific life to every judge who would listen.

Richard began collecting butterflies in kindergarten, he told the judges. When he was in the third grade, his mother volunteered him to tag monarch butterflies for a Canadian scientist who was studying their migration routes from Canada to the Gulf of Mexico. Catching large numbers of the big black-and-orange butterflies is difficult, so the best way to get insects to tag is to raise them yourself. For the next several years, Richard raised butterflies by the thousands and became thoroughly familiar with their development from egg to larva (caterpillar) to pupa (the cocoon stage) to flying adult.

A gold spot that appears on the mummylike pupae of monarchs and other butterflies intrigued Richard. He read everything he could find about the spot and learned that scientists, unable to determine its function, labeled it ornamental. Although then only 13, Richard found himself unable to accept this conclusion. "Everything in nature has some purpose," he said.

When he was in the ninth grade, he surgically removed the spot from pupae and discovered that the spotless pupae became adults with distorted wings. The dustlike scales that cover the wings were mal-formed and discolored. Richard proved that the gold spot is not just ornamental but is essential to the proper development of an adult butterfly. This work won for him the grand championship of his regional fair in 1975. At the ISEF in Oklahoma City, Okla., that year, he won third place in the zoology category; a summer job at the Walter Reed Army Medical Center in Washington, D.C.; and a trip to U.S. Navy facilities in San Diego. In addition, he received cash prizes of $100 and $50.

The Army job turned out to be crucial to his research. "I had access to a good library in the entomology department of the hospital," he said. "I learned a great deal that I never could have learned in Reading, and this information permitted me to continue the project."

While at Walter Reed, Richard hit upon the idea that the gold spot might influence growth by secreting a hormone. Using techniques he learned during the summer, he cut several butterfly pupae in two. When this is done correctly, both parts continue to live and function. The part of a butterfly pupa containing the brain has all the hormones for normal growth, and it develops into an adult. The other part remains a pupa indefinitely. Richard implanted live cells from the gold spot in the nondeveloping part of the pupa. Ten days later, that part had grown perfectly formed and colored scales. At 16, Richard had discovered a previously unknown insect hormone that is responsible for the growth of scales and wing-support structures. This work earned him first place in zoology at the 1976 ISEF, $300 in cash awards, another summer job, and a trip to the Nobel prize award ceremonies in Stockholm, Sweden.

As a senior, Richard began isolating the hormone from other substances in the gold spot, purifying it, and determining its chemical composition and structure. His 1977 ISEF exhibit dealt with this work. He explained to the judges that his research might lead to the production of an effective, environmentally safe insecticide. Chemicals designed to render the growth hormone ineffective could be sprayed on plants eaten by unwanted butterflies or other insect pests. Those that ate the chemical would not reproduce normal offspring, thereby reducing that pest's population. Richard hoped this work would bring him another first place in Cleveland.

The judges were impressed with Richard's work, but they had doubts, too. "I don't know if it's something new or not; some of the work is well above my head," one judge confessed. "It looks significant," said another, "but I want to compare it with other students' work and talk with other judges."

After a break for dinner, judging continued on Wednesday evening when some of the fair participants began to receive an indication of how they were doing. "If you get swarmed with judges who ask a lot of questions, you know you did well," Richard said. "If no one comes back in the evening, you know you didn't do well."

In the engineering section, a few judges drifted back to Paul Embree's booth. Assured and unflappable, he answered their questions. Experience at three previous ISEF's had developed his confidence and ability to talk with judges. Paul's projects, unlike those of Lori and Richard, were different each year. His main interest is electrical engineering, and he said that he chose "any good project I could think of that would produce some conclusive results." Although Paul's father is an electrical engineer by profession, the senior Embree says he gave his son no direct help with any of his projects.

In his freshman year, Paul devised a way to send voice and code messages over power-transmission lines. He believes the system could be used to read electric meters at local sites and send the information back to billing computers. This work won the grand championship of the Reading regional fair, first place in engineering at the 1974 ISEF, and a first prize from the American Patent Law Association. As a sophomore, he developed a scanning device that could record pictures on audio cassette tapes. When the tape was replayed, the picture could be reproduced. "It worked, but not too well," Paul remarked. "This was one of my poorer efforts." However, judges thought enough of it to name Paul champion of the regional fair, and award him a third

Lori Rhodes, *above left,* and Richard Ebright, *left,* explain their projects to judges at regional fairs, the last hurdle before the international fair. Paul Embree, *above,* accepts the award that sent him on to the ISEF.

A judge inspects Paul Embree's project at the ISEF in Cleveland.

place in engineering at the ISEF. The Institute of Electrical and Electronics Engineers added their first-place special award.

When he was a junior, Paul developed a magnetic suspension bearing with much less friction than the best ball bearings. He won the regional-fair championship, third place in engineering at ISEF, and seven special awards. These included a summer job at the U.S. Coast Guard's electronic engineering center in Wildwood, N.J., where he received $100 a week for doing things he would have gladly done for nothing—working with microcomputers and microprocessors. "It was very fulfilling," he said, "the best award I ever won."

At the 1977 ISEF, Paul explained to judges how he developed a method to detect and eliminate signal distortion in AM radio and television receivers. Distortion results from a broadcast signal traveling over dozens of paths that interfere with one another and produce undesirable effects on television, short-wave, and AM broadcast reception. Paul did computer analyses of signals from a local and a distant station, then designed a device to remove this interference. The AM receiver he built for the fair has excellent wideband reception and automatic fine-tuning. One judge was impressed enough to remark, "It's like listening to FM. There is no static or overlapping stations—very pleasing to the ear." Paul plans to patent his system.

Most students left the Convention Center about 8:30 P.M. on Wednesday, but the judges kept working. In a large room near the exhibits, they sat around tables littered with papers, cardboard cups, and soft-drink cans. They argued, discussed, debated, and thought. "It's a

Judges question students in detail about projects ranging from complex medical and biological research to computerized control systems before making final ratings.

difficult task," one judge summed up, "because every one of those kids is a winner." By 10 P.M., the judges had made their choices, but the students would not find out what they were for another 20 to 45 hours.

On Thursday, the ISEF was opened to the public. Participants with odd-numbered booths spent the morning explaining their projects to local students, teachers, and other visitors. Those with even-numbered booths manned them in the afternoon. On their free half day, Paul and Richard eased the tension of waiting by taking one of the 38 available tours. These included visits to museums, medical facilities, an automobile-assembly plant, and a steel plant; a bird walk; archaeology, botany, and geology field trips; and tours to various points of interest in the Cleveland area. Paul went to the computing facility at the National Aeronautics and Space Administration's (NASA) Lewis Research Center. Afterward, he told a companion, "I saw equipment I would not have seen otherwise, but they could have given us more information about it."

Richard chose to attend a medical-research seminar at St. Luke's Hospital, part of the Case Western Reserve University system. He and 19 other students heard lectures on neurophysiological aspects of learning and behavior, and on the latest research into lung diseases. They looked at sections of lung tissue under electron microscopes and watched open-heart surgery in which a heart valve in a dog was replaced with an artificial valve. "In three years at the ISEF, this is by far the most valuable of the tours that I've gone on," Richard commented. "It was on a level you could relate with – not too much above you or, as most of the tours are, too much below you."

On Thursday night, 42 special awards were presented by such organizations as the American Dental Association and American Speech and Hearing Association for projects dealing with various aspects of human and animal health. These prizes included cash awards of up to $100, certificates of achievement, plaques, trips to scientific meetings, and magazine subscriptions. Among the winners was Nanda Victorine Duhé of Houston, who won a first-prize citation and plaque from the American Medical Association for her attempt to develop a poison ivy preventive. Forced to seek frequent treatment for this unpleasant malady, Nanda extracted the toxic agent – urushiol – from poison ivy plants, then experimented with various detoxifying chemicals until she found a cheap, effective medication. Nanda persuaded members of her family to serve as subjects for her research, testing medications on them both before and after she exposed them to the toxin. Everyone was happy when the project turned out successfully.

When 278 additional special awards were announced on Friday morning, Lori Rhodes sat in the audience. "I had to come to see all my friends and feel the excitement again," she said. Richard did well in this part of the competition. The Entomological Society gave him a $100 U.S. savings bond and Eastman Kodak Company awarded him $100. He also received a certificate of merit, a $75 savings bond, and a

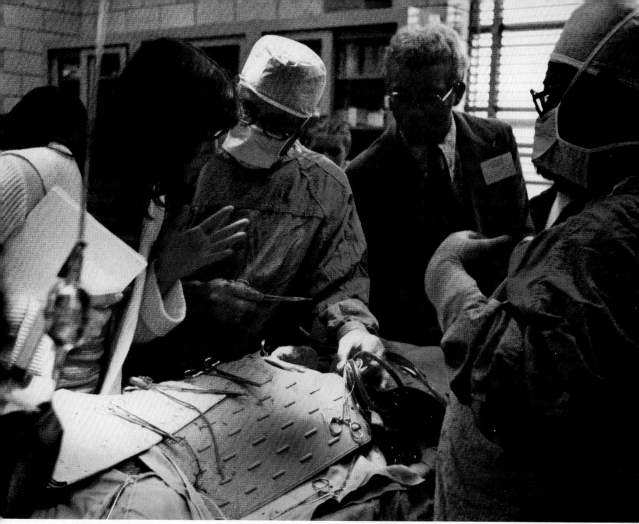

Students watch closely, *above,* as open-heart surgery is performed on a dog during a tour of St. Luke's Hospital. Paul Embree and other tour goers, *left,* listen to explanation of the computer system at the Lewis Research Center.

Judges raise hands to vote on choices for the ISEF awards, *above.* Paul Embree receives a gold medal from the U.S. Army, *above right,* one of several special awards his project won.

summer job offer from the U.S. Department of Agriculture. The Army added a gold medal and the choice of a summer job or an expense-paid trip to an Army research facility. Paul made as many trips to the stage as Richard, receiving the same Army award as well as $100 from General Motors, $50 from Motorola, Incorporated, and a certificate from NASA.

Highlight of the ceremony was the awarding of two-week trips to London and Japan to attend the London International Youth Science Fortnight and the annual Japan Student Science Awards Fair in To-kyo. John Dodge Hayes, 17, of Preston-Fountain High School, Preston, Minn., won the London trip for his development of a self-oscillating chemical system that can be used as a model for such biological processes as transmitting nerve impulses in humans. Paul Embree won the trip to Japan. His parents reached Cleveland in time to see him win, and they shared his delight at such an honor. "It's more than I expected," Paul said.

No official events were scheduled for Friday afternoon, so students used the time to rest, take tours, or study for tests coming up when they returned to school. No matter how they passed the time, all 416 contestants thought often about their chances of receiving one of the important prizes to be given that evening. These were the major awards in 11 categories presented by Science Service of Washington, D.C., the nonprofit institution that conducts the ISEF and administers the Westinghouse Science Talent Search.

At the ISEF award banquet, Lori sat at a table with Richard and Paul, Paul's parents, and Richard and Paul's sponsor—an official of the Reading regional fair. As they ate, the students discussed their future plans. Lori has been accepted into the medical education honors program at Northwestern University in Evanston, Ill. If she satisfactorily completes the first two years of accelerated work, she will be admitted to medical school. "If everything goes right, I'll be a doctor in six years," Lori said. Paul plans to major in electrical engineering at Lehigh University in Bethlehem, Pa. Richard intends to enroll at Harvard University in Cambridge, Mass., to study under a noted insect endocrinologist, Carroll Williams. He received a research grant to continue work on the hormone he discovered, and a National Merit Scholarship to help pay his tuition.

The laughter and talking during dinner were regularly punctuated by groups of students cheering their schools or states. The students tried to hurry the start of the awards by clapping, but officials stuck to

Nanda Duhé reacts with joy, *below,* at learning she has won a trip to the Nobel prize award ceremonies in Stockholm, Sweden. Lori Rhodes joins Paul Embree, Richard Ebright, and other friends at the final banquet, *bottom.*

After the fair is over, students dismantle their projects and pack for the journey home, *below.* Before leaving, *right,* Jeffrey Rice sends his robot ALPHA on final swing through the exhibit area.

their schedule. After the inevitable introductions, speeches, recognitions, and expressions of thanks, they began to announce the names of fourth-place winners. There were four in every category, and each received $25. These students stood up at their tables as their names were announced, and the audience was asked to hold applause until all were called. Next came third-place winners—three in every category, and each received $50. They went to the front of the dais, but the audience again was asked to hold its applause.

Paul and Richard now felt fairly sure they would capture a first or second place. The two second-place winners in each category received $75 each. The excitement grew steadily, and the audience now applauded as each student walked to the dais. Paul and Richard were not called, and a tinge of doubt crept into their minds. There was a good possibility they had won first places, but there was also a chance that they had not won anything.

Although judges selected two, and even three, top winners in some categories, most categories had only one $100 first-place winner. The officials announced the categories in alphabetical order. When they reached engineering, Paul's category, Arthur O. Grantz, 15, of Newport News, Va., took first place for designing a wind tunnel to test airfoil efficiency. Paul visibly stiffened. Another first place went to Gregory A. Dale, 16, of Arlington, Va., for developing a photographic technique using infrared light and a pulsed laser. Paul began to look grim. Then the officials announced an unexpected third first-place award. Paul won it. Richard shook his hand and Lori squeezed his arm as he got up to head for the dais.

For Richard, the wait was even longer. He sat, with barely concealed anxiety, until the officials reached the last category, zoology, but the wait was worth it. He won the only first prize.

Finally, officials announced the two winners of expense-paid trips to Sweden to attend the Stockholm International Youth Science Seminar and the Nobel prize ceremonies. Nanda Duhé won a trip for her work on poison ivy, and Wesley Alden, 17, of Wichita, Kans., won the other trip for his study of the interactions of hormone-receptor sites in normal and cancerous cells.

Richard had won the Stockholm trip at the 1976 ISEF, and he described it as "one of the best experiences I'll ever have." The high point of the trip for him was a reception during which he met nine Nobelists and the king and queen of Sweden. "What surprised me," he said, "was that the Nobel laureates were really friendly. I hadn't expected this. They talked to us as though they were our perfect equals, and never seemed impatient to end a conversation. It made you feel that there are important people interested in what you are doing—people you could call on later for help and information. The experience gave me a lot of encouragement to continue my work."

The ISEF award ceremonies were over at 9 P.M., but not the work. Students returned to the Convention Center to disassemble and crate their exhibits. Some worked in evening dresses or suits and ties; others changed to jeans and work shirts. Parents and sponsors helped pack the displays for transport by plane, truck, or family car. Everyone laughed, shouted, and shook hands; there was a warm comradeship fostered by sharing a challenging experience.

Not every student can do work on the same level as Paul and Richard. Not everyone can win a first, or even a fourth-place award. But they all shared a valuable experience.

"Some kids just think of the awards and prizes," Lori Rhodes said, "but one of the great benefits is the people you meet. I've made friends from all over the United States and from other countries. We correspond, and when I start college in the fall, I know I'll meet friends from the science fairs."

Jeffrey Rice summed up the feelings of many of those who must try again next year for the big prizes. The 16-year-old from Elkins, W. Va., designed and built a computer-controlled robot named ALPHA. It was one of the most popular exhibits at the ISEF, and Jeff was letting other students take turns operating ALPHA's controls.

"I didn't win much," he said as ALPHA glided toward a pretty girl, "but I sure had a lot of fun."

First-Place ISEF Winners

Behavioral and Social Sciences:
Jack A. Yanowski, 16, Wyncote, Pa.

Biochemistry:
Neil E. Goodman, 17, Columbus, Ohio.

Botany:
Eileen C. Villafane, 16, Ponce, Puerto Rico.

Chemistry:
John D. Hayes, 17, Preston, Minn.

Earth and Space Sciences:
Richard A. Sanger, 17, Coronado, Calif.;
Wayne R. Moyle, 18, Ogden, Utah

Engineering:
Arthur C. Grantz, 15, Newport News, Va.;
Paul M. Embree, 17, South Temple, Pa.;
Gregory A. Dale, 16, Arlington, Va.

Mathematics and Computers:
William M. Rojas, 17, Daytona Beach, Fla.

Medicine and Health:
Nanda V. Duhé, 16, Houston;
Wesley Alden, 17, Wichita, Kans.

Microbiology:
William H. Cork, 17, Louisville, Ky.;
David S. Kaplan, 17, Brooklyn, N.Y.

Physics:
Patty M. Sandborn, 15, Fort Collins, Colo.

Zoology:
Richard H. Ebright, 17, South Temple, Pa.

Samuel C. C. Ting, *left,* and Burton Richter, *right,* shared the 1976 Nobel prize in physics for their independent discovery of a new particle.

Awards And Prizes

A listing and description of the year's major awards and prizes in science, and the men and women who received them

Earth and Physical Sciences

Chemistry. Major awards in the field of chemistry included:

Nobel Prize. Professor William N. Lipscomb, Jr., of Harvard University was awarded the Nobel prize for chemistry in 1976. Lipscomb received the prize, which carries a cash award of $160,000, for his studies on the structure and bonding mechanisms of compounds known as boranes. Lipscomb explained how these compounds were built, why they exist, and the nature of their chemical bonds.

Lipscomb studied at the University of Kentucky and the California Institute of Technology (Caltech). He joined the Harvard faculty in 1959.

Cope Award. Elias J. Corey, professor of chemistry at Harvard University, won the Arthur C. Cope Award in 1976. The award is presented by the American Chemical Society (ACS) and includes a gold medal and a $10,000 research grant. Corey was honored for his "sustained record of contribution," particularly his syntheses of complex natural products and his pioneering use of computer analysis in this work.

Corey has synthesized such natural products as hormones, toxins, antibiotics, terpenes, and prostaglandins, hormonelike substances that occur in most mammal tissues.

Perkin Medal. Paul J. Flory, Stanford University professor emeritus, received the 1977 Perkin Medal. The award is presented in the United States by the American section of the Society of Chemical Industry for outstanding work in applied chemistry.

Flory won the Nobel prize in chemistry in 1974 for his research into the physical chemistry of macromolecules. He pioneered research on the chemical and physical properties of giant natural and synthetic molecules, including rubbers, fibers, plastics, proteins, and nucleic acids (DNA and RNA). The practical application of his findings is seen in the development of many new polymeric compounds used in plastics, fibers, paints, automobile tires, film, clothing, and many other products.

Priestley Medal. Henry Gilman, professor emeritus at Iowa State University, received the Priestley Medal in 1977 for his distinguished service to chemistry. The award, given by the ACS, is the highest in U.S. chemistry.

Gilman is credited with more than 1,000 scientific publications. Much of organometallic chemistry, which is important in synthesis, catalysis, and enzyme research, is built on his achievements. His work contributed to the development of the silicone polymers and other industrial products. Gilman taught at the University of Illinois and Iowa State.

Physics. Awards recognizing major work in physics included:

Nobel Prize. Physicists Burton Richter and Samuel C. C. Ting won the Nobel prize in physics in 1976 for their independent discovery of a new type of elementary particle, the psi, or J, particle. Richter is a professor at Stanford University and Ting at the Massachusetts Institute of Technology (M.I.T.).

Richter did his work at the Stanford Linear Accelerator in Palo Alto, Calif., and Ting made his discovery at the Brookhaven National Laboratory on Long Island, N.Y. The psi, or J, particle is unique, showing no kinship to other groups or families of particles. The discovery opens a new field of research that may reveal the nature of quarks, which are thought to be the smallest building blocks of matter.

Richter was born in New York City and studied at M.I.T. Ting was born in Ann Arbor, Mich., spent his youth in China and Taiwan, then attended the University of Michigan.

Bonner Prize. Nuclear physicists Stuart T. Butler of the University of Sydney in Australia and G. Raymond Satchler of the Oak Ridge (Tenn.) National Laboratory shared the Tom W. Bonner Prize in 1977. They were honored for their discovery that direct nuclear reactions can be used to determine angular moments of discrete nuclear states.

Buckley Prize. Leo P. Kadanoff, professor of physics and engineering at Brown University, received the 1977 Oliver E. Buckley Solid State Physics Prize. Kadanoff was honored "for his contribution to the conceptual understanding of phase transitions and to the theory of critical phenomena."

Kadanoff, a native of New York City, taught at the University of Illinois and at Cambridge University in England before going to Brown.

Henry Gilman

Earth and Physical Sciences

Continued

William N. Lipscomb, Jr., received the 1976 Nobel prize in chemistry from Sweden's King Carl XVI Gustaf in ceremonies in Stockholm in December.

E. O. Lawrence Memorial Award. Five scientists were awarded Lawrence Memorial Awards in 1976 for meritorious contributions to the development, use, or control of atomic energy. The scientists honored are:

A. Philip Bray, General Electric Company, San Jose, Calif.; James W. Cronin, University of Chicago; Kaye D. Lathrop, Los Alamos (N. Mex.) Scientific Laboratory; Adolphus L. Lotts, Oak Ridge National Laboratory; and Edwin D. McClanahan, Battelle Pacific Northwest Laboratories, Richland, Wash.

Geosciences. Awards for important work in the geosciences during the year included:

Bowie Medal. James A. Van Allen, head of the University of Iowa Physics and Astronomy Department, received the William Bowie Medal in 1977. Van Allen was honored for his outstanding contributions to geophysics and his unselfish cooperation in research. Van Allen discovered the Van Allen radiation belts that surround the earth.

Bucher Award. Bruce C. Heezen, oceanographer at the Lamont-Doherty Geophysical Observatory in New York City, received the Walter Bucher Award in 1977. Heezen was honored for his original contributions to the basic knowledge of the earth's crust.

Heezen is noted for his work on underwater geologic features and has drawn physiographic maps illustrating the continuity of the mid-ocean ridge.

Day Medal. Professor Akiho Miyashiro of the Department of Geology, State University of New York at Albany, received the Arthur L. Day Medal in 1977. The medal is awarded by the Geological Society of America (GSA). Miyashiro is best known for his investigations of progressive metamorphism of the Abukuma Plateau in Japan.

Penrose Medal. Earth scientist Robert P. Sharp was awarded the Penrose Medal by the GSA in 1977. Sharp was one of the first geologists to study the surface of Mars, and he is noted for the originality of his research. He served as chairman of the Division of Geological Sciences at Caltech from 1952 to 1967.

Life
Sciences

William L. Russell

Medicine. Major awards in medical sciences included the following:

Nobel Prize. Two American researchers shared the Nobel prize for physiology or medicine in 1976. They are Baruch S. Blumberg, 51, associate director of the Institute for Cancer Research in Philadelphia, and D. Carleton Gajdusek, researcher for the National Institute of Neurological and Communicative Disorders and Stroke in Bethesda, Md.

Blumberg was honored for research in which he discovered the Australia antigen. This led to a test for hepatitis viruses and an experimental vaccine against the disease. Gajdusek discovered the role of slow viruses in kuru, a mysterious ailment that killed many members of a cannibalistic New Guinea tribe. Slow viruses are now suspected of causing several disorders of the nervous system, such as multiple sclerosis and Parkinson's disease.

Blumberg, a native of New York City, studied at Union College and Columbia University, and at Oxford University in England. Gajdusek was born in Yonkers, N.Y., and studied at the University of Rochester, Harvard, and Caltech. A bachelor, he has adopted 16 boys during research trips to New Guinea and other Pacific islands and is raising them in the United States.

Enrico Fermi Award. William L. Russell, a principal geneticist at the Oak Ridge National Laboratory, received the $25,000 Fermi Award for 1976. Russell was honored for his outstanding contributions to the evaluation of the genetic effects of radiation in mammals. His studies serve as a major scientific base for national and international standards established for protecting human beings from radiation. The Fermi Award is given by the U.S. Energy Research and Development Administration.

Gairdner Awards. Keith J. R. Wightman, president of the Gairdner Foundation, received the first Gairdner Foundation Wightman Award, which is a $25,000 prize, in 1976. The award was established to honor Canadians for outstanding leadership in medicine and medical science.

Geneticists Seymour Benzer, *left,* of the California Institute of Technology and Charles Yanofsky, *above,* of Stanford University received the Louisa Gross Horwitz prize for their research on gene structure and function.

Life Sciences

Continued

Godfrey N. Hounsfield of the Central Research Laboratories in England received a $25,000 Award of Merit for his development of computerized X-ray tomography. Other awards:

George Klein of the Karolinska Institute in Stockholm, Sweden, $10,000 for his contributions to the understanding of the biology of neoplastic cells and distinguished work in tumor immunology.

Eugene P. Kennedy of the Harvard Medical School, $10,000 for his explanation of the biochemical pathways involved in triglyceride and phospholipid synthesis.

George D. Snell of Jackson Laboratory, Bar Harbour, Me., $10,000 for identifying the major system of graft tissue compatibility in mice and for establishing study methods basic to immunogenetics.

Thomas R. Dawber of Boston University School of Medicine and William B. Kannel of the Framingham (Mass.) Public Health Service shared a $10,000 award for studies revealing risk factors in cardiovascular disease.

Horwitz Prize. Seymour Benzer and Charles Yanofsky, pioneering geneticists, were awarded the 1976 Louisa Gross Horwitz prize. The $20,000 prize is given for outstanding research in biology or biochemistry. Benzer is a professor at Caltech and Yanofsky is a professor at Stanford University.

The two geneticists shared the award for their studies of gene structure and function. Their work helped to explain how genes code instructions for life, molecule by molecule.

Benzer showed in bacteria that each gene in a cell's chromosome has a unique function. Yanofsky showed how other parts of the cell "read" genetic instructions and use them to produce the chemical substances that are needed for life.

Biology. Among the awards presented in biology were the following:

Carski Award. Margaret Green, chairman of the University of Alabama's Department of Microbiology in Tuscaloosa, was awarded the 1977 Carski Foundation Distinguished Teaching

D. Carleton Gajdusek, *above,* and Baruch S. Blumberg, *right,* shared the 1976 Nobel prize in physiology or medicine.

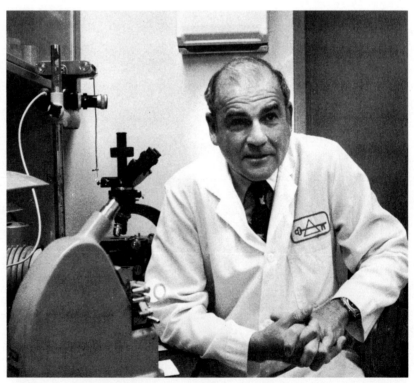

Life Sciences

Continued

Keith J. R. Wightman

Award. The $1,000 award is given annually "for distinguished teaching of microbiology to undergraduate students and for encouraging them to subsequent achievement."

Green joined the Alabama faculty in 1952. She became department chairperson in 1969.

Lilly Award. Alice S. Huang, associate professor of microbiology and molecular genetics at Harvard Medical School, received the $2,000 Lilly Award in 1977. Huang was honored for her contributions to the understanding of defective interfering particles in ribonucleic acid (RNA). The award is presented annually by Eli Lilly and Company.

NAS Award for Environmental Quality. Miron L. Heinselman, adjunct professor in the University of Minnesota's Department of Ecology and Behavioral Biology, received the $5,000 National Academy of Sciences Award for Environmental Quality. The award is given for outstanding contributions in science or technology to improve the environment.

Heinselman was honored for his contributions in clarifying the natural role of wildfire in virgin conifer forests, which provided strong reasons for public acceptance of wilderness protection.

Tyler Ecology Award. Eugene P. Odum, professor of ecology and director of the Institute of Ecology at the University of Georgia, received the John and Alice Tyler Ecology Award in 1977. Odum was granted the $150,000 award for his educational contributions to basic and applied ecology.

Odum established the University of Georgia's Institute of Ecology, and his research has spanned many facets of community and ecosystem ecology. The award is administered by Pepperdine University in Malibu, Calif.

U.S. Steel Foundation Award in Molecular Biology. Aaron J. Shatkin of the Cell Biology Department at the Roche Institute of Molecular Biology in Nutley, N.J., was awarded the $5,000 U.S. Steel Foundation Award for "his contributions to the understanding of eucaryotic, viral, and cellular messenger RNAs."

Space Sciences

James S. Martin, Jr.

Aerospace. The highest of awards granted in the aerospace sciences included:

Collier Trophy. General David C. Jones and Robert Anderson were designated to receive the Robert J. Collier Trophy for 1976. They received it on behalf of the Air Force-industry team that successfully produced and demonstrated the B-1 strategic bomber. General Jones is chief of staff of the United States Air Force and Anderson is president and chief executive officer of Rockwell International Corporation.

Goddard Astronautics Award. James S. Martin, Jr., manager of the National Aeronautics and Space Administration's Viking project, was awarded the Goddard Astronautics Award for 1977. Martin was honored for his "brilliant leadership of the Viking project to land an instrumented, automated spacecraft on the planet Mars. . . ."

Reed Aeronautics Award. William C. Dietz, vice-president for research and engineering of the General Dynamics Corporation, received the Reed

Aeronautics Award in 1977. Dietz was cited for his leadership in the development of military aircraft.

Astronomy. Major awards in astronomy during the year included:

Bruce Medal. Bart Bok, professor emeritus of astronomy at the University of Arizona, was awarded the Catherine Wolfe Bruce Gold Medal in 1977. Bok was honored for his distinguished services to astronomy. He is noted primarily for his investigations of the structure of our Galaxy, the Milky Way. He has also made valuable studies of small dark nebulae known as Bok Globules.

Draper Medal. Arno Penzias, director of the Bell Telephone radio research laboratory at Holmdel, N.J., and Robert W. Wilson, head of the Bell Telephone radio physics research department, shared the Henry Draper Medal in 1977. Penzias and Wilson were honored "for their discovery of the cosmic microwave radiation, a remnant of the very early universe, and their leading role in the discovery of interstellar molecules."

General Awards

Science and Humanity Awards for outstanding contributions to science and humanity during the past year included the following:

National Medal of Science. Fifteen scientists were awarded the National Medal of Science, the highest United States award for distinguished scientific achievement. President Gerald R. Ford presented the awards to:

John W. Backus, International Business Machines San Jose (Calif.) Research Laboratory, computer programming

Manson Benedict, professor emeritus, M.I.T., nuclear engineering

Hans A. Bethe, professor emeritus, Cornell University, nuclear research

Shiing-shen Chern, University of California, Berkeley, mathematics

George B. Dantzig, Stanford University, computer science

Hailowell Davis, Washington University, research on ear, nose, and throat diseases

Paul György, professor emeritus, University of Pennsylvania, vitamin and nutrition research (posthumous)

Sterling Brown Hendricks, formerly of the U.S. Department of Agriculture, soil and protein research

Joseph O. Hirschfelder, University of Wisconsin, quantum mechanics

William H. Pickering, director emeritus of Caltech's Jet Propulsion Laboratory, for planet and solar system exploration

Lewis H. Sarett, Merck & Company, Incorporated, research on chemotherapeutic agents

Frederick E. Terman, Stanford University, electronics

Orville Alvin Vogel, Department of Agriculture, agronomy

E. Bright Wilson, Jr., Harvard University, molecular research

Chien-hsiung Wu, Columbia University, physics.

Oersted Medal. H. Richard Crane, chairman of the University of Michigan Physics Department, was awarded the Oersted Medal in 1977. The award is made annually by the American Association of Physics Teachers for outstanding contributions to the teaching of physics.   [Joseph P. Spohn]

President Gerald R. Ford presented the National Medal of Science to University of Wisconsin chemist Joseph O. Hirschfelder, one of 15 U.S. scientists honored in White House ceremony.

Major Awards and Prizes

Award winners treated more fully in the first portion of this section are indicated by an asterisk (*)

AAAS-Rosenstiel Award (oceanography): Gordon A. Riley

Acoustical Society of America Gold Medal (physics): Raymond W. B. Stephens

A. Cressy Morrison Award (natural sciences): John F. Dewey

Adams Award (chemistry): William A. Johnson

Alexander von Humboldt Award (agriculture): Wendell Roelofs, Harry H. Shorey

American Physical Society High Polymer Physics Prize: Samuel Krimm

American Physical Society International Prize for New Materials: Francis Bundy, H. Tracy Hall, Herbert Strong, Robert Wentworf, Jr.

*Arthur Cope Award (chemistry): Elias J. Corey

*Arthur L. Day Medal (geophysics): Akiho Miyashiro

*Bonner Prize (nuclear physics): Stuart T. Butler, G. Raymond Satchler

*Bowie Medal (geophysics): James A. Van Allen

*Bruce Medal (astronomy): Bart Bok

*Bucher Award (geophysics): Bruce C. Heezen

*Buckley Solid State Physics Prize: Leo P. Kadanoff

*Carski Foundation Award (teaching): Margaret Green

*Collier Trophy (astronautics): General David C. Jones, Robert Anderson

Davisson-Germer Prize (optics): Walter Kohn, Norton Lang

*Draper Medal (astronomical physics): Arno Penzias, Robert W. Wilson

Earle K. Plyler Prize (physics): Charles H. Townes

*Enrico Fermi Award: William L. Russell

Fisher Award (microbiology): David Gottlieb

Fleming Medal (geophysics): Francis S. Johnson

*Gairdner Awards (medicine): Keith J. R. Wightman, Godfrey N. Hounsfield, George Klein, Eugene P. Kennedy, George D. Snell, Thomas R. Dawber, William B. Kannel

Garvan Medal (chemistry): Marjorie Groothuis Horning

*Goddard Astronautics Award: James S. Martin, Jr.

Haley Space Flight Award: William H. Dana

Harvey Human Health Prize (Israel): Seymour Benzer

Harvey Science and Technology Prize (Israel): Freeman John Dyson

Heineman Prize (American Physical Society): Steven Weinberg

*Horwitz Prize (biology): Seymour Benzer, Charles Yanofsky

Jeffries Medical Research Award (aerospace medicine): Harald J. von Beckh

Klumpke-Roberts Prize (astronomy): Sir Fred Hoyle

Langmuir Prize (chemical physics): Aneesur Rahman

Lasker Awards (medical research): Raymond P. Ahlquist, James W. Black, Rosalyn S. Yalow, World Health Organization

*Lawrence Memorial Award (atomic energy): A. Philip Bray, James W. Cronin, Kaye D. Lathrop, Adolphus L. Lotts, Edwin D. McClanahan

Lawrence Sperry Award (aeronautics, astronautics): Joseph L. Weingarten

*Lilly Award (microbiology): Alice S. Huang

Macelwane Award (geophysics): Paul Richards, Ignacio Rodriquez-Iturbe, Chris Russell

Maurice Ewing Award (oceanography): Henry Stommel

Mervin J. Kelly Award (telecommunications): Alton C. Dickieson

NAS Applied Mathematics Award: Chia-chiao Lin, George B. Dantzig

*NAS Award for Environmental Quality: Miron L. Heinselman

*National Medal of Science: John W. Backus, Manson Benedict, Hans A. Bethe, Shiing-shen Chern, George B. Dantzig, Hailowell Davis, Paul György, Sterling Brown Hendricks, Joseph O. Hirschfelder, William H. Pickering, Lewis H. Sarett, Frederick E. Terman, Orville A. Vogel, E. Bright Wilson, Jr., Chien-hsiung Wu

Nehru Award for International Understanding: Jonas Salk

New York Academy of Sciences Award: Alvin Weinberg

*Nobel Prize: chemistry, William N. Lipscomb, Jr.; physics, Burton Richter, Samuel C. C. Ting; physiology or medicine, Baruch S. Blumberg, D. Carleton Gajdusek

*Oersted Medal (teaching): H. Richard Crane

Oppenheimer Memorial Prize (physics): Sheldon Glashow, Feza Gürsey

Parsons Award (chemistry): William O. Baker

*Penrose Medal (geology): Robert P. Sharp

*Perkin Medal (chemistry): Paul J. Flory

Pfizer Award (enzyme chemistry): Michael S. Brown, Joseph L. Goldstein

*Priestley Medal (chemistry): Henry Gilman

*Reed Aeronautics Award: William C. Dietz

Roussel Prize (steroids): Elwood V. Jensen, Etienne Vaulieu

Space Science Award: Bruce Murray

Space Systems Award: Walter O. Lowrie

Trumpler Award (astronomy): John Harry Black

*Tyler Ecology Award: Eugene P. Odum

U.S. Steel Foundation Award (molecular biology): Aaron J. Shatkin

Walcott Medal (geology, paleontology): Preston B. Cloud, Jr.

Wright Brothers Memorial Trophy (aviation): William A. Patterson

Wyeth Award (microbiology): Erwin Neter

Deaths of Notable Scientists

Notable scientists and engineers who died between June 1, 1976, and June 1, 1977, include those listed. An asterisk (*) indicates that a biography appears in *The World Book Encyclopedia*.

Bailey, Pearce, (1902-June 23, 1976), neurologist, first director of the Institute of Neurological and Communicative Disorders and Stroke at the National Institutes of Health from 1956 to 1959.

Beecher, Henry K. (1904-July 25, 1976), physician who helped to make anesthesia a specialty in medicine and chairman of the Harvard Medical School's Committee to Examine the Definition of Brain Death.

Bramlette, Milton N. (1896-March 30, 1977), geologist and professor emeritus at Scripps Institution of Oceanography. He received the Distinguished Service Medal in 1963 for his study of transatlantic deep-sea cores.

Chamberlin, Clarence D. (1893-Oct. 30, 1976), first aerial photographer and first pilot to fly a passenger on a nonstop transatlantic flight, two weeks after Charles A. Lindbergh's historic solo flight to Paris in 1927.

Curme, George O., Jr. (1888-July 28, 1976), industrial chemist whose research led to such widely used products as ethyl alcohol, synthetic rubber, and antifreeze for automobiles.

Dufek, Rear Admiral George J. (1903-Feb. 10, 1977), commander of U.S. naval forces in Operation Deep Freeze in the Antarctic from 1957 to 1959 and the first American to set foot on the South Pole.

Emerson, Alfred E. (1896-Oct. 3, 1976), professor of geology at the University of Chicago from 1929 to 1962 and an authority on termites.

Fishbein, Morris (1889-Sept. 27, 1976), physician and outspoken editor of the *Journal of the American Medical Association* from 1924 to 1949.

Flint, Richard F. (1902-June 5, 1976), geologist and leader in the radiocarbon method of dating glacial periods. His many books included the standard *Physical Geology* (1932).

Fowler, Herbert G. (1918-Jan. 2, 1977), genetic psychiatrist, noted for his study of the inheritance patterns of mental illness in American Indians. He was to have received the Lenin Prize from Russia in May.

Friis, Harald T. (1893-June 15, 1976), Danish-born engineer who pioneered in radio and telephone systems in nearly 40 years with the Bell Telephone Laboratories. He also designed the rabbit-ear television antenna.

Gorini, Luigi (1903-Aug. 13, 1976), Italian-born microbiologist whose research in 1964 provided insights into the role of tiny organelles called ribosomes in the synthesis of protein.

Gurevich, Mikhail I. (1892-reported in Moscow Nov. 25, 1976), Russian aircraft designer who co-designed the MIG fighter plane in 1940.

Haagen-Smit, Arie J. (1900-March 17, 1977), Dutch-born biochemist who pioneered in fighting air pollution in an almost single-handed battle against the automotive and oil industries in the 1950s. His prizes included a $50,000 Tyler Ecology Award in 1973.

Harman, Harry H. (1913-June 8, 1976), Polish-born statistician and director of research for the Educational Testing Service. He developed the technique of factor analysis used in psychology and other fields.

Ilyushin, Sergei V. (1894-reported in Moscow Feb. 10, 1977), Russian aircraft designer who produced more than 50 planes– from World War II bombers to modern passenger jets.

Johnson, Clarence L. (1910-Oct. 7, 1976), aviation pioneer and early designer of outboard motors for boats.

Kupchan, S. Morris (1922-Oct. 19, 1976), chemist and cancer researcher, best known for his work with plant-derived drugs.

Lazarsfeld, Paul (1901-Aug. 30, 1976), Austrian-born sociologist who pioneered in research on the effects of mass communication. His studies of U.S. voting patterns became classics.

Levy, David M. (1892-March 1, 1977), psychiatrist who coined the term "sibling rivalry" and introduced the Rorschach test for diagnosing personality to the United States in 1926.

Levy, Milton (1903-Oct. 30, 1976), biochemist who helped to design the Lang-Levy pipette, a basic tool in biochemistry. He was also noted for his work on the structure of collagen.

Lilly, Eli (1885-Jan. 24, 1977), pharmaceutical chemist, former chairman of Eli Lilly and Company, and founder of the Lilly Endowment. He helped

Morris Fishbein

Harald T. Friis

Eli Lilly

Deaths of Notable Scientists

Continued

Trofim D. Lysenko

Lars Onsager

Leopold Ružička

develop such drugs as insulin, penicillin, and Salk polio vaccine.

Logan, Myra (1908-Jan. 13, 1977), the first black woman to be elected to the American College of Surgeons, was noted for her research on the use of antibiotics. She worked in Harlem in New York City for more than 35 years.

*Lysenko, Trofim D. (1898-Nov. 20, 1976), Russian biologist whose faulty theory that environment could cause hereditary changes in plants dominated Russian agriculture for 20 years.

Mackenzie, Locke L. (1900-May 10, 1977), gynecologist who helped George N. Papanicolaou develop the Pap test for early detection of cervical cancer.

Marine, David (1880-Nov. 26, 1976), pathologist who discovered the iodine treatment for goiter 70 years ago.

Marrack, John R. (1886-June 13, 1976), British chemist who revolutionized the study of immunology in 1934 with the proposal that the specific affinity of antibodies for antigens is determined by the same factors that determine the specific binding of molecules to form crystals.

Menzel, Donald H. (1901-Dec. 14, 1976), theoretical astrophysicist and director of Harvard Observatory from 1954 to 1966. He was a leading authority on the sun and its corona.

Miller, Harry W. (1880-Jan. 1, 1977), surgeon, nutritionist, and medical missionary known as the "China Doctor." He developed a process of extracting milk from soybeans and predicted that the world would be a vegetarian society by the year 2000.

Miller, Julian M. (1922-Dec. 14, 1976), nuclear chemist and educator whose research on the kinetic energies of atoms led to a better understanding of how nuclear reactions change one element into another.

Müller, Erwin W. (1911-May 17, 1977), German-born physicist who invented the powerful field ion microscope in 1951 and became the first person to see and photograph an atom.

Muschenheim, Carl (1905-April 27, 1977), physician, co-winner of the 1955 Lasker Award for developing a drug treatment for tuberculosis that reduced the death rate in the United States by almost 70 per cent.

Olds, James (1922-Aug. 21, 1976), behavioral biologist who in 1953 discovered "pleasure centers" in the brains of rats, a significant step toward understanding the basic physiological events underlying motivation.

Onsager, Lars (1903-Oct. 5, 1976), Norwegian-born theoretical chemist who won the 1968 Nobel prize for chemistry for his work on thermodynamics, in which he used complex equations to establish cause and effect in chemical reactions.

Ružička, Leopold (1887-Sept. 26, 1976), Swiss chemist, co-winner with Adolph Butenandt of the 1939 Nobel prize for chemistry for his work on polymethylenes. He was the first to synthesize the hormone testosterone.

Shipton, Eric (1907-March 28, 1977), British explorer who photographed the tracks of the "abominable snowman" near Mount Everest in 1951 and inspired several expeditions.

Sparks, William J. (1904-Oct. 23, 1976), former president of the American Chemical Society and winner of the 1965 Priestley Medal. He co-invented synthetic butyl rubber, which played a major role in World War II.

Spence, Sir Basil (1907-Nov. 18, 1976), British architect who, in 1962, designed Coventry Cathedral, which replaced the medieval cathedral destroyed by bombing in World War II.

Thornton, Sir Gerard (1892-Feb. 6, 1977), British microbiologist whose work laid the foundation for later research on nitrogen fixation in plants.

Vereshchagin, Leonid F. (1909-reported in Moscow Feb. 23, 1977), Russian physicist who developed a way to produce artificial diamonds.

Vickers, Harry F. (1898-Jan. 13, 1977), pioneer in hydraulic engineering. He devised such products as power steering for automobiles and trucks and the Vickers balanced vane type pump.

Wheeler, Sir R. E. Mortimer (1890-July 22, 1976), Scottish archaeologist who discovered relics in England believed to be from the legendary King Arthur's Camelot.

Wiener, Alexander S. (1907-Nov. 6, 1976), physiologist and immunohematologist who co-discovered the Rh blood factor.

Wilson, John A. (1899-Aug. 30, 1976), authority on ancient Egypt. His books include *The Burden of Egypt* (1951).   [Irene B. Keller]

Science and the Decision Makers

By Daniel M. Singer

Public involvement in the debate on recombinant DNA has taught us that nonscientists can learn enough about science to make competent judgments about the course of its progress

A powerful new experimental technique is revolutionizing biological science and is casting both scientists and nonscientists in unfamiliar roles. This technique, called recombinant DNA research, has generated enormous excitement and serious concern. The excitement comes from the realization that this experimental tool will permit molecular biologists to ask and perhaps answer a broad range of questions heretofore beyond their capability. The concern arises from a recognition that the technique makes possible the development of organisms with novel properties whose behavior may not be wholly predictable. See GENES: HANDLE WITH CARE.

The new technique will almost certainly vastly expand our understanding of the detailed biochemical processes of all living things—how they are nourished and grow, how they reproduce and pass informa-

The new recombinant DNA techniques will expand our understanding of how living things function and, perhaps, how we may affect those functions.

tion to their descendants, and how they grow old and die. We may learn what happens, biochemically, when a living thing gets sick. In short, recombinant DNA research will increase our understanding of how our biosphere functions.

Doubtless some of this increased knowledge will be applied in medical and agricultural technologies and in the manufacture of products that can be sold commercially. One can imagine in a few decades—perhaps even in this century—new ways to cure or to alleviate human genetic diseases. One can also imagine genetically altered corn and wheat that can take nitrogen directly from the air rather than from expensive fertilizers that require large quantities of energy to produce. Well-informed scientists have speculated about many such technologies based on the knowledge gained from recombinant DNA research.

No one denies that recombinant DNA techniques will dramatically expand our detailed understanding of living organisms—not only of plants and lower animals, but of human beings as well. However, some scientists and nonscientists have expressed reservations about using recombinant DNA techniques, both to acquire knowledge and to convert that knowledge into technology.

These concerns were first widely publicized following the famous conference in Asilomar, Calif., in February 1975, where 139 people, predominantly scientists, debated the potential dangers of the new technique and agreed to refrain from doing certain types of experiments and to exercise caution with others. During the following 18 months, the United States National Institutes of Health (NIH) developed research guidelines in response to the request of the Asilomar conferees. At the same time, a broader community of scientists, as well as articulate nonscientists, were making their concerns known in their local communities and throughout the United States.

These efforts took a unique turn in July 1976, following publication of the NIH guidelines. The City Council of Cambridge, Mass., voted a moratorium on all P3 recombinant DNA research in the city and called upon City Manager James L. Sullivan to establish a citizens' review board to study the issue before such research would be permitted to proceed. P3 refers to a level of physical containment of organisms produced through recombinant DNA research as described in the NIH guidelines. A P3 laboratory is designed to handle experiments such as the transfer of genes from birds into a crippled strain of the bacteria *Escherichia coli*. This crippled strain is a derivative of the already weakened form of *E. coli* called K-12. K-12 has been the form commonly used in genetic studies for the past 30 years.

Cambridge is the home of Harvard University and of the Massachusetts Institute of Technology (M.I.T.), both active centers of recombinant DNA research. The City Council's action was prompted by a variety of factors: the swirling public debate over the potential biohazards of recombinant DNA research, the long-standing and open hostility between Harvard and Alfred Vellucci, the mayor of Cam-

The author:
Daniel M. Singer is a partner in the law firm of Fried, Frank, Harris, Shriver & Kampelman in Washington, D.C. He organized the layman discussions at the 1975 Asilomar Conference.

bridge, and Harvard's application to Cambridge for permits to build a P3 research laboratory on the Harvard campus.

In August 1976, Sullivan appointed the Cambridge Experimentation Review Board (CERB) "to consider whether research on recombinant D.N.A. which is proposed to be conducted at the P3 level of containment in Cambridge may have any adverse effect on public health within the City." Sullivan considered and decided against appointing to CERB "both proponents and opponents to the experimentation and some neutral citizens," as he wrote in his appointment letter. He also considered appointing "knowledgeable scientists, biologists and geneticists who would approach the problem scientifically." "I rejected this approach as well," Sullivan said, "for the issue is before us because of a dispute within the scientific community as to the hazards involved and it would be extremely difficult to find knowledgeable scientists who did not have preconceived views on the subject." He concluded: "Since the issue has been raised by many who have expressed concern about the potential hazards of experimentation to the citizens of Cambridge, and the problems that can be generated by scientists who have a self interest in experimentation controlling the experimentation, it seemed only reasonable to create a committee of Cambridge citizens who could approach the subject in an unbiased manner and insure that the public safety is at all times the foremost consideration."

Daniel J. Hayes, a businessman and former mayor of Cambridge, served as chairperson of the board. Other members included: Cornelia Wheeler, a former city councillor; William J. LeMessurier, a structural engineer; Constance Hughes, a public health nurse and social worker; Sheldon Krimsky, a social scientist from Tufts University; Sister Mary Lucille Banach, a hospital administrator; Dr. Joseph L. Brusch, a physician specializing in infectious disease; and Mary Nicoloro, a person active in Cambridge community affairs. LeMessurier was the only board member who had formal ties with either Harvard or M.I.T.; he had previously taught courses in structural engineering at both universities. Only three members had ever met one another previously. And none had any prior association with recombinant DNA research. In short, CERB was probably as neutral a group as could have been assembled from the Cambridge citizenry.

How could a group of lay people, however intelligent, hope to thread its way through what appeared to be a highly technical and obscure subject? How could nonscientists make sound decisions that would affect the course of scientific research locally and perhaps lay the groundwork for similar groups to make research policy across the United States or even throughout the world? The members of CERB did it by long hours of hard work and a determination to succeed. They first educated themselves in the biochemical concepts involved. They insisted that scientists talk to them in language free of jargon and found that they could understand much of the science and the range of potential hazard. They proved to themselves and to others that, if they tried, nonscientists could understand the science even if they could not design or perform experiments. While they learned some science, they did not thereby become scientists. They remained intelligent laymen, exercising their common sense and independent judgment.

They visited the laboratories at Harvard and M.I.T. and observed one mock experiment that showed some stages of the recombinant process. As they stated in their report, the visit, "helped the Board members concretize many of the specifications found in the NIH guidelines relating to physical containment." CERB devoted more than 100 hours to its deliberations and to hearing testimony from scientists and cross-examining them and others holding widely different views.

The CERB proceedings took place in an atmosphere that was highly charged by newspaper and television coverage and by a mayor who has made a political career out of baiting Harvard University. Mayor Vellucci had publicly made clear on several occasions that he wanted no recombinant DNA research done in Cambridge, not even those experiments classified at the P1 and P2 levels—in which there is less potential danger than in experiments requiring P3 containment.

In January 1977, CERB issued its findings and recommended that the research be permitted to continue. Basically, the board's conclusions endorsed the NIH guidelines and called upon the City Council to establish a Cambridge Biohazards Committee to develop reporting and monitoring procedures. In addition, CERB urged the City Council, "on behalf of this Board and the citizenry of the country," to recommend uniform national licensure or regulation by Congress, "to insure conformity to [the NIH] guidelines in all sectors, both profit and non-profit." One month later the City Council adopted the board's recommendations.

Perhaps the most meaningful portion of the CERB report is the introduction. This not only defined CERB's task, but also recognized and clearly sketched the broad range of philosophical and practical issues that have surfaced in the recombinant DNA debate. CERB properly viewed those broader issues as being beyond its immediate

Scientists and laymen have come together in forums and conferences to debate the dangers and develop guidelines for research with recombinant DNA.

charge, but nonetheless recognized their importance and urged that they be considered by the general public. CERB invited the public and the U.S. Congress to repeat the self-education that the committee had undergone: to try to understand the risks and the benefits beyond the narrow safety considerations. The members also urged that efforts be made to develop a public consensus not only about safety standards but also about the much more difficult social issues.

Scientists were impressed with the performance of CERB. For example, David Baltimore, a Nobel laureate and microbiologist from M.I.T., wrote, "It proved that even complex scientific issues can be understood by lay people who devote the necessary time and energy to the problem."

CERB did not shrink from making ethical judgments about the safety of all citizens of Cambridge. While recognizing that "absolute assurance [of no hazard] was an impossible expectation," CERB found in the NIH guidelines "a sufficient number of safeguards...to protect the public against *any reasonable likelihood* of a biohazard." In essence, CERB judged that the people of Cambridge, including the scientists and laboratory workers, should accept the known and unknown risks of injury that are possible even with scrupulous adherence to the NIH guidelines.

Local citizens' groups have asked questions and found their own answers to whether the new research can be conducted with safety.

The board recognized that few, if any, human activities are completely free from the risk of injury to other persons and that we nonetheless continue to subject others to risks. We build bridges, tunnels, pipelines, and subways, and operate complicated machines in highly organized factories. We drive cars and motorcycles, and play football and basketball. We engage in many activities–including war–knowing that some people will be injured or killed.

Even though we rarely talk or even think about risks, casualty and liability insurance carriers do. They have an exquisitely articulated and detailed understanding of risk, including both the frequency of injury-causing events and the intensity of the injuries. Insurance carriers accept the fact that injuries will occur–for instance, in the construction of a bridge. They expect a certain number of accidents, ranging from minor to fatal. Construction companies are required by state laws to buy workmen's compensation insurance to protect employees who are injured on the job, whether or not as a result of negligence. The companies buy insurance that will cover them if non-employees are injured negligently.

This exposure of others to risk seems to be a necessary cost of living in an organized society in which each of us depends on the efforts of others in our work and recreation. While we permit ourselves to expose others to risk, we also try to prevent injury. We encourage or require people to act carefully. We punish those who injure others carelessly or intentionally. And we buy insurance to protect ourselves and others from the consequences of the risks. Where inflicting intentional injury is the purpose of the activity–as in fighting a war–we

require broad justification when we compel someone to suffer high risk of either injury or death.

When one person exposes another to risk of injury–whether or not the injury actually occurs–that person must somehow claim a right to do so. Society, for its collective survival, may require the exposure, as in conscription for military service. Society may license or permit the exposure, as with motor vehicles. Or the "victim" may consent, as in surgery or other medical treatment and in scientific experiments involving human subjects. In each instance, there is some procedure to assure that the event is treated seriously and cautiously. In other words, an orderly society needs to legitimize the conscious exposure of another human being to the possibility of harm.

As a public body, CERB, and subsequently the Cambridge City Council, provided that legitimacy for continuation of recombinant DNA research at Harvard and M.I.T. CERB recognized that the issue was not exclusively scientific. The benefits promised or speculated upon–including the increase of knowledge–were benefits to the public, and the risks were risks to the public also. The public was obligated to determine whether the risks were worth the benefits–however speculative–or whether it wanted those benefits at all.

There is also implicit in the CERB report a very high value placed upon the acquisition of knowledge for its own sake. I believe that there is widespread public acceptance of this goal and support for undertakings that pursue it. State legislatures support their state universities; Congress appropriates billions of dollars annually for basic scientific research; and charitable donations from individuals and private foundations have supported our private universities for many years.

Some of the risks people take are voluntary. Other risks are imposed by society so that it is able to function in the people's interest.

CERB recognized its responsibility to act even without complete data about both of the elements inherent in any calculation of risk: first, about the probability that each of a series of chance events will occur and will lead to injury; and second, about the magnitude of injury if each chance event in that series does in fact occur. All human judgments about the future are made on the basis of imperfect or incomplete information. In most ordinary matters we act on the basis of information or misinformation that is readily available. In purchasing a television set or an automobile, we respond largely to advertising and personal whim. In more serious matters, we marshal the available data, assess their importance and relevance, and then try to predict good or bad outcomes of various courses of action. In contemplating major surgery, for example, we are urged to seek advice from more than one physician. Ultimately, we exercise judgment by selecting one course of action; deciding whether to have the surgery, or not. This process of judgment and openness distinguished CERB's report and recommendations. In analyzing the safety issue in recombinant DNA research, CERB rendered a singular public service.

With few exceptions, scientists and laymen now accept the NIH guidelines as conservatively reasonable for most recombinant DNA research, and such dispute as exists about their "toughness" should no longer be central to the debate. If the research is performed according to the guidelines, the probability of human beings or the biosphere being adversely affected seems exceedingly small.

Most of those who have taken the trouble to learn the facts in a disciplined way—whether lay groups like CERB or scientists not immediately involved in the research—no longer seriously dispute the efficacy of the NIH guidelines. The small group opposing the research on safety grounds challenges neither the underlying scientific and philosophical rationales of the guidelines nor their generally reasonable approach to specific safety issues. This group will likely diminish in size as further research expands our knowledge of the genetic recombination that occurs naturally through evolution.

Even the most aggressive of the legislative proposals before Congress in the spring of 1977 were adopting the NIH guidelines as the basis on which the acquisition of knowledge might continue. Indeed, the main thrust of those proposals is to extend the guidelines to cover all research, including private industry and other work not supported by federal grants or contracts.

The really difficult public issue goes well beyond whether the NIH or any other guidelines for research safety are tough enough. That issue is what we should do with knowledge once it is acquired—to what technological purposes knowledge should be applied. This is an intensely practical as well as political issue. All parts of society have a right to be heard and an obligation to participate in the decision.

An earlier example was the debate in 1970 about the supersonic transport (SST). With the SST, there was no question that United

States airplane manufacturers could build a technically successful airplane. At the same time, it was clear that public funding would be required. Congress heard testimony, debated, and then, quite reasonably, chose not to continue spending public funds for the SST. After evaluating the costs in money and in noise and other environmental disturbances, not enough people were persuaded that cutting in half the time of transoceanic flights was important enough to warrant the investment. The British and French, also quite reasonably, reached the opposite conclusion and made the investment. Now, they are trying to persuade us to reverse our earlier environmental decision and allow their SST to serve United States airports. This points up a new complication in scientific and technological decisions that touches on the problems of recombinant DNA research as well. People in different nations will not always come to the same conclusions.

I think that the opponents of recombinant DNA research are unduly pessimistic that laymen will be able to make reasonable decisions on recombinant DNA research. They therefore offer sometimes silly and sometimes dangerous smoke screen arguments in opposition to the NIH guidelines and to the acquisition of scientific knowledge in general. Or they loudly demand "public participation" in the process. Yet, since its beginnings in 1973 recombinant DNA research has been open to public view, comment, and control to a far greater extent than any similar issue in recorded history. Again, in my judgment, the "public participation" argument conceals the genuine uneasiness of the many people who doubt that the democratic process can manage, in the public interest, the variety of technologies that may flow from genetic discoveries.

It is frightening to hear scholars, in the name of preserving individual freedom, urge a ban on recombinant DNA research because such research might lead to forbidden or dangerous knowledge. The history of Western civilization demonstrates that attempts to ban inquiry and knowledge are at best ineffective, and at worst the entering wedge of genuine tyranny. Neither papal edict nor humiliation could prevent the Italian astronomer Galileo from demonstrating that the sun—and not the earth—was the center of our "universe." A 20th-century example is the theories of Russian biologist Trofim Lysenko with which Russia tried to harness genetic research to political ideology. The Soviet politicians succeeded only in hamstringing their biologists and in making them the laughingstock of science.

Fortunately, creative inquiry is still flourishing in the United States in 1977. For instance, in May 1977, biochemists at the University of California, San Francisco, reported that they had isolated and reproduced the rat gene coding for the production of insulin. This is a necessary step in the eventual mass production in pharmaceutical laboratories of large quantities of insulin. As the number of identified diabetics in the world increases, the normal sources of clinical insulin—obtained from the pancreas of commercially slaughtered pigs and

The far-reaching
effects of recombinant
DNA research will
require carefully drawn
laws to protect the
public and permit the
science to go forward.

cows—are declining and becoming more expensive. An alternative, inexpensive source of insulin may bring treatment within reach of many more diabetics than ever before.

In addition, experiments are now underway to attempt to insert into non-leguminous plants, such as wheat and corn, the genetic capability of leguminous plants to use nitrogen from the air as fertilizer. This would sharply reduce the need for nitrogen-rich commercial fertilizers. If such experiments are successful, a revolution in agriculture and nutrition will be possible. Then, of course, it will fall to our institutions, including the agricultural industry, to determine whether to exploit them. The United States, because it is relatively well-nourished and comparatively rich in energy, may choose not to alter its agricultural economy so drastically. Other nations, such as those in southern Asia and Africa, may be virtually compelled to do so in order for the people to survive.

While enhancing our understanding of plant genetics and insulin production we may also learn to insert new genetic capabilities into human beings, and perhaps even acquire the capability of designed genetic change. This could be done either as treatment of human

genetic diseases, such as Tay-Sachs disease or sickle cell anemia, or as specific modification of certain human conditions now accepted as bad luck in the normal reproductive process.

The biologists involved in the recombinant DNA story can be proud of their achievements. Certainly they have performed important science and, to that extent, the public's investment in support of basic research has paid enormous dividends already. But beyond the scientific achievements, these biologists have taken an historically unique role in engaging the public from the beginning with their concerns about how recombinant DNA work will proceed. The scientific community recognized its responsibilities to inform the public about the potential hazards of such work, but did not stop there. Scientists have remained engaged as citizens in concert with the general public in attempting to reach a consensus for continuing the research.

The biologists' position stands in sharp contrast to the dilemma the nuclear physicists faced in the 1940s. The experiments in the late 1930s by the German chemists Otto Hahn and Fritz Strassmann—crucial to the development of atomic bombs—were reported to the American Physical Society in January 1939. For reasons of national security, scientists in the United States and England involved in such work imposed upon themselves a prohibition on publication or open discussion of such work. Within a year, all such work was highly classified and public discussion was forbidden. In addition, even the scientists were separated into working groups at different geographic locations and were barred from communicating with one another. The public knew nothing about the project until the basic science had been converted into atomic bombs, and World War II had ended.

It was unfortunate, but perhaps unavoidable, that the physicists were prevented from bringing the issue to public attention until 1946. After that, they threw themselves into the public debate with dedication and persistence, especially in their political activities supporting civilian control of the development of atomic energy. But they never quite overcame the legacy of wartime secrecy that barred the public from knowing about, and thus legitimizing, their scientific and technological efforts at the outset.

The molecular biologists of the 1970s have embarked on an open road with the recombinant DNA techniques. Henceforth the public— including biologists as a part of the public—will share responsibility for deciding how, if at all, our society shall utilize the results of the research. Will the public respond by shutting off creative inquiry into the biological nature of life through these techniques? In permitting the process of inquiry to continue, we are certain to enhance both our knowledge of our biosphere and our capability to alter it very specifically. Which, if any, portions of such enhanced knowledge shall we choose to convert into technology? Since these decisions will affect all of us, so all of us are obligated to do as CERB did. We must educate ourselves, and, once educated, exercise our good judgment.

Index

This index covers the contents of the 1976, 1977, and 1978 editions of *Science Year,* The World Book Science Annual.

Each index entry is followed by the edition year in *italics* and the page numbers:

   Muons, *78*-324, *77*-325, *76*-323

This means that information about Muons begins on the pages indicated for each of the editions.

An index entry that is the title of an article appearing in *Science Year* is printed in boldface italic letters: ***Archaeology.*** An entry that is not an article title, but a subject discussed in an article of some other title, is printed: **Plutonium.**

The various "See" and "See also" cross references in the index are to other entries within the index. Clue words or phrases are used when the entry needs further definition or when two or more references to the same subject appear in *Science Year.* These make it easy to locate the material on the page.

   Neurology, *78*-314, *77*-312, *76*-312; *Special Report,*
      76-54. See also **Brain; Nervous system.**

The indication *"il."* means that the reference is to an illustration only, as:

   Aphid, *il., 78*-269

Index

A

Aalto, Alvar, 77-394
Abdomen: surgery, 78-307
Aborigines, 78-231
Absorption lines, 78-246, 77-246
Accelerator, particle: Stanford Linear (SLAC), 76-107
Achievement, educational, 78-333
Acupuncture: electronics, 77-273
Additives, food: drugs, 78-266; public health, *Close-Up*, 77-336
Adenosine triphosphate, 78-228, 76-246
Adrenalin: allergies, 77-159
Adriamycin, 76-268
Aedui, 78-232
Aerosols: environment, 78-281, 76-278; meteorology, 77-305; ozone, *Close-Up*, 76-280
Aerospace: awards and prizes, 78-391, 77-391, 76-391; industry, 76-257
Africa: anthropology, 78-229; *Men and Women of Science,* 76-357; *Special Report,* 76-181
Agassiz Medal, 77-388
Aggression: among !Kung, 76-190
Aging: anthropology, 76-231; dentistry, 76-297
Agriculture, 78-226, 77-226, 76-226; *Books of Science,* 77-250; botany, 76-252; chemical technology, 76-254; disaster, *Special Report,* 76-168; *Essay,* 76-403; nitrogen *Special Report,* 76-94; Old World archaeology, 78-232, 76-232; space colonies, 76-38; *Special Report,* 76-181. See also *Botany; Chemical Technology; Climate; Food; Nutrition; Pesticide.*
Ainu: anthropology, 78-230
Air Mass Transformation Experiment, 78-311
Air pollution: agriculture, 77-226; energy, 77-274; environment, 78-279, 76-282, *Close-Up*, 78-282; meteorology, 77-305; wind engineering, *Special Report,* 78-200. See also *Automobile; Climate; Environment; Pollution.*
Airbags: transportation, 77-347
Airlines: meteorology, 77-306
Airship: *Special Report,* 77-190
Akuplas: electronics, 77-274
Alaska: archaeology, 76-234; wolves, *Special Report* 78-113
Albedo: meteorology, 76-306
Alberta Fireball, *il.,* 77-216
Alcator: plasma physics, 77-329
Alcohol: energy, 78-278; nutrition, 78-318; stellar astronomy, 76-240
Alfalfa, 78-226, 76-228
Algae, *ils.,* 77-253
Allende chondrite, 77-286, 76-286
Allergies: *Special Report,* 77-153

Allied Chemical Corp., 78-279
Alpha decay: nuclei, *Special Report,* 78-118
Alpha waves: biofeedback, *Special Report,* 78-81
Alpha-fetoprotein: immunology, 76-295
Alpine Snow and Avalanche Project: avalanche, *Special Report,* 77-110
Aluminum: geochemistry, 77-286
Alvin: FAMOUS, *Special Report,* 76-140; oceanography, 78-319
Amaranth: environment, 77-280; useful plants, *Special Report,* 77-68
Amazon River, 76-318
American Foundation for the Blind, 76-74
Amino-acid dating: mummies, *Special Report,* 77-93
Amino acids: allergies, *Special Report,* 77-161; genetics, 78-283; hormones, *Special Report,* 76-83; origins of life, *Special Report,* 77-124; stellar astronomy, 76-240
Ammonia: chemistry, 77-260; nitrogen, *Special Report,* 76-94
Amniotic fluid: *Special Report,* 76-49
Amoxicillin: drugs, 76-268
Amplification, gene: microbiology, 76-311
Amygdala: endogenous opiates, *Special Report,* 78-130
Analgesics: chemistry, 78-260
Andromeda Nebula: cosmology, *Special Report,* 77-102
Anemia: internal medicine, 78-303
Angina pectoris: internal medicine, 76-298; nutrition, 78-318
Angiogenesis: internal medicine, 78-303
Animal behavior: earthquake, *Special Report,* 76-158; wolves, *Special Report,* 78-103
Animal feed, 76-254
Ankle, artificial, 77-302
Ant: ecology, 76-271; zoology, 78-349, 77-349
Antacids: drugs, 76-268
Antarctica, *il.,* 78-320
Anteater: zoology, 78-352
Anterior pituitary, 76-82
Anthropology, 78-229, 77-230, 76-229; *Books of Science,* 78-252, 76-249; *Men and Women of Science,* 76-357; mummies, *Special Report,* 77-81; *Special Report,* 76-181. See also *Archaeology.*
Anti-antibodies, 77-293
Antibiotics: biochemistry, 78-251,77-247; chemical synthesis, 76-262; drugs, 78-267, 76-265, 76-268; microbiology, 76-311; public health, 78-337; science fair, 78-371
Antibodies: allergies, *Special Report,* 77-154; genetics, 78-284;

immunology, 78-297, 77-293; slow virus, *Special Report,* 76-54
Antigens: immunology, 78-299, 77-293; virus, *Special Report,* 76-62
Antihistamines: allergies, *Special Report,* 77-158
Anti-idiotype antibodies, 77-293
Antimicrobial drugs, 76-301
Antiquarks, 77-323, 76-323
Antitumor drug therapy, 77-297
Anxiety: biofeedback, *Special Report,* 78-84; psychology, 77-334
Apgar, Virginia, 76-394
Aphid, *il.,* 78-269
Aplastic anemia, 78-303
Apollo-Soyuz Test Project: space exploration, 77-343, 76-342; stellar astronomy, 77-242
Applications Technology Satellite: communications, 76-265; space exploration, 76-345
Aqualon: chemical technology, 77-257
Aquapulse gun: oil exploration, *Special Report,* 77-18
Aquatic weeds: *Special Report,* 76-14
Archaeology, 78-232, 77-232, 76-231; *Books of Science,* 78-252, 77-250, 76-249; Ebla, *Special Report,* 78-183; *Men and Women of Science,* 76-357; New World, 78-234, 77-234, 76-233, *Close-Up,* 78-236, 77-236; Old World, 78-232, 77-232, 76-231; *Special Report,* 76-181. See also *Anthropology; Geoscience.*
Archaeopteryx: dinosaurs, *Special Report,* 77-65
Archimède: FAMOUS, *Special Report,* 76-140
Architecture: wind engineering, *Special Report,* 78-196
Arctic Ice Dynamics Joint Experiment (AIDJEX): oceanography, 77-319
Arctowski Medal, 76-391
Argentina: archaeology, 76-233
Argo Merchant: environment, 78-283
Ariel 5: space exploration, 76-345
Arms Control and Disarmament Agency, U.S.: energy, 78-276
Arp, Halton C., 76-383
Arteriosclerosis, 76-315
Artery: microsurgery, *Special Report,* 77-49, *il.,* 77-44
Arthropods: paleontology, 76-294
Artificial gene: *Close-Up,* 78-249. See also *Recombinant DNA.*
Artificial insemination: agriculture, 76-227
Artificial joints, 77-302
Artificial organs: diabetes, *Special Report,* 78-57; surgery, 76-303
Artificial sweetener: drugs, 78-266
Artificial-vision device: *Special Report,* 76-68; electronics, 77-272
Ascites: surgery, 78-307

Index

Index

417

Index

Index

Index

Index

Index

Index

Acknowledgments

The publishers of *Science Year* gratefully acknowledge the courtesy of the following artists, photographers, publishers, institutions, agencies, and corporations for the illustrations in this volume. Credits should be read from left to right, top to bottom, on their respective pages. All entries marked with an asterisk (*) denote illustrations created exclusively for *Science Year.* All maps were created by the *World Book* Cartographic Staff.

Cover
Sigurgeir Jonasson

Advisory Board
7   Lee Boltin*; Stanford University; Roland Patry*; Dennis Galloway*; John Swanberg*; Kitt Peak National Observatory; State University of New York at Stony Brook

Special Reports
10   Rod Allin, Bruce Coleman Inc.; Thomas B. Kellogg; George Suyeoka*; Henry M. Cathey, U.S. Department of Agriculture; Handelan Pedersen, Inc.*
12-13   Jet Propulsion Laboratory
15   Jet Propulsion Laboratory; James Teason*
16-18   Jet Propulsion Laboratory
19   Jet Propulsion Laboratory; James Teason*; James Teason*
20-21   Jet Propulsion Laboratory
22   Martin Marietta Aerospace
23   Jet Propulsion Laboratory; James Teason*; James Teason*; James Teason*
24-26   Jet Propulsion Laboratory
29   Detail of *The Creation of Adam* by Michelangelo, Sistine Chapel, The Vatican, Rome (SCALA); Peter Menzel, Stock, Boston
31   Jack D. Griffith, Stanford University Medical Center; Stanley N. Cohen, Stanford University Medical Center
32-33   Handelan Pedersen, Inc.*
34   Bob Moody, University of Alabama Medical Center, Stanford University Medical Center; Virus Laboratory, University of California
36-37   Lou Bory*
39   Detail of *The Creation of Adam* by Michelangelo, Sistine Chapel, The Vatican, Rome (SCALA); Michael Foley*
41   Andrew A. Stern, National Academy of Sciences
42   National Institutes of Health
45   Paul Conklin, Academy Forum
47   United Press Int.
48   Bob Scott Studio*
52-55   George Suyeoka*
57   Herman Polet, M.D., Peter Bent Brigham Hospital
58   Mas Nakagawa*; Joe Friezer
60   United Press Int.
62   Sigurgeir Jonasson
64   R. S. Fiske, Smithsonian Institution
65-67   Marge Moran*
68-70   Hawaiian Volcano Observatory
71   Los Alamos Scientific Laboratory
72   Sigurgeir Jonasson
74   New Zealand Department of Scientific and Industrial Research; Marge Moran*
77-87   Dan Morrill*
94-100   Handelan Pedersen, Inc.*
102   Erwin A. Bauer
105   Illustration of Little Red Riding Hood and The Wolf by Louis Rhead in *Grimm's Fairy Tales,* Copyright 1917 by Harper & Row, Publishers, Inc. By permission of the publisher
106   Jack Couffer, Bruce Coleman Inc.; Douglas H. Pimlott; Rod Allin, Bruce Coleman Inc.
108-109   Joseph A. Erhardt*
110-111   Rolf O. Peterson
112   Rolf O. Peterson; Durward L. Allen
113   George B. Rabb, Brookfield Zoo
114   Brookfield Zoo
117   Gesellschaft für Schwerionenforschung mbH (GSI)
119-121   Herb Herrick*
122   Mas Nakagawa*

123   Robert V. Gentry, Oak Ridge National Laboratory
124   Mas Nakagawa*
125   Herb Herrick*
126   Lawrence Berkeley Laboratory, University of California
127   Los Alamos Scientific Laboratory, University of California
128-132   Lou Bory*
134   Lou Bory*; Johns Hopkins Medical School
136-138   Lou Bory*
140-141   (Background photo) © California Institute of Technology and Carnegie Institution of Washington, from Hale Observatories, (illus.) P. Kronberg, (left photo) Max-Planck-Institut für Radioastronomie, (right photo) WORLD BOOK photo*
144   Mas Nakagawa*; National Radio Astronomy Observatory
145   Mas Nakagawa*; National Radio Astronomy Observatory; National Radio Astronomy Observatory; National Radio Astronomy Observatory
147   National Radio Astronomy Observatory; WORLD BOOK photo*; National Radio Astronomy Observatory; Max-Planck-Institut für Radioastronomie
148   National Radio Astronomy Observatory
150   Kenneth I. Kellerman, National Radio Astronomy Observatory
151   © California Institute of Technology and Carnegie Institution of Washington, from Hale Observatories
155   Ric Ferro, Black Star; David Falconer, Black Star
157   George Suyeoka*
159   Thomas B. Kellogg; Rod Ruth*; Specimens courtesy of T. Saito of Lamont-Doherty Geological Observatory, and scanning electron micrographs by Dee Breger
161   CLIMAP, International Decade of Ocean Exploration, National Science Foundation from *Science,* Copyright © 1976 by the American Association for the Advancement of Science
162   Oregon State University
164   Rod Ruth
166   George Suyeoka*
168   Dan Morrill
171   Fritz Goro
173   Frank B. Salisbury, Utah State University; S. B. Hendricks
174-175   Henry M. Cathey, U.S. Department of Agriculture
176   S. B. Hendricks; Lee Pratt, Vanderbilt University; John Mackenzie, Jr., Stanford University, Richard Coleman, University of North Carolina, Winslow Briggs, Carnegie Institution and Lee Pratt, Vanderbilt University
178   Richard Coleman, University of North Carolina, and Lee Pratt, Vanderbilt University; Brookhaven National Laboratory
179   Brookhaven National Laboratory
180   Dan Morrill
182-189   Kinuko Craft*
190   David Noel Freedman
192-193   Kinuko Craft*
196-197   Jim Curran*
199   James A. Garrison, Fluid Dynamics and Diffusion Laboratory, Colorado State University; Jim Curran*; Wide World
200   Cary Wolinsky, Stock, Boston
202   Fluid Dynamics and Diffusion Laboratory, Colorado State University
203   Jim Curran*; Fluid Dynamics and Diffusion Laboratory, Colorado State University; Fluid Dynamics and Diffusion Laboratory, Colorado State University
204-205   Fluid Dynamics and Diffusion Laboratory, Colorado State University
206-207   James A. Garrison, Fluid Dynamics and Diffusion Laboratory, Colorado State University
210   Ted Thai; Peter Menzel
214   United Press Int.; United Press Int.; Center for Disease Control; Wide World; Keith Meyers, NYT Pictures; Center for Disease Control

215 Wide World; Michael Dressler, NYT Pictures; Center for Disease Control; Center for Disease Control
216-217 Peter Menzel
218 Center for Disease Control
219 Fred Murphy, Center for Disease Control
220 Center for Disease Control
221 Keystone
222 Center for Disease Control
224 U.S. Department of Agriculture; Santa Luca, Crompton and Charig from *Nature* vol. 264 (Peabody Museum, Harvard University); William J. Hammond III, University of California at Davis and Mary K. Seeley, Desert Ecological Research Unit, South West Africa; Mas Nakagawa*; CL Systems, Inc.

Science File
226-227 U.S. Department of Agriculture
228 E. A. Pillemer and W. M. Tingey, New York State College of Agriculture and Life Sciences. Copyright © 1976 by the American Association for the Advancement of Science
229 University of California at Davis
230 School of Dentistry, University of Michigan
231 Wide World
232 J. Kruk and S. Milisauskas excavations; W. Hensel, Project Director (Carole L. Crumley)
233 *The New York Times*
234 Carole L. Crumley
235 G. C. Frison, University of Wyoming
236 Michael E. Moseley
237 Al Kaufman, *Bulletin of the Atomic Scientists*
238 Cornell University
240 Mas Nakagawa*
241 S. Michael Scarrott, Cathleen White, and W. Stewart Pallister, University of Durham, Durham, UK and Allen B. Solinger, Massachusetts Institute of Technology
242 Australian Information Service; Mas Nakagawa*
243 Halton Arp, Hale Observatories
245 TRW Systems, Redondo Beach, California
246 J. C. Theys, State University of New York at Stony Brook and E. A. Spiegel, Columbia University; Kitt Peak National Observatory
248 Drawing by H. Martin; © 1977 The New Yorker Magazine, Inc.
250 James A. Lake
255 S. C. Ducker and R. B. Knox, University of Melbourne
256 Brookhaven National Laboratory
257 Larry L. Hench and David E. Clark, College of Engineering, University of Florida (Ron Franklin)
258 Minnesota Mining and Manufacturing Company
260 Paul F. Swan, *Bulletin of the Atomic Scientists*
263 Ernest Grunwald, et al., Brandeis University
265 Philips Forschungslaboratorium Aachen GmbH
267 Alan F. Hofmann, M.D., Mayo Clinic
268 Oak Ridge National Laboratory
269 L. R. Nault, M. E. Montgomery, Ohio Agricultural Research and Development Center and W. S. Bowers, New York Agricultural Experiment Station, from *SCIENCE*, June 25, 1976
270 Oak Ridge National Laboratory
272 Frank Aleksandrowicz, *Business Week*
273 Calspan Technology Products
274 CL Systems, Inc.
275 General Electric Research and Development Center
277 Sandia Laboratories
278 Energy Research and Development Administration
279 Keystone
280 Wide World
281 NASA; Wide World
284 Drawing by Lorenz; © 1976 The New Yorker Magazine, Inc.
286 *New Scientist,* London
288 Mas Nakagawa*; G. W. Wetherill; G. W. Wetherill
290 Wide World
294 Peter Sheehan, Princeton University
295 Santa Luca, Crompton and Charig from *Nature* vol. 264 (Peabody Museum, Harvard University)
296 Harry Clemmey, University of Leeds

297 Reprinted from *New England Journal of Medicine,* vol. 295: 1341 (Dec. 9), 1976
300 *Temporomandibular Joint Dysfunction and Occlusal Equilibration* by Nathan A. Shore, D.D.S. © 1976 J. B. Lippincott Company
302 Cyclotron Corporation; Mary Catterall, M.D., Hammersmith Hospital; Mary Catterall, M.D., Hammersmith Hospital; Mary Catterall, M.D., Hammersmith Hospital; Mary Catterall, M.D., Hammersmith Hospital
304 Robert Levy, M.D., National Institutes of Health; Robert Levy, M.D., National Institutes of Health; George Tames, *Medical World News*
306 Mentor Corporation
307 Dennis Shermeta, M.D., Johns Hopkins Hospital
309 Theodore Fujita, University of Chicago
310 Sovfoto
311 National Oceanic and Atmospheric Administration
313 Julius Adler
315 Mas Nakagawa*; S. J. Williamson, New York University
317 G. P. Gundlach & Co.
318 Du Pont
319 National Bureau of Standards
320 National Oceanic and Atmospheric Administration
322 Mas Nakagawa*; Oak Ridge National Laboratory
325 Fermi National Accelerator Laboratory; Mas Nakagawa*
326 Stanford University
327 Gordon Watt, *Bulletin of the Atomic Scientists*
329 David T. Attwood and Lamar W. Coleman, Lawrence Livermore Laboratory, University of California
331 International Business Machines Corporation
332 Bell Laboratories
336 Gordon Watt, *Bulletin of the Atomic Scientists*
337 Ministry of Agriculture, Fisheries & Food of Great Britain
340 Reprinted with permission from *SCIENCE NEWS,* the weekly news magazine of science, copyright 1977 by Science Service, Inc. (John H. Douglas); Wide World
341 Paul Conklin, Academy Forum
343-344 NASA
346 BRK Electronics Division of Pittway Corporation (Michael Foley*); Paul J. Fields
347 Muirhead Limited
348 Nick Hobart, *Bulletin of the Atomic Scientists*
350 William J. Hamilton III, University of California at Davis and Mary K. Seely, Desert Ecological Research Unit, South West Africa
351 D. J. Lyon, American Institute for Toxin Research, Inc.
353 United Press Int.

People in Science
354 Steve Hale*; Daniel D. Miller*; Daniel D. Miller*; Jim Freeman, Photo-Graphics*; Steve Hale*
356-357 Daniel D. Miller*
359-360 Gabriela Noris*
363-367 Daniel D. Miller*
368 Gabriela Noris*
370 Dale Stedman, Stedman Studio*; Steve Hale*; Steve Hale*; Steve Hale*; Steve Hale*
374 Jim Freeman, Photo-Graphics*
375 Dale Stedman, Stedman Studio*
377 Dale Stedman, Stedman Studios*; Tom Weigand, Photo-Graphics*; Tom Weigand, Photo-Graphics*
378-384 Steve Hale*
386 M.I.T. Historical Collections; Stanford University
387 American Chemical Society
388 Swedish Information Service
389 Oak Ridge National Laboratory; California Institute of Technology; Stanford University
390 Swedish Information Service; The Fox Chase Cancer Center
391 The Gairdner Foundation; Langley Research Center, NASA
392 Wide World
394 Karsh of Ottawa, Woodfin Camp & Assoc.; Bell Laboratories; Eli Lilly & Co.
395 Tass, Sovfoto; Keystone; Pictorial Parade
397-409 Richard Mlodock*

Typography
Display — Univers
Total Typography, Inc., Chicago
Text — Baskerville Linofilm
Total Typography, Inc., Chicago
Text — Baskerville Linotron
Black Dot Computer Typesetting
Corporation, Chicago

Offset Positives
Collins, Miller, & Hutchings, Chicago
Schawkgraphics, Inc., Chicago
Capper, Inc., Knoxville, Tenn.

Printing
Kingsport Press, Inc., Kingsport, Tenn.

Binding
Kingsport Press, Inc., Kingsport, Tenn.

Paper
Text
Childcraft Text, Web Offset (basis 60 pound)
Mead, Escanaba, Mich.

Cover Material
Oyster White Lexotone
Holliston Mills, Inc., Kingsport, Tenn.
White Offset Blubak
Holliston Mills, Inc., Kingsport, Tenn.